TOWARDS
X-RAY FREE
ELECTRON LASERS

TOWARDS X-RAY FREE ELECTRON LASERS

Workshop on Single Pass, High Gain FELs
Starting from Noise, Aiming at Coherent X-Rays

Garda Lake, Italy June 1997

EDITORS

Rodolfo Bonifacio
University of Milan

William A. Barletta
Lawrence Berkeley National Laboratory

AIP CONFERENCE
PROCEEDINGS 413

American Institute of Physics   Woodbury, New York

phys
Seplae

L.C. Catalog Card No. 97-77194
ISBN 1-56396-744-8
ISSN 0094-243X
DOE CONF- 9706161

Printed in the United States of America

CONTENTS

TA1693
W67
1997
PHYS

INVITED PRESENTATIONS

WORKING GROUPS SUMMARY

CONTRIBUTIONS

PREFACE

The Workshop on Single Pass, High Gain FELs Starting from Noise, aiming at Coherent X-Rays was held at the Palazzo Feltrinelli, Gargnano, on the Garda Lake, Italy, June 97. The Workshop has been sponsored by the Department of Physics of the University of Milan, DESY (Hamburg), SLAC (USA) and LBNL (USA). In attendance were 52 scientists from all over the word, including the most outstanding theoreticians and experimentalists of Free Electron Lasers.

Particular attention was given to the following topics: analytical and numerical modeling of starting from noise; longitudinal and transverse coherence; fluctuations, spiking and photon statistics; and experimental projects status and results. The Workshop was organized in Plenary Sections in the mornings and Working Groups in the afternoon, on three specific subjects:

I)   FEL Physics and Computer Modeling Codes, coordinated by W. M. Fawley (LBNL) and E. T. Scharlemann (LLNL);

II) Beam Production (guns, injectors, accelerators), coordinated by M. Cornacchia (SLAC) and L. Serafini (INFN);

III) FEL Components and Diagnostics, coordinated by R. Carr (SLAC) and J. Rossbach (DESY).

Divergent opinions emerged especially during the Working Group I meetings. Photon statistics, the relevance of quantum fluctuations and the interpretation of the UCLA experiment presented by M. Hogan are examples of the controversial topics discussed during the meeting. Some of the material published here on these and other subjects (WG I report and my contribution), as a matter of fact, was stimulated by the Workshop and worked out after it. We decided to publish this extended works to provide more complete and up-to-date information for the readers.

The meeting location, Gargnano, is a small old village, located among the olive-trees and the lemon houses at the Garda Lake. The Garda Lake is considered the "Botanical Garden of Europe" for its beauties and Gargnano is its heart. Lawrence, D'Annunzio and Churchill are among the most famous personalities who stopped in Gargnano. The meeting took place at the historical Palazzo Feltrinelli, now property of the University of Milan, facing the Garda lake and its far crown of the Dolomite mountains. The place has a perfect organization not only with respect to meeting rooms, and lodging, but also for the good food and wine for which we are indebted to Mr. Gardumi and his collaborators, who created a pleasant and comfortable atmosphere for the participants.

On behalf of the Organizing Committee, we would like to thank the sponsoring institutions for supporting the idea of the Workshop and its organization. In particular we are grateful to Prof. B. Wiik and Dr. J. Rossbach, from DESY, for the most generous support from their laboratory. We want to thank especially the Secretary staff, Franca Tempesta, Giovanna La Pietra and Susanna Mornati for their total and generous commitment to the Workshop. We thank the Scientific Advisory Committee and, in particular, the Speakers and the Working Group Coordinators, who have contributed to the scientific success of the Workshop.

R. Bonifacio
September 15, 1997

Scientific Advisory Committee

W. A. Barletta,	LBNL
A. Bienenstock,	SLAC
M. Cornacchia,	SLAC
C. Pellegrini,	UCLA
J. Rossbach,	DESY
B. Wiik,	DESY
H. Winick,	SLAC

Organizing Committee

R. Bonifacio,	University of Milan, **Chairman**
L. De Salvo,	University of Milan, INFM
P. Pierini,	INFN, Milan
N. Piovella,	University of Milan
L. Serafini,	INFN, Milan
F. Tempesta,	University of Milan, **Local Secretary**

LIST OF PARTICIPANTS

BONIFACIO Rodolfo	Milan Univ.	bonifacio@mi.infn.it
BRAU Charles A	Vanderbilt Un.	brauca@ctrvax.vanderbilt.edu
CARR Roger ·	SLAC	Carr@slac.stanford.edu
CLENDENIN James E	SLAC	clen@slac.stanford.edu
CORNACCHIA Max	SLAC	cornacchia@ssrl.slac.stanford.edu
FAATZ Bart	DESY	faatz@desy.de
FAWLEY William M	LBNL	Fawley@lbl.gov
FELDHAUS Josef	DESY	feldhaus@desy.de
FERRARIO Massimo	INFN-LNF	ferrario@lnf.infn.it
FLOTTMANN Klaus	DESY	mpyflo@mail.desy.de
GIANNESSI Luca	ENEA	giannessi@frascati.enea.it
GLUSKIN Efim	ANL	gluskin@aps.anl.gov
HOGAN Mark	UCLA	hogan@physics.ucla.edu.
JACKSON Alan	LBNL	ajackson@lbl.gov
JAROSZYNSKI Dino	Strathclyde Un.	d.a.jaroszynski@strath.ac.uk
LIMBERG Torsten	DESY	limberg@batman.desy.de
KAMADA Susumu	KEK	Susumu.Kamada@KEK.JP
KIM Kwang - Je	LBNL	Kwang-Je@lbl.gov
KINCAID Brian M	LBNL	BMKincaid@lbl.gov
MARCHESINI Stefano	CEA-Grenoble	marchesini@cea.fr
MAROLI Cesare	Milan Univ	Maroli@mi.infn.it
MASUDA Kai	Kyoto Univ	masuda@iae.kyoto-u.ac.jp
MICHELATO Paolo	INFN	michelato@mi.infn.it
MILTON Stephen	ANL	milton@aps.anl..gov
MULLER Ute Carina	DESY	umuller@mail.desy.de
NG Johnny	DESY	ng@desy.de
NUHN Heinz-Dieter	SLAC	Nuhn@slac.stanford.edu
OEPTS Dick	FOM	oepts@rijnh.nl
ORTEGA J M	LURE	ortega@lure.u-psud.fr
PALMER Dennis	SLAC	dtp@unixhub.slac.stanford.edu
PARMIGIANI F	Milan Polytech.	parmigiani@mail.polimi.it
PETRILLO Vittoria	Milan Univ.	Petrillo@mi.infn.it

PIERINI Paolo	INFN	Paolo.Pierini@mi.infn.it
PIOVELLA Nicola	Milan Univ.	Piovella@mi.infn.it
REICHE Sven	DESY	reichesv@x4u2.desy.de
RENIERI Alberto	ENEA	renieri@frascati.enea.it
ROBB Gordon	Strathclyde Un.	g.r.m.robb@strath.ac.uk
ROSSBACH Jorg	DESY	rossbach@desy.de
SALDIN Evgeny	ASC Samara	Yurkov@vxdesy.desy.de
SCHARLEMANN Ted	LLNL	ets@llnl.gov
SERAFINI Luca	INFN	serafini@vesta.physics.ucla.edu
SCHNEIDMILLER Evgeny	ASC Samara	Yurkov@vxdesy.desy.de
SHEPPARD John C	SLAC	JCSRL@slac.stanford.edu
SUZUKI Shinsuke	SPring-8	shin@haru01.spring8.or.jp
TOMIMASU Takio	FELI	HCC01112@niftyserve.or.jp
TRAVISH Gil	INFN	Gil.Travish@mi.infn.it
VINOKUROV N	Budker INP	vinokurov@inp.nsk.su
YAMAMOTO Shigeru	KEK	Shigeru@kekvax.kek.jp
YANAGIDA Kenichi	SPring-8	ken@haru01.spring8.or.jp
YEREMIAN A Dian	SLAC	Anahid@slac.stanford.edu
YOSHIKAWA Kiyoshi	Kyoto Univ.	kiyoshi@iae.kyoto-u.ac.jp
YURKOV Mikhail	LSVE Dubna	Yurkov@sunse.jinr.ru

xiii

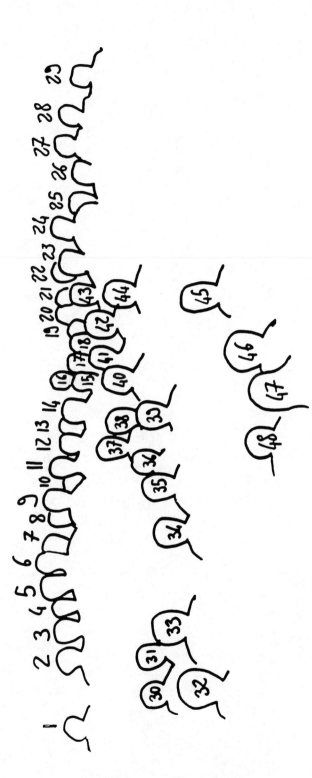

Alphabetical listing of participants in group photo:

48	Bonifacio, Rodolfo	8	Flottmann, Klaus	40	Marchesini, Stefano
5	Brau, Charles A.	31	Giannessi, Luca	17	Masuda, Kai
27	Carr, Roger	12	Gluskin, Efim	2	Milton, Stephen
4	Clendenin, James E.	16	Hogan, Mark	44	Muller, Ute Carina
9	Cornacchia, Max	28	Jackson, Alan	25	Johnny, Ng
41	Faatz, Bart	34	Jaroszynski, Dino	33	Nuhn, Heniz-Dieter
39	Fawley, William M.	11	Limberg, Torsten	24	Oepts, Dick
20	Feldhaus, Josef	14	Kamada, Susumu	1	Palmer, Dennis
6	Ferrario, Massimo	26	Kim, Kwang-Je	15	Piovella, Nicola

22	Reiche, Sven	42	Suzuki, Shinsuke
3	Renieri, Alberto	43	Tomimasu, Takio
35	Robb, Gordon	45	Travish, Gil
21	Rossbach, Jorg	38	Vinokurov, N.
36	Saldin, Evgeny	29	Yamamoto, Shigeru
7	Scharlemann, Ted	18	Yanagida, Kenichi
30	Serafini, Luca	10	Yeremian, A. Dian
37	Schneidmiller, Evgeny	23	Yoshikawa, Kiyoshi
13	Sheppard, John C.	19	Yurkov, Mikhail

47 ?

Invited Presentations

Temporal and Transverse Coherence of Self-Amplified Spontaneous Emission[1]

Kwang-Je Kim

Lawrence Berkeley National Laboratory
University of California
Berkeley, California 94720

Abstract. We review the coherence properties of the self-amplified spontaneous emission (SASE). Temporally, SASE is similar to the spontaneous undulator radiation except that the spectral bandwidth is about ten times narrower compared with typical undulator radiation. The situation is quite different in the transverse dimension, where SASE is fully coherent.

INTRODUCTION

Several laboratories are pursuing the R and D of the linac-based light source based on the SASE principle [1], [2] for its promise of extreme high brightness and time resolution [3]. It therefore is important to clearly understand the coherence properties of SASE, in particular, compared to the current generation of synchrotron radiation sources. This paper is a review of this topic.

In the temporal (longitudinal) dimension, the first order coherence of SASE sources is improved about ten times compared with typical undulator sources in current generation synchrotron radiation facilities. The second order coherence has to do with the intensity fluctuation, and SASE in this regard is very similar to undulator radiation, consisting of a random superposition of wave packets. The fluctuation property of such light, known as thermal or chaotic light, is well-known [4]: the fluctuation of intensity at a given time or frequency is 100%. However, the fluctuation for an integrated intensity is smaller by a factor $\sqrt{m_c}$, where m_c is the number of modes in the integration interval. The probability distribution of the integrated intensity is given by

[1] This work was supported by the Director, Office of Energy Research, Office of High Energy and Nuclear Physics, Division of High Energy Physics, of the U.S. Department of Energy under Contract No. DE-AC 03-76SF00098.

the Gamma probability distribution. The fluctuation properties of undulator radiation was discussed in [5], [6], and of SASE in [7], [8], [9].

In the transverse dimension, SASE is a linear superposition of a set of transverse eigenmodes, each having a different growth rate [10]. If the growth rate of one or more higher order modes were the same as the fundamental mode (degenerate), then SASE would be partially coherent transversely. However, it turned out that the higher order modes have significantly smaller growth rates than the fundamental mode [11]. Therefore, the transverse behavior of SASE is determined entirely by the fundamental mode, and SASE becomes fully coherent transversely. This situation is markedly different from the usual undulator radiation which is partially coherent [12], the degree of coherence being determined by the ratio of the coherent phase space area $\lambda/2$ and the electron beam phase space area ($2\pi \times$ emittance).

TEMPORAL COHERENCE

The temporal coherence characteristics of self-amplified spontaneous emission (SASE) is based on the 1-D analysis of Maxwell-Vlasov Equations [13], [14]. In the frequency domain, the evolution of the electric field in the exponential gain regime is given by

$$E_\omega(z) = G_\omega(z)\ E_\omega(0)\quad.\tag{1}$$

Here $E_\omega(0)$ is the initial amplitude

$$E_\omega(0) = \sum_{i=1}^{N_e} e^{-i2\pi\omega t_i{}^0}\quad,\tag{2}$$

where ct_i^0 is the initial coordinate of ith electron with respect to the bunch center, and N_e is the number of the electrons. The gain function $G_\omega(z)$, neglecting electron beam energy spread, is given by

$$G_\omega(z) = \frac{1}{3} e^{\sqrt{3}\rho k_u z}\ e^{-i\left(\frac{2}{3}\frac{\Delta\omega}{\omega_0} + \rho\right)k_u z}\ e^{-\frac{(\Delta\omega)^2}{4\sigma_\omega^2}\left(1+\frac{i}{\sqrt{3}}\right)}\tag{3}$$

Here ρ is the FEL scale parameter [1], $k_u = 2\pi/\lambda_u$, λ_u = undulator period length, z = distance along the undulator, $\Delta\omega = \omega - \omega_0$, ω_0 = central frequency, and σ_ω is the gain bandwidth

$$\sigma_\omega = \omega_0\sqrt{\frac{9\rho}{2\pi\sqrt{3}z/\lambda_u}} \approx \omega_0\sqrt{\frac{\rho}{z/\lambda_u}}\quad.\tag{4}$$

The field in the time domain is obtained by the Fourier transform:

4

$$E(z,t) = \frac{1}{\sqrt{2\pi}} \int d\omega\, E_\omega(z)\, e^{i\omega(t-z/c)}$$

$$= \frac{1}{9\sqrt{z}} e^{\sqrt{3}\rho k_u z} \sum_{i=1}^{N_e} e^{i\omega_0(t-z/c(1+\rho\Delta\beta)-t_i)}$$

$$\times\; e^{-\frac{\left(t-\frac{z}{c}\left(1+\frac{2}{3}\Delta\beta\right)-t_i^0\right)^2}{4\sigma_\tau^2}\left(1-\frac{i}{\sqrt{3}}\right)} \tag{5}$$

Here $\Delta\beta = 1 - \beta$, β = electrons average longitudinal speed, and

$$\sigma_\tau = \frac{1}{2\sigma_\omega} \approx \frac{1}{2\omega_0}\sqrt{\frac{z/\lambda_\mu}{\rho}} \tag{6}$$

is the coherence length. Equation (5) describes a sum of N_e wave packets of rms pulse length σ_τ, propagating with a group velocity $v_g = c/\left(1+\frac{2}{3}\Delta\beta\right)$, slower by $\frac{2c}{3}\Delta\beta$ than the speed of light in a vacuum [7]. As it propagates, the wave amplitude grows exponentially. The wave packets are randomly distributed with relative positions ct_i^0. The time domain picture was emphasized by Bonifacio, et al [7].

For our purpose here, the relevant features of the results in the above are the coherence length of the individual wave packet, which characterizes the first order coherence, and the random distribution of the center of the wave packets, which refers to the second order coherence.

The coherence length increases as given by Equation (6), or equivalently, the radiation bandwidth decreases, as given by Equation (4). The saturation takes place at about $z/\lambda_u = 1/\rho$, so that the bandwidth of SASE is about $\sigma_\omega/\omega_0 \approx \rho$. Since ρ is typically about 10^{-3}, the SASE bandwidth is about ten times narrower than the typical bandwidth of undulator radiation in current synchrotron radiation sources.

The random distribution of wave packets of SASE is similar to the undulator radiation. Figure 1 shows an example of random distribution of wave packets. This kind of light wave is very common in nature and is referred to as "chaotic light"; almost all light encountered in daily life, such as sunlight, as well as synchrotron radiation, is this kind. The properties of chaotic light have been discussed extensively in the literature, for example in [4]. The following is a brief summary of these results as applied to undulator radiation [5] [6] and SASE [7], [8], [9].

A simplified model of chaotic light (observed at a fixed position) can be represented in the time domain as a super position of Gaussian pulses

$$E(t) = E_0 \sum_{i=1}^{E_e} e^{-\frac{(t-t_i)^2}{4\sigma_\tau 2} - i\omega_0(t-t_i)} \quad, \tag{7}$$

or in the frequency domain (In fact, the undulator radiation is a truncated sinusoidal pulse. However, the Gaussian pulse is easier to analyze.)

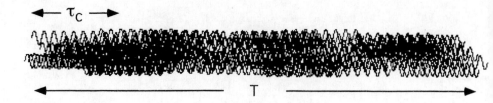

FIGURE 1. An example of chaotic light given by a random superposition of 100 sinusoidal wave packets each with six periods long. The total length of the pulse is T.

$$E(\omega) = E_0 \sum_{i=1}^{N_e} e^{-\frac{(\omega-\omega_0)^2}{4\sigma_\omega{}^2} - i\omega t} \quad . \tag{8}$$

Figure 2 shows an example of explicitly adding wave packets with $\lambda = 1$, $\sigma_\tau = 2$ ($\sigma_\omega = 0.25$), $N_e = 100$, assuming that t_i's are randomly distributed with equal probability in a bunch length $T = 100$. The remarkable feature of this figure is that the resultant wave is a relatively regular oscillation interrupted only a few times, much less than one might have naïvely expected from the fact that it is a superposition of 100 waves. In fact, the number of the regular regions has nothing to do with the number of wave packets. Each regular region is a coherent mode, the length of which is the coherent length. Therefore, the number of the regular region is the number of the coherent modes, which is the ratio of the bunch length to the coherence length,

$$m_c = T/2\sqrt{\pi}\sigma_\tau \approx T/4\sigma_\tau \tag{9}$$

The situation in the frequency domain is similar. Figure 3 shows the intensity spectrum $dn/d\omega \propto |E(\omega)|^2$ (n = number of photons), with $E(\omega)$ computed by Equation 8 with the same wave parameters as in Figure 2. The spectrum consists of sharp peaks of width $\Delta\omega_p \sim 2/T$ randomly distributed within the radiation bandwidth $\sigma_\omega \sim 1/2\sigma_\tau$. Thus the number of the spectral peaks is the same as the number of the coherent modes in the time domain.

The intensity fluctuation described here is a simple consequence of the central limit theorem: for a random superposition of a large number of waves, the probability distribution of the field amplitude in either frequency domain or time domain is given by a Gaussian function. This implies that the intensity distribution is given by the inverse exponential function with a variance the same as the average.

In the above, we have discussed the case in which the radiation is observed within a narrow interval. If the whole pulse is integrated, the fluctuation will be smoothed out, and the variance reduced roughly by $\sqrt{m_c}$. The probability distribution in this case is the so-called gamma probability distribution [4].

6

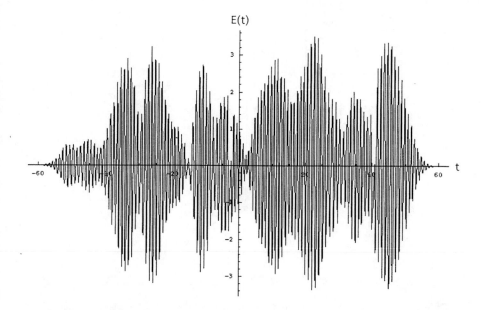

E(t)

FIGURE 2. The resultant wave of a random superposition of Gaussian wave packets as in Eq. (8). The parameters of each wave packet are $\lambda = 1$, $\sigma_\tau = 1/2\sigma_\omega = 2$. A total of 100 wave packets are distributed randomly with uniform probability in a length of 100.

This statement can be generalized for the partial intensity integrated over a finite frequency interval $\Delta\omega$. Thus, let us consider

$$n_\Delta\left(\omega\right) = \int_{\omega-\Delta\omega/2}^{\omega+\Delta\omega/2} d\omega' \, dP/d\omega' \quad . \tag{10}$$

As long as $\Delta\omega > \Delta\omega_p$, we define the number of the coherent modes within the integration (observation) interval

$$m_c = \Delta\omega/\Delta\omega_p \quad . \tag{11}$$

The probability distribution for $n_\Delta\left(\omega\right)$ is then given by the gamma distribution, peaked at $\langle n_\Delta\rangle$ with a variance $\sigma_{n_\Delta} = \sqrt{\langle n_\Delta^2\rangle - \langle n_\Delta\rangle^2} \approx \frac{1}{\sqrt{m_c}}\langle n_\Delta\rangle$. Figure 4 illustrates the reduction of fluctuation for the integrated intensity.

Taking into account the quantum fluctuation of photons, we have

$$\sigma_{n_\Delta}^{2} = \frac{\langle n_\Delta\rangle^2}{m_c} + \langle n_\Delta\rangle \quad . \tag{12}$$

The second term in Equation (12), due to the quantum fluctuation, can be written as

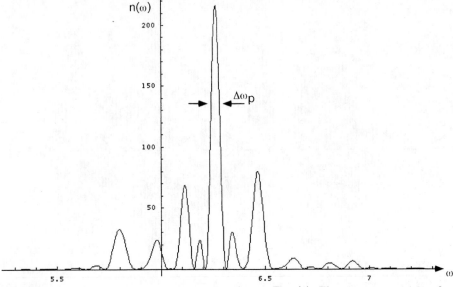

FIGURE 3. The intensity spectrum corresponding to Fig. (2). The spectrum consisits of sharp peaks of width $\Delta\omega_p \sim 1/T$, which are distributed within a Gaussian envelope or rms width σ_ω. The height of the spectral peaks fluctuates 100%.

$$\langle n_\Delta \rangle = \frac{\langle n_\Delta \rangle^2}{m_c \delta}, \quad \delta = \frac{\langle n_\Delta \rangle}{m_c} \quad . \tag{13}$$

Here δ is the number of photons per mode, known as the degeneracy parameter. Comparing Equation (13) with the first term of Equation (12), we find that the quantum fluctuation is negligible for $\delta \gg 1$. This is the case in SASE.

TRANSVERSE COHERENCE

Taking into account the 3-D effects, the field amplitude for SASE as a function of z can be written as

$$E_\omega(\mathbf{x}, z) = \sum_n e^{\mu_n z} C_n(z) E_n(\mathbf{x}) + \text{continuum modes} \tag{14}$$

Here $\mathbf{x}$ is the transverse coordinate, and μ_n and E_n are the discrete solutions for the eigen value and the eigen mode of the Maxwell-Vlasov equations. An example of the intensity profiles for the fundamental and second order mode are shown in Figure 5 [11]. The simplest way to determine the expansion coefficient C_n seems to be the Van Kampen method [10], [15].

As z increases, the field amplitude is dominated by the fundamental mode with the largest growth rate, the real part of μ_1, if there is no degeneracy, i.e., if the growth rate of the higher order modes is sufficiently smaller than the fundamental.

The growth rates of the fundamental, and the first and second higher order modes obtained by solving the FEL eigenvalue equation [11] are shown as a function of the ratio of the 1-D gain length to the Rayleigh length in Fig. 6. All SASE projects currently under discussion are in the region far away from the degenerate limit.

The fact that SASE is dominated by a single transverse mode means that it is completely coherent transversely. This is in marked contrast to the spontaneous emission in which the coherence criteria is, under the optimal matching condition, determined by the ratio of the electron beam emittance to the coherent radiation emittance $(\lambda/4\pi)$ [12]. In the case of SASE, the modes are non-degenerate even if this ratio is large, and therefore is transversely coherent. For temporal coherence, the case of SASE was similar to the spontaneous emission. For transverse coherence, the SASE is almost always fully coherent.

The difference of the transverse coherence characteristics of SASE from those of the spontaneous emission may be understood as follows [16]. In the case of the spontaneous radiation the total radiation phase space area is the incoherent sum of the electron phase space area. In the case of an electron beam bunched by the FEL action, on the other hand, the radiation angular

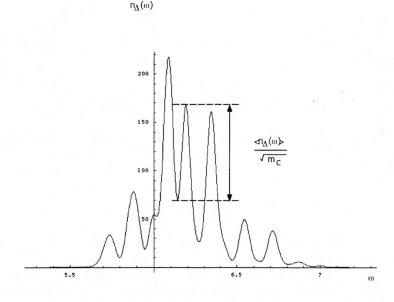

FIGURE 4. The spectrum of the integrated intensity defined by Eq. (10). Notice that the spectrum is more smooth, the fluctuation being reduced by $\sqrt{m_c}$.

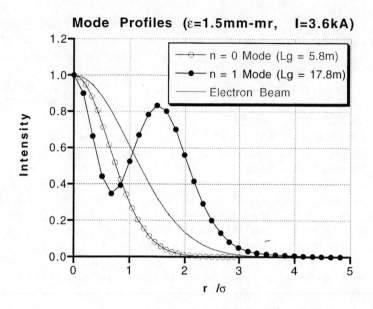

FIGURE 5. An example of mode profiles. (Courtesy of M. Xie)

divergence is the diffraction angle determined by the electron beam transverse size, implying that the radiation is fully coherent transversely. Although the bunching in SASE is not complete, the SASE radiation is dominated by the bunched part, and is therefore transversely coherent.

QUANTUM EFFECTS IN SASE

So far we have treated electrons as classical. There are two corrections due to the quantum nature of electrons. The first is a correction on the gain, which is small if the recoil is small compared with the electrons' energy spread [17]:

$$\frac{\hbar\omega}{E_e/N_u} \approx \frac{\hbar\omega}{\Delta E_e} << 1 \quad . \tag{15}$$

Here N_u is the number of the undulator periods (which for SASE is about ρ^{-1}) and ΔE_e is the e-beam energy spread.

The second is the effect on the statistics, such as the effective input signal in SASE. The quantum correction on SASE noise was first considered in [18]. Continuing the analysis in that paper, and taking into account the correct multi-particle electron wave functions, it is found that the correction is small [19] when

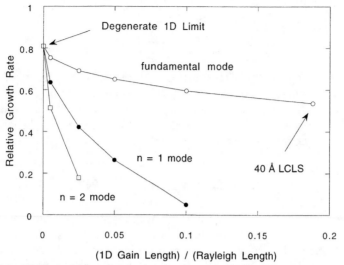

FIGURE 6. The growth rates for fundamental, and first and second higher modes, illustrating the dominance of the fundamental mode in most of the parameter space. (Courtesy of M. Xie)

$$\varepsilon_{\perp n}{}^2 \, \varepsilon_{\| \, n} < N_e (\lambda_c)^3 \quad . \tag{16}$$

Here $\varepsilon_{\perp n}$ is the normalized transverse emittance and $\varepsilon_{\| n}$ is the normalized longitudinal emittance ($\sigma_{\Delta \gamma} \, \sigma_z$). The inequality (16) could have been expected from quantum statistical mechanics.

In all cases of SASE under discussion at present times, the inequalities (15) and (16) are well satisfied.

We have remarked that quantum corrections in SASE are negligible. This may also be seen from the comparison of the effective noises in atomic lasers and in SASE. For this purpose, it is convenient to compare the brightness, i.e., photons per unit six dimensional phase space volume:

$$\frac{d^6 \, n_{ph}}{d^2 \mathbf{x} d^2 \phi \, (d\omega / \omega) \, dz} = B_{\text{noise}} \, e^{z/L_G} \quad . \tag{17}$$

Here L_G is one gain length.

For the atomic laser, it is well known that the effective noise is due to the vacuum fluctuation given by one photon per mode with phase space volume λ^3 [20]. Thus

$$B_{noise} \approx \frac{1}{\lambda^3} \quad . \tag{18}$$

The effective noise for SASE derived in Refs. [13] [14] [10] can be written approximately as

$$B_{noise} \cong \frac{\alpha K^2 N_G^e}{\lambda^3} \quad , \tag{19}$$

where α = fine structure constant, K = the undulator parameter, N_G^e is the number of electrons in a distance $\lambda \left(L_G / \lambda_u \right)$. The number of the noise photons per mode in the case of SASE is therefore $\alpha K^2 N_G^e$, which is much larger than one. This is because the start-up noise of SASE is the classical shot noise and not from quantum fluctuation.

ACKNOWLEDGMENTS

The author thanks M. Zolotorev for illuminating discussions on the topics of temporal coherence, fluctuations, and quantum effects. He also thanks E. Saldin for helpful discussion on temporal coherence and M. Xie on transverse coherence. He is also grateful to R. Bonifacio for asking presentation of an invited talk on this topic at the Lake Garda Workshop.

REFERENCES

1. Bonifacio, R., Pellegrini, C., and Narducci, L., *Opt. Commun.*, 50, 373 (1984).
2. Pellegrini, C., "A 4 to 1 mm FEL based on SLAC Linac", Proceedings of the Workshop on 4th Generation Light Sources, Feb. 24 - 27, 1992, SLAC, SSRL 92/02 (1992).
3. Winick, H., "Next Generation SR Sources", Proceedings of PAC '97.
4. Goodman, J., "Statistical Optics" (John Wiley & Sons, New York, 1985).
5. Tanabe, T., et al., NIM, A304, 97 (1991).
6. Zolotorev, M., and Stupakov, G. V., SLAC-PUB-7132 (March 1996); Proceedings of PAC '97.
7. Bonifacio, R., et al., *Phys. Rev. Lett.*, 74, 70 (1994).
8. Saldin, E. L., Schneidmiller, E. A., and Yurkov, M. V., "Statistical Properties of Radiation form VUV and X-Ray FEL," DESY Preprint TESLA-FEL97-02 (April 1997).
9. Kim, K.-J., talk at ICFA Workshop at Archidosso (Sept. 1996).
10. Kim, K.-J., *Phys. Rev. Lett.* 57, 1871 (1986).
11. Xie, M., to be published.
12. Kim, K.-J., "Characteristics of Synchrotron Radiation", pg. 585, AIP Conference Proceedings '84, M. Month and m. Dienes, eds. (1989).
13. Kim, K.-J., *Nucl. Instrum. Methods* A250, 396 (1986).
14. Wang, J.-M., and Yu, L. H., *Nucl. Instrum. Methods* A250, 484 (1986).
15. Van Kampen, N. G., Physica XXI, 949 (1955).

16. Kim, K.-J., "Emittance in Particle and Radiation Beam Techniques," Proceedings of the Advanced Accelerator Workshop, Lake Tahoe, CA, Oct. 12-18, 1996.

17. For a review, see G. Dattoli and A. Renieri, pg. 221, Laser Handbook, Vol. 6, W. Colson, C. Pellegrini and A. Renieri, eds., (North-Holland, 1990).

18. Bonifacio, R., Casagrande, R., Opt. Commun. 50, 25 (1984).

19. Kim, K. J., to be published.

20. Milonni, P. W., and Eberly, J. H., "Lasers" (John Wiley & Sons, New York, 1988).

On the Classical and Quantum Feature of Self Amplified Spontaneous Emission[1]

Rodolfo Bonifacio

Dipartimento di Fisica, Università di Milano, INFN, INFM, Sezione di Milano,
Via Celoria, 16, 20133, Milano, Italy

In this paper we shortly review the classical and the quantum theory of SASE to include the definition of a new SASE concept: the quantum SASE.

INTRODUCTION

After summarizing the definition and main properties of classical SASE, we describe the quantum theory and the photon statistics using a rigorous QED approach "alla" Glauber. We include both classical shot noise and quantum noise, as well as coherent field and coherent bunching initiation. Finally we define the concept of quantum SASE, i.e., when the seed is only due to quantum fluctuations from the uncertainty principle. That is the unique noise that, in principle, cannot be eliminated. The observation of quantum SASE would be a new quantum phenomena in which position-momentum single electron uncertainty fluctuations give rise to exponential amplification of the vacuum field.

CLASSICAL SASE

Let us start with some historical remarks and definitions.

SASE as a collective instability *starting from noise with exponential growth of radiated intensity, bunching and energy spread up to saturation values which are independent on the initial noise* level has been firstly described in ref.[1]. SASE scaling laws, the fundamental FEL parameter ρ and the shot noise onset as $1/N$ have

---

[1] The material of this paper has been partially stimulated by the discussion during this workshop.

CP413, *Towards X-Ray Free Electron Lasers*
edited by R. Bonifacio and W. A. Barletta
© 1997 The American Institute of Physics 1-56396-744-8/97/$10.00

been clearly spelled out in ref. [2]. In particular, in the Steady State regime, in which slippage is negligible, the radiation power at saturation scales as $N^{4/3}$ and the energy spread as $\Delta\gamma/\gamma \approx \rho$.

Some specifications are necessary. In classical SASE the onset is due only to classical shot noise, scaling as $1/N$, and coming from classical fluctuations of the electron's positions. Starting from noise means that no input radiation signal or coherent pre-bunching of the beam is provided. Furthermore, coherent bunching contribution coming from the Fourier components of the shape of the input current must be negligible. The coherent bunching contribution would give rise to coherent spontaneous contribution at the beginning of the wiggler, i.e., to contribution as z^2 and N^2, where z is the position on the wiggler and N is the number of electrons. This should not be confused with super-radiance [3]. In super-radiance the electrons start radiating independently as N, evolving spontaneously by radiation reaction to a correlated state in which they radiate as N^2. Note that in super-radiance the saturation is inhibited by slippage [3].

The word SASE came out in a private discussion between Claudio Pellegrini and myself to distinguish steady state regime starting from noise from super-radiance, and appeared shortly after in different contributions to the FEL conference in Rome, 1984 [4]. At the Garda Workshop the group of UCLA [5] presented what they claim is the first observation of High Gain SASE. However, looking at the experimental data presented in Fig. (1) and (2) of their contribution [5], one can see that the so called high gain claimed in the title is just a factor 5 or 6 above spontaneous emission and not 10^5 or 10^6 as it should be in the high gain exponential regime. Furthermore, the experimental data, with the corresponding error bars, can be fitted not only by the exponential presented by the authors, but also by almost any power law, including a square law characteristics of Coherent Spontaneous Emission (CSE), as it is pointed in the report of Working Group I in this proceedings. In this report it is suggested that the term SASE should stand for "Slightly-Amplified-Spontaneous-Emission" to describe the UCLA experimental results.

A many modes propagation theory of SASE has been firstly proposed in ref. [6]. In this paper it is shown that the integrated power grows exponentially as a function of z, up to a saturation value in agreement with the steady state theory of ref. [2]. Regarding the spectrum, the paper describes the *envelope* of the spectrum in what we shall call the Gaussian approximation.

A complete classical theory describing not only integrated power and spectrum envelope but also *temporal* and *spectral structure* of SASE pulses is contained in ref. [7], where also *shot to shot fluctuations* are described. The radiation pulse in the exponential regime is composed by super-radiant spikes, whose peak intensities are proportional to N^2 and whose width are of the order of $2\pi L_c$. Here L_c is the so-called cooperation or coherence length, defined as $L_c = \lambda_r/4\pi\rho$, where λ_r is the radiation wavelength. Neglecting transverse effects, the total number of super-radiant spikes in the exponential regime is ruled, on the average, by the number of cooperation lengths in the bunch length L_b, $m = L_b/2\pi L_c$.

On the average, since the spikes have a random amplitude, they occur at random times, and their configuration varies from shot to shot. Let us stress, however, that at the beginning of the undulator the number of spikes is much larger than m (of the order of L_b/λ_r). The reduction to the average number m has been called the phenomena of "cleaning up" of noise. Note that if m≈1 one has only a single super-radiant spike. These features have been confirmed by 3-D GINGER simulations [8]. The spectrum has the complementary feature, i.e., its total width in wave-numbers, scales as $1/L_c$ and is composed by random pulses of width $1/L_b$. Hence, if $2\pi L_c > L_b$ there is only one pulse. The integrated power increases exponentially but it does not stop at the steady state saturation value, as in ref. [6], but it continues to grow to higher values, due to the super-radiant character of the spikes. Also this feature has been confirmed by 3-D GINGER simulations in ref. [8]. Regarding the space time dependence of the shape of the spikes, it cannot be approximated by a Gaussian function of t-z/v_s as in ref.[6]. The shape is almost a *hyperbolic secant* shape [9, 10], with some ringing. The dependence on z and t is not a simple function of t-z/v_s, but is a the self similar pulse in the

variable y defined as $y = \sqrt{\dfrac{c(z - v_e t)}{v_e L_c}} * \dfrac{(ct - z)}{L_c}$ [9], where v_e is the average electron's

velocity. The growth, as a matter of fact, is not even simply exponential, but is described by $\exp(y^{2/3})$. Only the peak of the pulse grows exponentially as a function of

z and propagates at the characteristic velocity $v_s = \dfrac{3v_e}{2 + (v_e / c)}$ [9]. It is easy to verify

that $v_e < v_s < c$ as described in the Appendix. The Gaussian shape of ref. [6] arises from an approximation on the pulse spectrum which implies the disappearance of an *essential singularity* [9] in an inverse Laplace-Fourier transform. This essential singularity originates the *non exponential* and *non Gaussian* space-time shape of the super-radiant pulses. Regarding shot to shot fluctuations, the positions and the amplitude of each spikes vary randomly; what is almost fixed is the number of spikes, m. The integrated intensity increases exponentially as a function of z, as predicted in ref. [6]. However, its value, from shot to shot, at a given z, in the exponential regime presents fluctuations which goes as $1/\sqrt{m}$ [7]. This can be understood as fluctuations due to the radiation coming from m uncorrelated independent coherence length. The probability distribution of these fluctuations has been described in this proceeding [11], fitting numerical results, via the classical binomial distribution described in ref. [12]. In the next we give a first principle quantum mechanical derivation of the photon statistics of SASE, which in the particular case m>>1, becomes a binomial distribution.

QUANTUM THEORY OF PHOTON STATISTICS

Many authors use "intuitive" or semi-classical arguments to deal with quantum fluctuations and photon statistics. These arguments are a statistical mixture of intuitive, classical and quantum arguments. This is the so-called quasi classical approach, which, sometimes, can reproduce, up to some degree, some quantum results on the basis of more or less hand-waving arguments. However, since the *concept of photon **does not** exists classically*, there is only one *rigorous* way of dealing with photon statistics: the QED way, originated by Roy John Glauber in 1963, in a famous series of papers ref. [13]. Since then the field of quantum optics has been very successfully applied to atomic laser physics giving results which, sometimes can be, and sometimes cannot be reproduced semi-classically. In this section, generalizing and summarizing the results of ref. [14], we will briefly describe the photon statistics of a high gain FEL in the framework of Glauber QED theory and coherent state formalism.

Fermions or Not Fermions

First of all we should answer the following question: in a FEL should the electrons be treated as fermions? If yes, the result would be dramatic, since there is not any classical analog of the Pauli exclusion principle. In the case of the FEL this would imply that there cannot be more than one electron per each momentum state, i.e., the energy spread would be enormous. Furthermore it would be compulsory to use anti-symmetric wave functions or second-quantization formalism. In the following we show that for any reasonable values of the current, in agreement with ref. [15], the fermion character can be ignored. Let us remember that one should take into account the symmetry or anti-symmetry of the wave function only if the particles are indistinguishable, i.e., if the single particle wave packets overlap. It is easy to show that this happens if, and only if [16],

$$\varepsilon_{//}\varepsilon_{\perp}^2 < N\lambdabar_c^3 \qquad (1)$$

where $\lambdabar_c$ is the Compton wavelength, $\varepsilon_{//}$ and $\varepsilon_{\perp}$ are respectively the parallel and transverse normalized emittance.

In fact, in 1-D theory one must impose that the classical phase space area $\Delta z \Delta p_z$ divided by the minimum wavepacket quantum uncertainty area $\hbar$ must be smaller than the number of electrons, such that the wavepackets overlap. Since $p=\gamma mc$ and the longitudinal emittance is $\varepsilon_{//}=\Delta\gamma\Delta z$, one has $\varepsilon_{//} < N\lambdabar_c$. The extension to 3-D leads to Eq.(1). Note that, since $\Delta z = L_b$ and $N = I.L_b/ec$, Eq.(1) can be written as $I > ec\Delta\gamma\varepsilon_{\perp}^2/\lambdabar_c^3$. For all reasonable values of the parameters, inequality (1) is violated, so that the wavepackets *do not* overlap. Hence, the electrons can be considered as *distinguishable particles* and not as fermions so that the total wave function can be taken as a *simple product* of wave functions. The fact that the electrons cannot be considered as fermions, *does not* imply at all that quantum fluctuations are negligible as argued in ref. [16]. They are negligible, in principle, only when one neglects the *uncertainty principle*.

In fact, all kind of classical fluctuations are, in principle, negligible, and can be taken to be *arbitrarily small* even if this can look impractical due to technological problems. What *cannot* be taken arbitrarily small are the fluctuations due to uncertainty principle. To be specific, the classical source of SASE comes from shot noise, i.e., from classical fluctuations of the electron position. However, in principle, one can imagine a "crystal beam" in which the electrons come almost exactly at the same distance one from the other, at a regular distance. In this case, the classical shot noise bunching is zero, i.e.,.there is not any classical source for the field. Spontaneous emission from one electron, in this case, would cancel the spontaneous emission of another electron that would have opposite phase. What cannot be ignored is the *quantum shot noise*, i.e., the one induced by the quantum single electron's position-momentum fluctuations due to the *uncertainty principle*.

We define as **Quantum SASE** the self-amplification and self-bunching when the *initiation* of the process is *due to quantum fluctuations*. In the next we will give a precise condition for observing quantum SASE *even in presence of classical shot noise*.

Quantum Statistics

In the classical theory the field amplitude is a complex number a. In the quantum theory field amplitude is represented by an operator â such that $[\hat{a}, \hat{a}^+] = 1$. Hence the field amplitude can be thought as random quantum variable whose statistic can be described in terms of the so called Glauber P function. The P function can be thought as a quasi probability function for the field operator to take the value α. Precisely, the P function is the diagonal expansion of the field density operator $\hat{\rho}$ in terms of the projector on coherent states $\hat{\rho} = \int d^2\alpha P(\alpha)|\alpha\rangle\langle\alpha|$, where $d^2\alpha = d(\text{Re}\alpha)d(\text{Im}\alpha)$. Here α are the so called Glauber coherent state defined as the eigenstate of â, correspondent to the complex eigenvalue a, i.e., $\hat{a}|\alpha\rangle = \alpha|\alpha\rangle$. $f(\alpha)$ is the eigenstate of the photon number $|n\rangle$ where $|\alpha\rangle = e^{-\frac{|\alpha|^2}{2}} \sum \frac{\alpha^n}{\sqrt{n!}}|n\rangle$. The existence of the P function in the space of tempered distribution has been shown in ref. [17].

The one mode calculation of photon statistics in the linear regime of a high gain FEL operating in steady state was performed many years ago in ref. [14]. In these papers we calculated the Glauber P function [13], i.e., the normally ordered quasi probability function for the complex field amplitude α. The main properties of the P as a quasi probability function are listed below [13]:
the ensemble average for the field value, $< \hat{a} >$ is given by

$$< \hat{a} >= \int d^2\alpha P(\alpha)\alpha ; \qquad (2)$$

where the average value of the photon number is given by

$$<n> = \int d^2\alpha |\alpha|^2 P(\alpha) \tag{3}$$

and the probability distribution for the photon number n is

$$P(n) = \langle n|\hat{\rho}|n\rangle = \int d^2\alpha e^{-|\alpha|^2}\frac{|\alpha|^{2n}}{n!}P(\alpha) = \int dI \frac{I^n}{n!}e^{-I}P(I) \tag{4}$$

The last expression is valid when P is a function of $|\alpha|^2 = I$, i.e., when the phase of the field is random. Equation (4) is the rigorous QED derivation of the semi classical Mandel formula [12]. In ref. [14] we calculated the P function under the following assumptions:

1) the initial state of the electrons is assigned by the product of N identical Gaussian wave packets for the phase $\theta_j = kz_j$ of the particles; each packet is centered on the value $<\theta_j>_0$ and has a width σ_θ^2. The center of the wave packet corresponds to the classical phase, whereas θ_j is a quantum variable which, in proper units, obeys the minimum uncertainty principle $\sigma_\theta\sigma_p = \frac{1}{2}$, since the canonical momentum is normalized to $\hbar k$. The assumption of minimum uncertainty insures a description as close as possible to the classical one. Note that the centers of the packets are real numbers and not quantum observables.

2) $\overline{\theta}_j = \theta_j - <\theta_j>_0$ and p_j are small quantities, so that we can linearize the theory.

3) the "classical bunching" b_c is defined as

$$b_c \equiv \frac{1}{N}\sum_j e^{-i<\theta_j>_0} = 0 \tag{5}$$

In this way we have obtained the following expression for the Glauber P function [13]

$$P(\alpha) = \frac{1}{\pi <n_s>}e^{\frac{-|\alpha - f_3\alpha_0|^2}{<n_s>}} \tag{6}$$

where α_0 represents a seed field.

Hence, the average photon number is given by a coherent contribution proportional to $|\alpha|^2$ plus the spontaneous photon number $<n_s>$, for $\overline{z} >> 1$, is given by

$$\frac{<n_s>}{N\overline{\rho}} \approx \frac{1}{9}e^{\sqrt{3}\overline{z}}\left[\frac{1}{N}\left(\sigma_\theta^2 + \frac{\sigma_p^2}{\overline{\rho}^2} + \frac{1}{2\overline{\rho}}\right)\right] \tag{7}$$

20

where $\bar{z} = z/L_g$, $L_g = \dfrac{\lambda_w}{4\pi\rho}$ is the gain length, λ_w is the wiggler wavelength and

$\bar{\rho} = \dfrac{\gamma_0 mc^2}{\hbar\omega}\rho$. In ref. [14] $\bar{\rho}$ is indicated as c. The term into the square brackets represents the "bunching" induced by the quantum noise fluctuations for phase and momentum. By the definition of ρ, as the energy efficiency at saturation [2], it is easy to see that $\bar{\rho}$ is the photon number per electron at saturation. Furthermore, $\bar{\rho} = \Delta E/\hbar\omega$ is the ratio between the classical energy spread $\Delta E = \rho E$ at saturation and the electron recoil energy $\hbar\omega$. Hence the condition $\bar{\rho} \gg 1$, which is necessary to neglect the last term of Eq.(7), coincides with the condition that the electron recoil energy is smaller than the classical energy spread [16]. This condition is very easily satisfied and it allows, eventually, to say that the last term of Eq.(7) is small.

However this does not allow to say that quantum fluctuations are negligible, as argued in ref. [16], unless the first two terms of Eq(7) are also negligible. These terms represent the contribution to bunching due to the quantum width of the single electron wavepacket σ_θ and σ_p related by the uncertainty relation $\sigma_p\sigma_\theta = 1/2$. The term $\sigma_p/\bar{\rho}$, keeping in mind that $p = \gamma mc/\hbar k$, represents the ratio between the quantum energy spread and the classical spread at saturation $\Delta\gamma = \rho\gamma$. Clearly, this term must also be negligible, to neglect quantum fluctuations. As we shall see later, this term can become very relevant. If $\alpha_0 = 0$, the initial state of the field is the vacuum state, and Eq. (6) reduces to the exponential distribution

$$P(I) = \frac{1}{\bar{I}} e^{-\frac{I}{\bar{I}}} \tag{8}$$

where $I = |\alpha|^2$ corresponds to the field intensity and $\bar{I} = <n_s>$. Using Eq. (4) and (6) one obtains the Bose-Einstein photon distribution

$$p(n) = \left(\frac{\langle n_s \rangle}{1 + \langle n_s \rangle}\right)^n \frac{1}{1 + \langle n_s \rangle} \tag{9}$$

Note that the most probable value of the intensity I and the photon number n is zero, while the photon expectation value is $<n_s>$. This photon statistics has been experimentally observed for a single mode atomic laser during its transient approach to the steady state regime [18].

Up to now we have assumed a "one mode theory" or, better to say, there is one coherence length in the bunch length L_b. More precisely, neglecting transverse effects, we have assumed that $m = L_b/2\pi L_c$ is of the order of unity. A rigorous many mode theory with $m > 1$ would be very complicated. Here we just extend the previous treatment using a simple intuitive argument, i.e., rephrasing in quantum mechanical terms the classical theory of ref. [12]. Let us assume that there are no coherent contributions to Eq.(18), i.e., $\alpha_0 = \bar{b}_c = 0$, so that the P function depends only on I. Let us now define the characteristic function

$$f(\xi) = \int_0^\infty dI e^{iI\xi} P(I) \tag{10}$$

After performing the integral one obtains

$$f(\xi) = \frac{1}{1 - i\xi\bar{I}}. \tag{11}$$

Let us now assume that we have m independent longitudinal "modes". The intensity for each mode is $\bar{I}_m = \dfrac{\bar{I}}{m}$ and the characteristic function is the product of single mode characteristic functions. In this way one obtains

$$f_m(\xi) = \frac{1}{\left(1 - i\xi\dfrac{\bar{I}}{m}\right)^m} \tag{12}$$

Note that for m going to infinity $f_m(\xi) = e^{i\xi\bar{I}}$, so that, after inverting Eq. (10) $P(I) = \delta(I - \bar{I})$. Hence, in this limit, using Eq.(4), one has the Poisson distribution

$$p(n) = \frac{\bar{I}^n}{n!} e^{-\bar{I}}$$

i.e., the field is a coherent field in the Glauber sense. On the contrary, for m ≈ 1 one has again the exponential Bose-Einstein distribution of Eq.(8) and (9). In general, performing the inverse Fourier transform of Eq.(12) one obtains the gamma distribution function

$$P(I) = \left(\frac{m}{\bar{I}}\right)^m \frac{I^{m-1} e^{-mI/\bar{I}}}{(m-1)!} \tag{13}$$

where for simplicity we have assumed m to be an integer. This corresponds to the classical intensity distribution for polarized thermal light of ref. [16]. If the light is umpolarized *m* must be replaced by 2m [16]. Inserting expression (13) into Eq.(4) one obtains the negative binomial distribution for the photon number, which is identical to the classical distribution of ref. [11,12]:

$$P(n) = \frac{(n+m-1)!}{n!(m-1)!}\left(1+\frac{m}{\bar{I}}\right)^{-n}\left(1+\frac{\bar{I}}{m}\right)^{-m}. \tag{14}$$

However, here we have given a first principle quantum mechanical derivation of the photon-statistics. We stress that P(I) of Eq.(13) gives the Glauber $P(\alpha)$ function replacing I with $|\alpha|^2$. This function acts as a *probability function* only for *normally ordered* field creation and annihilation operators a and a^+ [3]. For example, if one wants to calculate the variance of the intensity, the normal ordering requirement [2] gives

$$\sigma^2(I) = \frac{\bar{I}^2}{m} + \bar{I} \tag{15}$$

The first term is the one which originates relative fluctuations going as $1/\sqrt{m}$ described in ref. [7].

One can argue that this term has complete classical origin so that if the number of photons per mode n/m is much larger than unity quantum fluctuations are negligible [16]. This is misleading for two reasons. First, the first term of Eq.(12) can be reproduced semi classically but its rigorous origin is quantum mechanical. As a matter of fact it is responsible of the bunching effect of thermal or chaotic light, which originated long time ago the so called Hanbury-Brown Twiss effect [13]. This effect is at the origin of the Glauber quantum theory of coherence. Second, it is not true that the field having a large photon number per mode is a classical field; for example, a field in a eigenstate of the photon number (n-th state) is not a classical field, since, for quantum reasons has a completely random phase and its Glauber P function is highly singular and negative (it contains derivative up to the order n of the Dirac delta function) [13]. Another example are the well known "squeezed states" which can have an arbitrary number of photons per mode but are completely non classical.

The photon statistics we have described above ignores all classical contributions, due to classical bunching or to a seed signal. This is the definition of Quantum SASE. In fact, the average photon number given by Eq.(7) arises from the exponential amplification of position and momentum quantum fluctuation related by the uncertainty principle.

Note that the assumption implicit in Eq.(5), which implies the absence of coherent or shot noise classical bunching, is *in principle*, possible. In fact, we are just assuming that *the centers* of the quantum wavepackets, which corresponds to the "classical" electron position, are initially equally spaced as in a "crystal beam" with arbitrarily small fluctuations. This can be assumed even, if not practical today. However, some

---

[2] $\langle a^+aa^+a\rangle = \langle a^+a^+aa\rangle + \langle a^+a\rangle = \langle|\alpha|^4\rangle + \langle|\alpha|^2\rangle$

years ago, just to think to crystal of atoms cooled at nanokelvin effective temperatures was absolutely impractical. Today it is a common laser cooling technique.

Quantum SASE would have a *fundamental* meaning, since it would represent one of the few phenomena in which *microscopic* quantum fluctuations would be amplified at a *macroscopic* level via a *collective instability*.

CLASSICAL CONTRIBUTIONS

The classical contributions can be taken into account easily since assumption (4) is not necessary. In fact, it is straightforward to generalize the theory to take into account the contribution of a classical bunching, as defined in Eq.(5), without taking b_c=0 [19]. The detailed calculations will be published elsewhere [20]. Let me just quote the simple results. The expression of n_s as given by Eq.(7) must be substituted by <n> given by

$$\frac{<n>}{N\bar{\rho}} = 9e^{\sqrt{3}z}\left[\frac{1}{N}\left(\frac{|\alpha_0|^2}{\bar{\rho}} + N|b_c|^2 + 1 + \sigma_\theta^2 + \frac{\sigma_p^2}{\bar{\rho}^2} + \frac{1}{2\bar{\rho}}\right)\right] \qquad (16)$$

where the term in square brackets represents the electron bunching factor including classical and quantum contributions. The $|\alpha_0|^2$ term represents the contribution of a seed signal, b_c represents the so called coherent classical bunching coming, for example, from a prebunching of the beam or simply from the Fourier transform of the current shape. The third term, as 1/N, is the so called shot noise contribution, which is there if and only if the *center* of the wavepacket are themselves *random variables* whose value is undetermined in the wavelength. The last three terms are the origin of the *Quantum SASE*, whereas the shot noise term is the origin of *classical SASE*.

When Quantum SASE is dominant on classical SASE? First of all a seed signal can be taken to be zero as well as the classical bunching contribution, including the shot noise, as discussed before for a "crystal beam". However, in any case, for a Gaussian bunch much longer than the wavelength the Fourier component of the current is zero. Regarding the shot noise contribution one sees very easily that taking as σ_z the De Broglie wavelength, keeping in mind that p=γmc/$\hbar$k, and using the uncertainty relation, one obtains that the momentum fluctuation contribution in Eq.(16) is dominant on the shot noise contribution if

$$\gamma\rho<1. \qquad (17)$$

This condition can be easily satisfied in a FEL design.

CONCLUSIONS

In conclusion, we have given a rigorous quantum theory of SASE, in the exponential regime, including classical and quantum noise as well as many modes effects. The photon statistics goes from the Bose-Einstein distribution to a Poisson distribution as the number of cooperation length in the bunch becomes very large. In this paper we have for the first time defined the concept of quantum SASE, i.e., the one originated by purely quantum noise due to the single electron position-momentum uncertainty fluctuations. This is the only noise which cannot be eliminated, since, in principle, for a crystal beam shot noise does not exist. However even taking into account classical shot noise, we have found that quantum SASE becomes dominant on the classical process as the quantity $\gamma\rho$ becomes smaller than one.

ACKOWLEDGEMENTS

We are indebted to Dr. W. A. Barletta and Dr. W. M. Fawley for helpful discussions. We would like to thank Dr. N. Piovella for suggestions on the Appendix.

APPENDIX

The linear equations for collective variables [2] can be written

$$\frac{\partial^2 b}{\partial \bar{z}^2} = iA$$

$$\frac{\partial A}{\partial \bar{z}} + \frac{\partial A}{\partial z_1} = b$$

where b is the bunching factor, A is the adimensional amplitude, $\bar{z} = z/L_g$, $z_1 = \dfrac{z - v_e t}{(1-\beta)L_g}$ is the normalized retarded time ($\beta = v_e/c$). Defining a Laplace-Fourier transform in $\bar{z}$ and z_1 $f(k,\omega) = \int\limits_{-\infty}^{\infty} d\omega \int\limits_{0}^{\infty} dk e^{i(\omega z_1 - kz)} f(z, z_1)$ one obtains easily [7]

$$A(k,\omega) = b_0(\omega)G(k,\omega) \tag{1A}$$

where $b_0(\omega)$ is the transform of the $b_0(z_1) \equiv b(\bar{z}=0, z_1)$. Also the other initial values at $z=0$ have been taken to be zero. The Green function $G(k,\omega)$ is given by

$$G(k,\omega) = -\frac{k}{k^3 - \omega k^2 + 1} \tag{2A}$$

Hence,

$$A(z, z_1) = \int_{-\infty}^{\infty} dz'_1 \, b_0(z_1 - z'_1) G(\bar{z}, z'_1) \tag{3A}$$

where $G(z, z_1)$ is the inverse Laplace transform of $G(k, \omega)$ and it represents the response function to the δ-function $b_0(z_1)$. In case of a linear superposition of quasi δ-functions occurring at random times, the field A will be a linear superposition of G-functions centered at random times.

$G(z, z_1)$ is given by the inverse Laplace-Fourier transforms $G(k, \omega)$, i.e.,

$$G(z, z_1) = \frac{1}{2\pi} \int d\omega e^{-i\omega z_1} \frac{1}{2\pi} \int dk e^{ikz} G(k, \omega) \tag{4A}$$

where the path integral must be properly chosen to include all singularities. One can see that $G(z, z_1) = 0$ for $z_1 > z$.

There are two ways of going on:

i) the Gaussian approximation [6, 16]. Calculate $G(z, \omega)$ as:

$$G(z, \omega) = \frac{1}{2\pi} \int dk e^{ikz} G(k, \omega) = \sum_{j=1}^{3} c_j e^{ik_j(\omega)z} \tag{5A}$$

where $k_j(\omega)$ are the simple poles of $G(k, \omega)$ i.e., the roots of the "classical cubic" [2] $k^3 - \omega k^2 + 1 = 0$. Taking the divergent root $k_1(\omega)$, in the approximation $k_1(\omega) \approx \frac{\sqrt{3}}{2}\left(1 - \frac{\omega^2}{9}\right)$, one obtains a Gaussian in ω for $G(z, \omega)$ which goes as $1/\sqrt{z}$. This is the procedure of ref. [6, 16] Hence $G(z, t)$ will also be a Gaussian function on $(t - z/v_s)$ [16] centered on random times with some characteristics velocity v_s (see text). However, as shown below, this turns out not to be the correct behavior of $G(z, t)$.

ii) Exact calculation [9]. In Eq.(4A) let us perform first the integral in ω. Observing that $G(k, \omega)$ has a single pole at $\omega = (k + 1/k^2)$, one finds easily that $G(z_1, k) = \frac{1}{2\pi}\int_{-\infty}^{\infty} d\omega e^{-i\omega z_1} G(k, \omega) = i \frac{e^{-iz_1(k+1/k^2)}}{k}$. Inserting into Eq.(4A) one obtains, with some algebra,

$$G(z,z_1) = \frac{i}{2\pi} \int \frac{dk}{k} e^{ik(\bar{z}-z_1)-iz_1/k^2} = \frac{i}{2\pi} \int \frac{dk'}{k'} e^{i\left(k'-\frac{1}{k'^2}\right)y^{2/3}} = G(y). \qquad (6A)$$

Note that in Eq.(6A) has a **essential singularity** at k'=0. Here $z_2 = z - z_1 = \frac{ct-z}{L_c}$, where

$L_c = L_g \frac{c-v_e}{v_e}$ is the so called cooperation or coherence length, $k' = k\left(\frac{z_2}{z_1}\right)^{1/3}$ and the "self-similar" variable y is defined as

$$y = \sqrt{z_1 z_2} = \frac{(z-v_e t)^{1/2}(ct-z)}{L_c^{3/2}} \qquad (7A)$$

Evaluating the integral (6A) with the method of stationary phase, one finds [9] $G(y) = a\exp(ay^{2/3})$ where $a = 3/2(\sqrt{3}+i)$. A non linear theory shows that G(y), as discussed before, represents the linear exponential tale of an hyperbolic secant [9,10]. Hence the correct solution is quite different from the Gaussian approximation described before and used in ref. [6, 16].

REFERENCES

1. Bonifacio R., Casagrande F., Casati G., *Opt. Commun.*, **40**, 219 (1982).
2. Bonifacio R., Pellegrini C., Narducci L. M., *Opt. Commun.*, **50**, 373 (1984).
3. Bonifacio R., Mc Neil B. W. J., Pierini P., *Phys. Rev.* **A40**, 4467 (1989).
4. Proceedings of the FEL Conference, Castelgandolfo, Italy (1984).
5. Hogan M., Anderson S., Bishofberger K., Frigola P., Murokh A., Osmanov N., Pellegrini C., Reiche S., Rosenzweig J., Travish G., Tremaine A., Varfolomeev A., this proceedings.
6. Kim K. J., *Phys. Rev. Lett.*, **57**, 1871 (1986).
7. Bonifacio R., De Salvo L., Pierini P., Piovella P. and Pellegrini P., *Phys. Rev. Lett.*, **73**, 70 (1994).
8. Fawley W. M., SPIE Proc. 2988, FEL Challenges (1997).
9. Bonifacio R., Maroli C., Piovella N., *Opt. Commun.*, **68**, 369 (1983) and Bonifacio R., Casagrande F., Cerchione G., De Salvo Souza L., Pierini P., and Piovella N., *Riv. Nuovo Cimento*, vol. **13**, n.9 (1990).
10. Piovella N., Opt. Commun., **83**, 92 (1988).
11. Saldin E. L., Schneidmiller E. A., and Yurkov M. V., DESY Print TESLA-FEL 96-07, Hamburg, DESY, (1996), and in this proceedings.
12. Goodman J. W., in *Statistical Optics*, Ed. John Wiley & Sons (1984).
13. Glauber R. J., *Phys. Rev.* **131**, 2766 (1963).
14. Bonifacio R., Casagrande F., *Opt. Commun.* **50**, 251 (1984), and Bonifacio R., Casagrande F., *Nucl. Instr. Meth.* **A 237**, 168 (1985)
15. Becker W. and Mc Iver J. K., *Phys. Rev.* **A27**, 1030 (1983)
16. Kim K. J, this proceedings.
17. Bonifacio R., Narducci L. M. and Montaldi E., *Phys. Rev. Lett.* **16**, 1125 (1966).
18. Arecchi F. T. and Degiorgio V., in *Laser Handbook*, edited by F. T. Arecchi and E. S. Schulz - Du Bois, North-Holland, Amsterdam, 191 (1972).

19. We are indebted to K. J. Kim for this observation.
20. Bonifacio R., to appear in *Nucl. Instr. Meth.* A.

Numerical Study of X-Ray FELS Including Quantum Fluctuation

S. Reiche[1], E.L. Saldin[2], E.A. Schneidmiller[2], M.V. Yurkov[3]

[1] Deutsches Elektronen Synchrotron, 22607 Hamburg, Germany
[2] Automatic Systems Corporation, 443050 Samara, Russia
[3] Joint Institute for Nuclear Research, Dubna, 141980 Moscow Region, Russia

Abstract. One of the fundamental limitations towards achieving very short wavelength in a self amplified spontaneous emission free electron laser (SASE FEL) is connected with the energy diffusion in the electron beam due to quantum fluctuations of undulator radiation. Parameters of the LCLS and TESLA X-ray FEL projects are very close to this limit and there exists necessity in upgrading FEL simulation codes for optimization of SASE FEL for operation at a shortest possible wavelength. In this report we describe a one-dimensional FEL simulation code taking into account the effects of incoherent undulator radiation. Using similarity techniques we have calculated universal functions describing degradation of the FEL process due to quantum fluctuations of undulator radiation.

I INTRODUCTION

It has been realized more than ten years ago that single-pass free electron laser (FEL) can provide the possibility to generate powerful, coherent VUV and X-ray radiation [1–3]. Several projects of such FEL amplifiers are under development at present [4–6]. The FEL process is critical with respect to the quality of the driving electron beam and the minimal achievable wavelength λ_{cr} is given by the expression [7]:

$$\lambda_{\mathrm{cr}} \simeq 18\pi\epsilon\frac{\sigma_{\mathrm{E}}}{\mathcal{E}_0}\left[\frac{\gamma I_A}{I}\frac{1+K^2}{K^2}\right]^{1/2} \quad , \tag{1}$$

where ϵ is emittance of the electron beam, σ_{E} is the energy spread in the electron beam, $\mathcal{E}_0$ is the energy of the electrons, I is the beam current, $K = eH_{\mathrm{w}}\lambda_{\mathrm{w}}/2\pi m_e c^2$ is the undulator parameter, H_{w} is the undulator field, λ_{w} is the undulator period, $(-e)$ and m_e are the charge and the mass of the electron, respectively, c is the velocity of light, γ the Lorentz factor, and

CP413, *Towards X-Ray Free Electron Lasers*
edited by R. Bonifacio and W. A. Barletta
© 1997 The American Institute of Physics 1-56396-744-8/97/$10.00

$I_A = mc^3/e \simeq 17$ kA is Alfven's current. When designing an FEL amplifier operating at the wavelength around 1 Å one should also take into account the effect of energy diffusion in the electron beam due to quantum fluctuations of undulator radiation. Recent study of performance limitations of an X-ray free electron laser has shown that this effect leads to the growth of the energy spread in the electron beam, σ_E (see eq. (1)), when the electron beam passes the undulator. This effect imposes fundamental limit towards achieving very short wavelength given by the following estimation [7] (for the case of zero energy spread at the undulator entrance):

$$\lambda_{\min} \simeq 45\pi \left[\lambda_c r_e\right]^{1/5} L_w^{-7/15} \left[\epsilon_n^2 \frac{I_A}{I}\right]^{8/15} , \tag{2}$$

or, in practical units

$$\lambda_{\min}[\overset{\circ}{A}] \simeq 4 \frac{\pi \epsilon_n[\text{mm mrad}]}{\sqrt{I[\text{kA}]L_w[\text{m}]}} ,$$

where $\lambda_c = \hbar/mc$, $\hbar$ is Planck constant, $r_e = e^2/m_e c^2$, L_w is the length of the undulator and $\epsilon_n = \gamma\epsilon$ is normalized emittance.

All the existent FEL simulation codes do not take into account the effect of the energy diffusion in the electron beam due to quantum fluctuations of undulator radiation. On the other hand, design parameters of existent projects of X-ray FELs (LCLS at SLAC [5] and X-ray FEL at linear collider TESLA [6]) are very close to this limit and there exists an urgent necessity in more rigorous simulations of their parameters.

In this report we describe a one-dimensional FEL simulation code taking into account the effects of incoherent undulator radiation. Using similarity techniques we have calculated universal functions describing degradation of the FEL process due to quantum fluctuations of undulator radiation.

II   NUMERICAL SIMULATION ALGORITHM

In this section we present brief description of the one-dimensional simulation code upgraded with the equations taking into account the effects of incoherent undulator radiation. The self-consistent FEL equations are identical to those described in paper [8] (section 3). To describe the influence of incoherent undulator radiation on the FEL process we have included two physical effects into the FEL code. The first one is additional energy loss which is given by well known classical expression:

$$d < \mathcal{E} > /dz = -2r_e^2 \gamma^2 H_w^2(z)/3 . \tag{3}$$

Another effect is energy diffusion in the electron beam due to quantum fluctuations of the undulator radiation. The rate of energy diffusion is given by the expression:

$$\frac{d<(\delta\mathcal{E})^2>}{dt} = \int d\omega\hbar\omega\frac{dI}{d\omega} \ , \tag{4}$$

where $dI/d\omega$ is the spectral intensity of an undulator radiation. Explicit expression for the rate of the energy diffusion has the following form [9]:

$$\frac{d<(\delta\gamma)^2>}{dt} = \frac{14}{15}c\lambda_c r_e \gamma^4 \kappa_w^3 K^2 F(K) \ , \tag{5}$$

where $F(K)$ is given by the following fitting formulae:

$$F(K) = 1.42K + \frac{1}{1+1.50K+0.95K^2} \qquad \text{for helical undulator} \ , \tag{6a}$$

$$F(K) = 0.60K + \frac{1}{2+2.66K+0.80K^2} \qquad \text{for planar undulator} \ . \tag{6b}$$

The simulation algorithm is organized as follows. Equations of motion of macroparticles and the field equations (see ref. [8], section 3) are integrated by means of Runge-Kutta scheme. An additional loss of the electron energy is calculated with eq. (3) and additional energy spread is introduced by means of random generator after each integration step in accordance with eq. (5). For the latter procedure to be physically correct, one should care about suppression of numerical noise connected with finite number of macroparticles. In other words, all the moments of the distribution function $f(\Psi, P, z)$ (here $P = \mathcal{E} - \mathcal{E}_0$):

$$a_k = \int P^k \exp(-i\Psi)f(\Psi, P, z)dPd\Psi \ , \tag{7}$$

must have the same values before and after performing the procedure of introducing an additional energy spread. Otherwise, an additional (unphysical) bunching due to numerical noise will appear which will produce an error in the results of calculations. The necessity of compensation of the numerical noise can be explained in the following way. Suppose one has a problem to prepare an initial ensemble of the particles corresponding to unmodulated electron beam:

$$a_0(0) = \int \exp(-i\Psi)f(\Psi, P, 0)dPd\Psi = 0 \ , \tag{8}$$

and some arbitrary distribution in the energy. Let us consider the evolution of such a distribution function $f(\Psi, P, z)$ in a drift space (no FEL process):

$$\frac{df(\Psi, P, z)}{dz} - \alpha P f(\Psi, P, z) = 0 \ . \tag{9}$$

31

It follows from this equation that distribution functions at coordinates z and $z + \Delta z$ are connected by the relation:

$$f(\Psi, P, z + \Delta z) = f(\Psi, P, z) \exp(\alpha P \Delta z) \ . \tag{10}$$

Using eqs. (10) and (8) the evolution of the bunching factor a_0 is as follows:

$$a_0(z) = a_0(0) + \sum_{k=1}^{\infty} a_k \frac{\alpha^k}{k!} z^k \ . \tag{11}$$

It can been seen in eq. (11) that although starting with bunching factor $a_0(0) = 0$, bunching can occur if higher moments exist non equal to zero. In the presence of the FEL process this can introduce an error in the results of the calculation.

For the correction of the moments the phase space is divided up into N stripes with a limited range of the momentum ($P_{min} < P < P_{max}$), where the correction scheme is applied for each stripe. A larger number of stripes improves the results because a correlation between P and $\exp(-i\Psi)$ is reduced. In general all higher moments are suppressed in the limit of $f(\Psi, P) = f_1(\Psi) \cdot f_2(P)$ for $N \to \infty$. Numerical simulations show that for practical calculations it is sufficient to compensate only the first two moments of the distribution function, a_0 and a_1. In this case the accuracy of calculations is better than 1 %.

III SIMULATION RESULTS

We consider simplified situation of a "cold" electron beam at the undulator entrance and neglect the influence of the space charge field. It is assumed that the FEL amplifier is tuned to the resonance frequency. Under these approximations operation of the conventional FEL amplifier is described in terms of the gain parameter Γ and the efficiency parameter ρ (see, e.g. refs. [2,8,10]):

$$\Gamma = \left[\frac{2\pi^2 j_0 K^2}{I_A \lambda_w \gamma^3} \right]^{1/3} \ , \qquad \rho = \frac{\lambda_w \Gamma}{4\pi}, \tag{12}$$

where j_0 is the beam current density (for the case of a helical undulator).

We assume that the mean energy loss (see eq. (3)) are compensated by an appropriate undulator tapering and study pure effect of the energy diffusion in the electron beam due to quantum fluctuations of undulator radiation (see eq. (5)). When simulating SASE FEL with steady-state simulation code, one should set the value of the "effective" power of input shot noise which is given approximately by the relation [11,12]:

$$W_{sh} \simeq \frac{3\sqrt{4\pi}\rho^2 P_b}{N_\lambda \sqrt{\ln(N_\lambda/\rho)}} \ , \tag{13}$$

where $P_b = \gamma m_e c^2 I/e$ is the power of the electron beam and $N_\lambda = I\lambda/(ec)$. For the case of the X-ray SASE FEL at TESLA [6] the value of the reduced input power is of about $\hat{W}_{sh} = W_{sh}/\rho P_b \simeq 10^{-8}$. This value has been used in the simulations.

Under accepted approximations the value of the reduced power at saturation, $\hat{\eta} = P_{out}/\rho P_b$, and the saturation length, $\hat{L}_{sat} = L_{sat}\Gamma$ are universal functions of the parameter of quantum fluctuations $\hat{q}$ (see eq. (5)):

$$\hat{q} = \frac{7}{15}\frac{\lambda_c r_e}{\rho^3}\gamma^2 k_w^2 K^2 F(K) . \tag{14}$$

Fig. 1 presents the plots of these universal functions. It is seen that operation of the FEL amplifier degrades significantly when the value of the parameter of quantum fluctuations is increased.

Operating point of 1 Å FEL at TESLA (50 GeV energy of the electron beam [6]) corresponds to the value of the parameter of quantum fluctuations $\hat{q} = 9 \times 10^{-3}$. Fig. 2 illustrate the degradation of the FEL performance of the 1 ÅFEL at TESLA. It is seen that parameters of the project have been chosen correctly and there is only a slight degradation of the FEL performance due to quantum fluctuations of undulator radiation.

REFERENCES

1. Derbenev, Y.S., Kondratenko, A.M., and Saldin, E.L., *Nucl. Instrum. and Methods* **193** 415 (1982).
2. Bonifacio, R., Pellegrini, C., and Narducci, L., *Opt. Commun.* **50** 373 (1984).
3. Murphy, J.B., and Pellegrini, C., *Nucl. Instrum. and Methods* **A237** 159 (1985).
4. "A VUV Free Electron Laser at the TESLA Test Facility: Conceptual Design Report", *DESY Print TESLA-FEL 95-03*, Hamburg, DESY, (1995).
5. Tatchyn, R., et al., *Nucl. Instrum. and Methods* **A375** 274 (1996).
6. Linear Collider Conceptual Design Report, *DESY print 97-048*, Hamburg, DESY, (1997).
7. Rossbach, J., Saldin, E.L., Schneidmiller, E.A., and Yurkov, M.V., *Nucl. Instrum. and Methods* **A374** 401 (1996).
8. Saldin, E.L., Schneidmiller, E.A., and Yurkov, M.V., *Phys. Rep.* **260** 187 (1995).
9. Saldin, E.L., Schneidmiller, E.A., and Yurkov, M.V. *Nucl. Instrum. and Methods* **A381** 545 (1996).
10. Bonifacio, R., et al., *Riv. Nuovo Cimento*, Vol.13, No.9 (1990).
11. Kim, K.J., *Phys. Rev. Lett.* **57** 1871 (1986).
12. Saldin, E.L., Schneidmiller, E.A., and Yurkov, M.V., *DESY Print TESLA-FEL 96-07*, Hamburg, DESY, (1996).

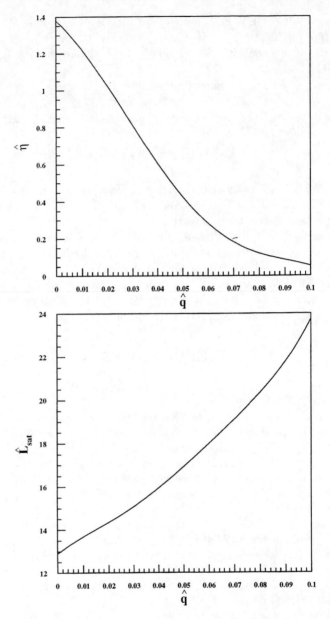

FIGURE 1. Dependency of the saturation efficiency $\hat{\eta} = P_{\text{out}}/\rho P_{\text{b}}$ and saturation length $\hat{L}_{\text{sat}} = L_{\text{sat}}\Gamma$ on the value of the parameter of quantum fluctuations $\hat{q}$.

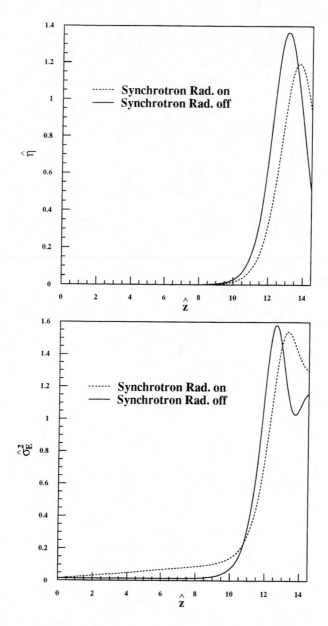

FIGURE 2. Reduced efficiency $\hat{\eta} = P_{\text{out}}/\rho P_{\text{b}}$ and induced energy spread $\hat{\sigma}_E^2 = \langle \mathcal{E}^2 - \langle \mathcal{E} \rangle^2 \rangle / \rho^2 \langle \mathcal{E} \rangle^2$ for different position $\hat{z}$ in the TESLA 50 GeV FEL. Results including quantum fluctuations are drawn by a dashed line ($\hat{q} = 9 \times 10^{-3}$), without – by a solid line.

Where Do We Stand with High Gain FEL Simulations?

Gil Travish

Istituto Nazionale di Fisica Nucleare (INFN), Sezione di Milano. Via Cervi, 201. 20090 Segrate (MI), Italy.

Abstract. Computer technology improvements have allowed for more complete and detailed free electron laser simulations, yet the demands of the large number of new experiments and proposed projects has outpaced the capability and availability of present codes. This paper, based on a talk given at the conference of these proceedings, presents a brief assessment of Free Electron Laser (FEL) codes, their availability and features, as well as some opinions on what direction the FEL code community should take for the near future. The discussion of FEL codes is restricted here to ones for high gain amplifiers: no codes for oscillators, waveguides or exotic configurations are considered.

INTRODUCTION

Free Electron Lasers (FELs), as with many other fields, have benefited from the availability of computers and computer codes: the ability to study the collective behavior (many particle effects) in an FEL have allowed theorists to confirm existing analytic work and discover new effects (1); the ability to predict FEL performance has allowed experimentalists to design and propose challenging machines. While a number of general characteristics of FELs (such as gain before saturation, the effect of smooth focusing, and the effect of energy spread) are well modeled by simple analytic formulas, a number of issues (such as saturation, undulator errors, and slippage) can only be addressed carefully by simulations.

There are three areas where high gain experimental FEL research is most active, and where simulations are, perhaps, most needed: physics test systems, short wavelength facilities, and industrial applications. Physics test systems are usually small scale efforts with rapid and numerous reconfigurations. As such, test systems require frequent use of simulations and comparison with theory. Short wavelength facilities, on the other hand, are complex, large scale systems which require numerous simulations to verify the designs and lend credence to the extrapolation to short wavelengths. Industrial applications generally require integrating detailed engineering considerations such as heat load, reliability, and cost along with physics modeling.

In the following sections we discuss some basics of an FEL code, present some opinions on the impact of computer technology on codes, review some past directions the community has taken, and consider some new directions including the needs of the above mentioned three areas. Finally, a challenge to the community is issued to develop and support a new code available to all, and benchmarked against a wide range of cases.

CP413, *Towards X-Ray Free Electron Lasers*
edited by R. Bonifacio and W. A. Barletta

AN FEL AS SEEN BY A CODE

An FEL simulation is one link in a chain of tools available to the FEL researcher: the chain begins with quantum mechanics, next follows the Maxwell and Lorentz equations, which lead to the FEL (pendulum) equations, averaging over a wiggle period and applying eikonal approximations leads to a set of simulation equations, and the code provides a means of solving this model. Continuing with the analogy, the experiment or user facility may be viewed as the final link of the chain.

An FEL code can be understood in terms of a simplified box diagram (see Figure 1). The **Beam Input** includes user input detailing the beam's initial six-dimensional phase space (i.e., energy spread, particle distribution model, prebuncher, etc.). The **Radiation Input** contains parameters such as the power, wavelength, and beam size of the initial radiation, and information on how the startup should be modeled (i.e., whether it is spontaneous or from a coherent source). The **Beam Radiation Interaction** is actually the FEL equation solver. Various flavors of the solver exist including 1D, 2D, 3D, and time dependent models. The **Optics** section allows for feedback or manipulation of the output radiation to better model experimental realities (such as a long drift of the radiation to a far away detector). Finally, the **Output** section provides numerical versions of radiation and beam diagnostics as well as graphics support.

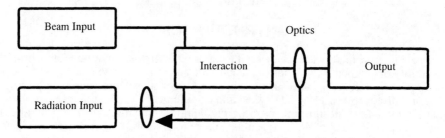

FIGURE 1. A box diagram for a generic FEL code.

As an illustration to the reader, we review the single frequency amplifier equations, highlighting some of the salient parameters. The energy equation can be expressed as

$$\frac{d\gamma_n}{dz} = -\frac{\omega_r}{2c}\frac{a_r a_w}{\gamma_n}\sin\theta_n + \text{Space Charge}, \tag{1}$$

where ω_r and a_r are the radiation wavelength and normalized field, a_w is the wiggler normalized field, c is the speed of light, z is the distance along the undulator, and γ_n and θ_n are the particle (n) energy and phase. As indicated, space charge forces can be added to the above equation. The phase equation is given by

$$\frac{d\theta_n}{dz} = k_w - \omega_r \frac{1 + \mathbf{p}_\perp^2 + a_w^2 + 2a_r a_w \cos\theta_n}{2c\gamma_n^2}, \tag{2}$$

where k_w is wiggler wavenumber and $\mathbf{p}_\perp$ is the perpendicular particle momentum given by

$$\frac{d\mathbf{p}_\perp}{dz} = -\frac{1}{\gamma_n}\frac{\partial a_w^2}{\partial \mathbf{r}_\perp} + \text{Focusing}. \tag{3}$$

Above, the external focusing terms can be added as needed and $\mathbf{r}_\perp$, the perpendicular particle position, is given by

$$\frac{d\mathbf{r}_\perp}{dz} = -\frac{\mathbf{p}_\perp}{\gamma_n}. \tag{4}$$

Finally, the evolution of the radiation is expressed as a wave equation:

$$\left[2ik_r\frac{\partial}{\partial z} + \nabla_\perp^2\right]a_r^2 \propto I\left\langle\frac{a_w e^{-i\theta_n}}{\gamma_n}\right\rangle, \tag{5}$$

where k_r is the radiation wavenumber, I is the beam current and the term in the angled brackets represents the average "bunching" of the particles.

To the above equations, one must add the initial particle phase space, the initial radiation profile and any relevant boundary conditions (such as those imposed by waveguides). Additionally, the model may include choices of undulator type (i.e., planar, helical, tapered, etc.), spatial modes of the radiation, higher harmonics of the radiation, time dependent effects, etc..

With a basic model of an FEL code in hand, one can go on to consider the type of computing facilities one should have to model an FEL. Instead, we turn to the corollary issue of the impact of computers on FEL simulations.

OPINIONS AND FACTS ON COMPUTERS

It is known that, at least by certain types of benchmarks, computing power is growing exponentially with time (see Figure 2) (2). Some estimates state that CPU speed doubles every 18 months. Perhaps what is more significant to the majority of the scientific community (those not working on Grand Challenge problems) is the common *availability* of computing power: "home" computers now exceed the computational speed of the previous generation of workstations. Indeed, desktop computers are within an order of magnitude of the fastest (single processor) machines, and researchers are often conducting work on machines sold to the consumer market. Finally, it is important to note that unlike the situation in the past, the physics community now has little influence over the computer market due to the overwhelming size of the business and personal sectors. Thus, it has become desirable (even necessary) to take advantage of commercial applications, software tools and standards in performing scientific computing. The question then arises, what can one do in FEL simulation with all the available computing power?

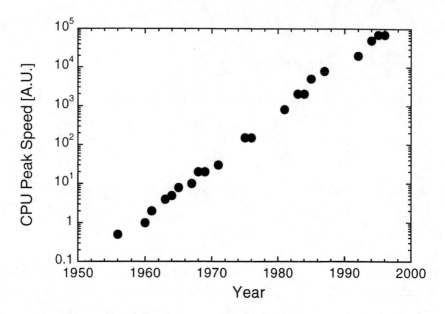

FIGURE 2. The peak available CPU speed (in arbitrary units) is plotted on a log scale against the calendar year. Note the exponential growth.

There are a number of interesting directions, of which we will only list a few here, including fully three (3D) and four dimensional (4D) codes (in both the beam and radiation), unaveraged codes (where the wiggle motion is included), startup modeling, and the inclusion of more details of the beam-radiation interaction including realistic undulator fields, and particle distributions obtained from multiparticle beam dynamics codes. Fully 3D and 4D Codes can yield more exact models of the FEL interaction but remain CPU intensive and require complex codes. "Unaveraged" codes are also CPU intensive, and produce results which are often difficult to interpret, and hard to compare to analytical work. Startup modeling, while urgently needed for experimental comparisons, is extremely CPU intensive, may require complex 4D models and a large number of particles.

It is almost an axiom of computational physics that a given problem will grow to consume all available CPU time. However, while more computing power leads to the ability to solve new problems, it can do so at the cost of neglecting older but still relevant issues. A balance needs to be found between solving new problems slowly, and solving old problems quickly. The use and availability of codes is, obviously, heavily influenced by the way codes are developed. Thus, we next turn to the past and present trends in FEL code production.

SUCCESSES AND FAILURES OF THE PAST

The history of FEL codes follows, to a large extent, the history of FELs themselves. The forces that effected FELs such as the large funding by the United States (US)

Defense programs of the Eighties, and the almost total lack of funding (at least in the US) following the end of the Strategic Defense Initiative (SDI), also effected the production and availability of FEL simulations (3).

In the early days of FEL research, many codes where developed by many authors. Most of these codes where for personal use by one researcher or internal use by a small group. A few "big" codes where developed primarily by the SDI projects. Their well developed, more capable codes where not generally available to the community either because of national security laws, or because there was no incentive to do so. The notable exception to this lack of distribution was, and still is, Tran and Wurtele's TDA code (4). As a consequence of the private nature of the codes, and perhaps for other reasons, most early codes where not well documented or supported (i.e., there was no installed user base). After the cancellation of the SDI FEL programs, many researchers left the field, and the original authors of the codes went on to other research.

SUCCESSES AND FAILURES OF THE PRESENT

Where as individual and small group efforts marked the early days of FEL research, and defense related work dominated the field after that, today the field is marked by a mix of small scale groups doing basic research and large groups considering big user facilities and light sources. The "big" groups are producing their own codes, which is a good thing. However, this code development seems to be for internal use, resembling more the efforts of the SDI groups than those of open scientific efforts. On the other hand, some of the SDI "big" codes are now available, and the authors are providing support on a volunteer basis. Through the efforts of such authors and a few dedicated users, documentation is slowly becoming available. Sadly, there is still no serious high-gain FEL code available on a personal computer, and users must still contend with archaic and cumbersome interfaces.

Despite the aforementioned problems, willing users have a set of codes available with a respectable list of capabilities (5). A summary of the present state of FEL simulations can be given in the form or a laundry list of well modeled parameters: beam current, steering errors, (external) focusing, multiple undulators (sections), saturation, fluctuations, slippage, and harmonics. We must also include a list of not so well modeled parameters: emittance, energy spread, startup, superradiance, and optical beam profile. As we discuss next, in addition to improving our modeling of the previously listed parameters, each area of FEL work has specific demands from future simulations.

NEW APPLICATIONS AND NEW FEATURES

We can first consider some of the modeling characteristics and features needed by the three areas of FEL research, and then we can sample some specific considerations. Perhaps the most demanding of the three areas are the short wavelength facilities, which we examine first. Careful modeling of the beam trajectory, including correctors, beam position monitors, and a realistic undulator field are central to understanding these long devices (6). In addition, external focusing and any attendant errors should be included. A realistic beam phase space, such as might be provided by a multiparticle beam dynamics code (i.e., Parmela), also raises the confidence level in the numerical models of short wavelength devices. Startup modeling, if feasible, will also provide a boost to projects trying to extrapolate from infrared experiments to x-ray proposals. Finally, effects such as harmonics and coherent undulator radiations may have to be included for a realistic and accurate simulation.

The physics test systems may (depending on the configuration) share some of the demands of the short wavelength facilities, but would benefit immediately from simulations which more closely resemble experimental realities. The ability to interface to the accelerator control system would improve the ease and speed of testing a larger parameter space. Related to the control system interface is the need to have the code accept inputs of measured beam parameters, as opposed to theoretical parameters. Finally, a two-dimensional transverse profile of the output radiation (including drifts and simple optics) would greatly aid in comparing experimental results to simulation.

Industrial systems have received the least attention, perhaps because FELs have not matured sufficiently. Nevertheless, some systems are under consideration and their computational needs should be addressed. In particular, industrial systems tend to be high duty factor devices with high average and/or peak powers. Thus, space charge and beam "halo" effects may need to be included in the model.

Having listed some demands that present and near future FELs have placed on simulations, we can also list some specific considerations that future code writers ought to take into account. The language, platform, operating system, structure and user interface play a significant role in the usability and accessibility of a code, and these issues have been unappreciated in the past. Along with the hardware improvements, software and programming techniques have matured since the development of early FEL codes. High level programing techniques including memory management, data structures and modular coded are now standard practice in industry. A code writer today must choose between several approaches, including weighing the benefits of staying with the well known FORTRAN language or changing to the more (industry) standard C series of languages. Proper choice of the language and good programing techniques can also assure portability of the code across a number of platforms and operating systems. Software engineering is no longer merely good programing style.

A new simulation must also offer the FEL community the ability to easily modify the source code. Again, modularity and documentation provide the means for users to readily accomplish code alterations. And, while well structured codes and intuitive user interfaces consume a good deal of time to produce, they provide users with the ability to rapidly learn the code, make fewer mistakes, and more easily make changes to the source code. Interfaces, documentation and code comments should not be viewed as features or "niceties", but rather as core components of a complete simulation.

Finally, a new author needs to consider details of the numerics such as the type of grid (mesh), solver and integration scheme. For example, a radial mesh requires less CPU time and is numerically more stable than a rectangular mesh (which explains why most codes use the radial mesh). On the other hand, a rectangular mesh does not require a modal decomposition, and is easier to compare to experiments.

Future FEL codes will surely be more capable than past efforts, and may include many new physics and engineering models. However, progress in FEL research is dependent on well tested and widely available simulations in addition to the highly specialized, state of the art, private codes.

A CHALLENGE TO THE COMMUNITY

We need a new set of codes. The codes should address the shortcomings of previous efforts. Namely, we need a set of widely distributed, well tested, easy to modify codes which operate on a number of platforms, and integrate/interface well with existing codes. Surely groups proposing multi-million dollar projects should verify their proposals on well tested, and community-wide benchmarked codes.

ACKNOWLEDGMENTS

A survey was sent to a large number of people (as well as posted on the web) prior to presenting this work. The author thanks the following people for responding and hence influencing the content of this paper: J. Wurtele, W. Vernon, D. Nguyen, J. Schmerge, W. Fawley, B. Faatz, S. Reiche, M. Yurkov, and M. Hogan.

REFERENCES

1. Orzechowski, T. J., *et al.* "Free-electron laser results from the advanced test accelerator," in the 1988 Linear Accelerator Conference Proceedings (CEBAF-Report-89-001). 1988. Newport News, VA, USA.
2. Special Issue: 50 years of Computers and Physics, Physics Today (Oct. 1996).
3. Warren, R., *Star Wars and the FEL.* Unpublished: 1995: 92.
4. Tran, T. M. and Wurtele, J. S., *Computer Physics Communications*, 54(2-3): p. 263-72 (1989).
5. Travish, G., "The How, What, Why, Where and When of Free Electron Laser (FEL) Simulation." Invited Talk presented at the 1993 Computers in Accelerator Physics (CAP93) Conference. (Pleasanton, CA: 1993).
6. Travish, G. "Performance simulation and parameter optimization for high gain short wavelength FEL amplifiers," Invited Talk presented at the Sixteenth International Free Electron Laser Conference (FEL94) (Stanford, CA, USA: 1994). And published in *NIM A* 358 48-51 (1995).

High-Gradient Acceleration of Electron Beams in a Plasma-Loaded Wiggler

V. Petrillo and C. Maroli

Department of Physics, University of Milano, INFN and INFM
Milano, Via Celoria 16, 20133-Milano, Italy

Abstract. The presence of a static periodic magnetic field on a plasma-based system for the acceleration of electrons has the effect of enhancing both acceleration gradient and dephasing length.

INTRODUCTION

Plasma-based devices for high gradient acceleration of electron beams are considered with great interest in these years. Strong longitudinal waves excited in nearly cold and spatially uniform plasmas have been found to accelerate electron bunches to high energies in short distances, the extrapolated values of the acceleration rates obtained in actual experiments being of the order of several GeV and in some cases even of several tens of GeV per meters[1] . The present beat-wave and wake-field experiments consider a plasma with no external magnetic fields. We have recently demonstrated [2] that if the plasma is in presence of a stationary magnetic field which is also periodic in space, like the wiggler magnetic field of a free-electron-laser, the acceleration rate and the dephasing length of the process can be increased considerably. The external magnetic field also leads to the possibility of using plasmas with decidedly lower densities and accelerating waves with correspondingly longer wavelengths.

If we may think that the positive ions of the plasma do not move over the time scales that characterise the acceleration process, the longitudinal accelerating part of the electric field of a plasma wave is estimated to be

$$E_{//} \approx 10^2 \, \frac{\omega_p}{\omega} \, \frac{\omega}{ck} \, (\frac{\delta n}{n})_{peak} \sqrt{n_p (cm^{-3})} \qquad (\frac{Volt}{m}) \qquad (1)$$

where n_p is the density of the undisturbed plasma, $\omega_p = (4\pi e^2 n_p/m)^{1/2}$ the plasma frequency of the electrons of the plasma, $\delta n/n_p$ the relative density perturbation

CP413, *Towards X-Ray Free Electron Lasers*
edited by R. Bonifacio and W. A. Barletta
© 1997 The American Institute of Physics 1-56396-744-8/97/$10.00

associated to the partially longitudinal wave, k and $\omega(k)$ the wave-number and frequency of the accelerating wave.

In a non-magnetic plasma the accelerating wave is the purely longitudinal Langmuir wave whose frequency $\omega = \omega_p$. It follows from (1), in this case, that to increase the value of $E_{//}$ one must necessarily use very high-density plasmas and correspondingly short wavelength electrostatic waves ($ck=\omega$). When, instead, the plasma is in the presence of a magnetic field, its dispersive properties change and the plasma itself usually carries (diamagnetic) currents which tend to screen the plasma from the external field. As the external magnetic field is increased the currents become relativistic and the corresponding increase of the electron mass leads to lower values of all plasma eigenfrequencies. If the wave that is now used as the accelerating wave has a frequency ω (much) smaller than ω_p, we gain a factor $\omega_p/\omega \gg 1$ in the strength of the accelerating field $E_{//}$. At the same time, the strong periodic magnetic field forces the electrons of the beam to follow the characteristic "wiggling" motion in a plane transverse to the original direction of propagation and this fact, in turn, leads to much longer dephasing lengths and to higher values of the energy gained by the electron beam at the saturation of the process.

1D TREATMENT

We consider a cold, 1D plasma with immobile ions distributed in space with the (constant) density n_p. The behaviour of the electrons of the plasma is described by the conservation equation of the number density $n(z,t)$ and the relativistic equation of motion for the momentum $\mathbf{p}(z,t)$.

In one dimension the electron beam is conveniently described by the charged sheet model that has been used, for instance, by O'Neil, Winfrey and Malmberg [3] in their treatment of the beam-plasma instability. Finally, the electromagnetic radiation fields are given in terms of a transverse vector potential $\mathbf{A}(z,t)$ and the axial component $E_z(z,t)$ of the electric field that comprises the parallel accelerating component of the wave field and the Coulomb repulsion between the electrons of the beam.

As we said, the plasma in equilibrium with the strong wiggler field carries diamagnetic currents which flow perpendicularly to the z-axis. The waves that propagate inside this system satisfy a cubic dispersion relation which shows clearly that all the frequencies of the plasma waves have been lowered as a result of the relativistic increase of the electron mass. In particular, the waves belonging to the lower branch of the dispersion relation are nearly longitudinal waves. The wave whose phase velocity is very near to the velocity of light c is the wave that must be used in the acceleration process.

46

We do not give, here, any detail about the deduction of the basic set of partial differential equations that describe the process and simply lay them down as follows. If we write the relative plasma density $\delta n/n_p$ in the form of a wave-packet, i. e.,

$$\frac{\delta n}{n_p}(z,t) = M_L(z,t)e^{i(kz-\omega t)} + cc \qquad (2)$$

with the carrier wave having the wave-number k and frequency $\omega(k)$ and $M_L(z,t)$ the (complex) slowly varying envelope of the packet, we can write the propagation equation

$$(\frac{\partial}{\partial t} + v_g \frac{\partial}{\partial z})M_L(z,t) = -iC_1 \sum_j e^{-i\theta_j(t)}\delta(z-z_j(t)) - iC_2 \sum_j \frac{1}{\gamma_j(t)}e^{-i\theta_j(t)}\delta(z-z_j(t)) \qquad (3)$$

In this equation $z_j(t)$ and $\theta_j(t)=kz_j(t)-\omega(k)t$ are the instantaneous positions and phases, respectively, of the charged sheets of the beam model, j ranges from 1 to N, the total number of sheets, v_g is the group velocity of the accelerating wave-packet and C_1 and C_2 are two constants.

To this equation we must obviously add the equations that describe the axial motion of the "electrons" of the beam, namely

$$\frac{dz_j}{dt} = c\beta_j(t) \qquad (4)$$

$$\frac{dp_j}{dt} = -i\frac{\omega_b^2}{2ck}(be^{i\theta_j} - cc) - i\left[\frac{\omega_p^2}{ck} + \frac{\gamma_p^2}{ck\gamma_j(t)}\left(\omega^2 - \frac{\omega_p^2}{\gamma_p}\right)\right](M_L e^{i\theta_j} - cc) \qquad (5)$$

where $\omega_b=(4\pi e^2 n_b/m)^{1/2}$ is the plasma frequency of the electrons of the beam ($n_b \ll n_p$ is the volume density of the bunch of electrons), $p_j(t)=\beta_j(t)\gamma_j(t)$, $\gamma_j(t)$ is the relativistic factor of each beam "electron" and

$$b = \frac{1}{N}\sum_{j=1}^{N} e^{-i\theta_j(t)}$$

is the bunching factor of the whole beam.

CONCLUSIONS

From the last equation (5) one can obtain the following estimate of the acceleration rate. If we admit that the process has already appreciably increased the average energy of the beam defined as $<\varepsilon>=mc^2<\gamma>= mc^2 \dfrac{1}{N}\sum\limits_{j=1}^{N}\gamma_j(t)$, dropping the first term in the r.h.s. of (5) which represents the Coulomb electrostatic interactions between the electrons of the beam and also the term proportional to $1/\gamma_j(t)$, we obtain directly in MeV per meter

$$G = \frac{d}{dz} <\varepsilon> = 10^{-4}\frac{\omega_p}{\omega}\sqrt{n_p(cm^{-3})}|M_L||b| \left(\frac{MeV}{m}\right). \qquad (6)$$

In particular, from the analysis of the factor

$$g = 10^{-4}\frac{\omega_p}{\omega}\sqrt{n_p(cm^{-3})} \qquad \left(\frac{MeV}{m}\right) \qquad (7)$$

as a function of the ambient plasma density n_p, for a fixed value of the wiggler periodicity λ_w and different increasing values of the wiggler field B_w, one can see that with strong periodic magnetic fields the acceleration can be very effective at values of np much lower than those used in current experiments. On the other hand, one can also see that really high values of the acceleration rates, of the order of several tens of GeV per meter require correspondingly very high values of the applied magnetic field of the order of several hundreds Tesla.

REFERENCES

1. Esarey,E., Sprangle, P., Krall, J., andTing, A., *IEEE Trans. on Plasma Science* **24**, 252(1996).
2. Maroli, C., Petrillo, V., and Bonifacio, R., *Phys.Rev.Letters* **76**, 3578(1996).
3. O'Neil, T.M., Winfrey, J.H., and Malmberg, J.H., *Phys.of Fluids* **14**, 1204(1971).

Measurements of High Gain and Noise Fluctuations in a SASE Free Electron Laser[*]

M. Hogan, S. Anderson, K. Bishofberger, P. Frigola, A. Murokh, N. Osmanov[✣], C. Pellegrini, S. Reiche[†], J. Rosenzweig, G. Travish, A. Tremaine, A. Varfolomeev[✣]

UCLA Department of Physics and Astronomy, Los Angeles, California

Abstract. We report measurements of large gain for a single pass Free Electron Laser operating in Self Amplified Spontaneous Emission (SASE) at 16 μm starting from noise. We also report the first observation and analysis of intensity fluctuations of the SASE radiation intensity in the high gain regime. The results are compared with theoretical predictions and simulations.

INTRODUCTION

The measurements have been done using the Saturnus linac (1), consisting of a 1 1/2 cell BNL photocathode RF gun, and a PWT accelerating structure (2), followed by a beam transport line and a 1.5 cm period, 0.75 Tesla peak field, Undulator Parameter of 1, 40 period undulator built at the Kurchatov Institute (3), (4). The undulator provides focusing in both planes. The linac operates at 5 Hz, with 2.5 μs long macropulses, and one 13 MeV electron bunch per macropulse. Steering magnets control the beam trajectory and align the beam in the undulator, slits measure the emittance (5), an Integrating Current Transformer (ICT) and Faraday cups measure the beam charge, and phosphor screens measure the beam transverse cross section. The beam can be propagated straight through the undulator, or bent through a momentum analyzer to measure the energy and energy spread.

The radiation produced by the undulator is focused by mirrors to a copper doped germanium detector cooled at liquid helium temperature. The detctor can measure the radiation produced by a single electron bunch and has a response time of about 5 ns, while our electron pulses are typically 4 to 6 ps long. The detector has been calibrated and the linearity of its response measured using the 10 picosecond long radiation pulses from the Firefly FEL at the Stanford Subpicosend FEL Laboratory. The noise level in the detector and its associated electronics is of the order of 10 mV. A detector signal of 20mV corresponds to 10^7 photons at 16 μm.

The experiment is done with an undulator of fixed length by changing the electron bunch charge from a low value (0.2nC), where we expect no or small amplification, and observe only spontaneous radiation, to a large value (0.6 nC) where we expect to

[✣]RRC Kurchatov Institute, Moscow, Russia

[†]DESY, Hamburg, Federal Republic of Germany

[*]Work supported by DOE Grant DE-FG03-92ER40793

see amplified spontaneous radiation. We produce one electron bunch, send it through the undulator, and measure the pulse charge and the intensity of the infrared radiation. This is repeated many times to accumulate statistics. We then repeat the experiment blocking the infrared radiation to measure the noise level due to background X-rays. The charge is measured non destructively with the ICT.

When changing the electron bunch charge other beam parameters (energy spread, emittance, pulse length, and beam transverse radius in the undulator) also change. Since all these quantities are important to understand the amplification and fluctuation properties, they have been measured independently as a function of charge. The energy spread changes from about 0.08 to 0.14% rms, when the charge changes from 0.2 to 0.58 nC, putting an upper limit to the rms bunch length of 0.64 to 0.84 mm, corresponding to a peak current (I) of 38 to 83 A. The normalized rms emittance changes from about 8 mm mrad at the lowest charge of about 0.2 nC to about 10 mm mrad at 0.58 nC. Beam losses in the 4mm inner diameter beam pipe, which can produce an X-ray background in our detector, were less than the resolution of our diagnostics. Beam transport and the IR signal were maximized with the beam focused to a spot size of about 0.4 mm (FWHM) at the undulator exit and about three times larger at the entrance.

In an FEL the undulator radiation emitted by the electron beam has a wavelength

$$\lambda = \lambda_0 (1 + K^2/2 + \gamma^2 \theta^2)/2\gamma^2 \tag{1}$$

where λ_0 is the undulator period, K the undulator deflection parameter, θ the angle with respect to the beam axis, and γmc^2 the beam energy. The FEL theory shows that the radiation intensity can grow exponentially along the undulator axis, z, as $I_{rad} \sim \exp(z/L_g)$. In the simple 1D theory (6), (7) neglecting diffraction and slippage, the gain length (L_g) is inversely proportional to the ratio of the beam peak current to the beam cross section, Σ, raised to the power 1/3.

GAIN

The gain is evaluated by comparing the SASE with the spontaneous (non amplified) undulator radiation which is linearly proportional to the charge in the bunch. Another effect which can increase the radiation intensity above the spontaneous undulator radiation level is coherent spontaneous emission, which gives an intensity proportional to the bunch form factor and to the square of the charge. Since our bunch is 1.5 - 2 mm long and our wavelength is 16μm, we expect this term to be small. Further, our intensity measurements at low charge, where we expect no FEL amplification, agree within the errors with the calculated spontaneous undulator radiation, with no discernible contributions from coherent spontaneous emission. The undulator infrared radiation (IR) is measured in the forward direction, within a solid angle Ω corresponding to an angle θ=7.7 mrad defined by the exit window of the beam line, and over all photon frequencies transmitted to the detector. The detector has a peak sensitivity between 2 and 32 μm. The KrS5 beamline exit window and detector window attenuate wavelengths shorter than 0.6 µm and longer than 30 µm, but have a transmission of 70% for wavelengths in between. Hence, we integrate the intensity over the undulator spectrum within Δf=2-30 µm and over Ω defined by the exit window.

The signal we expect from non amplified spontaneous radiation within Ω and Δf after reduction for the windows attenuation, is evaluated using undulator radiation formulas. The energy in a single IR pulse is calculated to be 4.9×10^6 eV, or about 8×10^{-13} J at 0.2 nC. The detector noise including its' amplifier is of the order of 10 mV, so we expect a signal to noise ratio of about 1 at 0.2 nC. X-rays hitting the detector have been minimized with lead shielding, measured while blocking the IR radiation, and have a mean value of 18 mV over our charge range. The almost constant X-ray background between 0.2 to about 0.6 nC indicates that the X-rays are mainly due to distributed background in the detector area, produced by the dark current from the electron source, and not to beam losses through the undulator.

ICT noise corresponds to a mean charge of 7 pC with a standard deviation of 2.3 pC. The measured IR intensities have been divided in bins, corresponding to a charge interval of +/- 2.5% of the central charge value. For each charge interval we accumulate 100 events or more, determine the mean IR intensity and the standard deviation, then subtract the mean X-ray background. The mean IR intensity and standard deviation is ploted vs. charge in Figure 1, where we have also plotted the calculated undulator radiation intensity, reduced by the windows attenuation. Again, our calculation does not include coherent spontaneous radiation, but our observations are consistent with this contribution being negligible. At 0.6 nC the measured intensity is about 2.5 times the calculated spontaneous intensity, thus showing SASE.

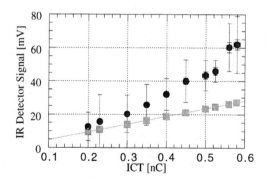

FIGURE 1. IR radiation intensity vs. charge. The vertical bars are the standard deviation for the intensity fluctuations due to starting from noise. For comparison, the effect of beam charge and transverse beam size uncertainties is 4 mV at 0.56 nC. The lower line is the calculated spontaneous emission intensity.

The IR intensity in Figure 1 contains photons in the third harmonic and outside the coherent solid angle Ω_c, a region where the FEL gain is very small compared with the gain in the first harmonic and within Ω_c. To establish the FEL gain for the coherent first harmonic we have measured at the lowest charge of 0.2 nC, the intensity of the third harmonic and the change in intensity when reducing the solid angle to Ω_c. The third harmonic has been measured using a CAF2 filter that does not transmit radiation above 10 μm; the filtered intensity was ~5/12 of the total intensity. The ratio of the intensity within the coherent solid angle, Ω_c and in the total solid angle, Ω, has been measured to be ~1/2 using an iris near the beamline exit window to reduce the solid angle. We have used this experimental information to evaluate the intensity in Ω at the third harmonic, plus that of the first harmonic outside Ωc. These radiation components can be extrapolated linearly with charge (if we assume that they are not amplified), and subtracted from the measured value leaving only the first harmonic within Ω_c. The result, along with the calculated value for the spontaneous first harmonic within Ω_c, is plotted in Figure 2. The ratio of the first harmonic intensity, 42.7 mV, measured at a charge of 0.58 nC, to the extrapolated spontaneous first harmonic at the same charge, 7.5 mV, is about 5.6.

The first harmonic experimental points in Figure 2 are fitted with a curve of the form:

$$I=\alpha Qexp(\Gamma(Q/\Sigma)^{1/3}) \tag{1}$$

which gives an intensity proportional to the charge (Q) for low beam brightness, when we expect to recover the spontaneous radiation limit, and growing exponentially with $Q^{1/3}$ for large electron beam brightness, as one would expect from a 1-D FEL theory (6), (7). The fit gives an exponent of 3.7 at 0.58 nC indicating that at the largest charge we have about 3.7 power gain lengths in our system.

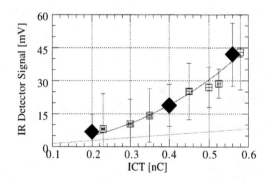

FIGURE 2. First harmonic coherent IR vs. charge. The vertical bars are the standard deviaion for the intensity fluctuations due to starting from noise. For comparison, the effect of beam charge and transverse beam size uncertainties is 4 mV at 0.56 nC. The lower line is the calculated spontaneous emission intensity. The curve fit to the data is IR = 1.85 x ICTexp(4.4ICT1/3). The three diamonds at 0.2, 0.4, 0.6 nC are the results of simulations with the code Ginger, normalized to fit the data point at 0.2nC.

The value of the FEL parameter ρ for the beam and undulator used in this experiment is $\rho \sim 0.01$, and the gain length evaluated from this value in the 1-D theory is about 7 cm. When including 3-D effects but no slippage this increases to about 11 cm. The larger gain length we observe is due to the slippage, defined for a bunch length σ_z as $\lambda N_u / \sigma_z$. The code Ginger (8), which includes both 3D effects and slippage, has been used to simulate three cases: 0.2. 0.4, and 0.58 nC (38, 64, and 83A), while keeping the same beam transverse cross sections. The results have been normalized to fit the experimental point at 0.2 nC to take into account experimental effects like the attenuation from the windows. The normalized results are shown in Figure 2 and fit the data well.

FLUCTUATIONS

We observe output intensity fluctuations due to starting from noise. In the case of no gain the IR power will scale linearly with charge, and does not depend on other beam parameters. When there is gain, a change in charge, Q, will lead to a change in output power which we can evaluate using (1). A maximum gain of 5 would give a maximum change in the IR intensity of 4% for $\Delta Q/Q = +/- 2.5\%$. The beam transverse area, Σ, at the undulator exit has been observed to change by about +/-10%. Since the gain changes as $\Sigma^{-1/3}$ in the exponent, we can expect a change in IR intensity of ~5%. The combined error due to uncertainties in Q and Σ is +/-7%. The much larger power fluctuations, are due to starting from noise. Following the work of (9), (10), (11) the intensity fluctuations are expected to follow a distribution with a relative standard deviation given by $1/M^{1/2}$, where M is qualitatively the number of degrees of freedom, or modes in the radiation pulse: $M = (L/L_c) \times (\Omega/\Omega_c)$, and L_c is the cooperation length. Following (24) when the observed frequency spectrum is larger than the FEL line width we have $L_c = \sqrt{\hat{z}} \lambda / 6\sqrt{2\pi} \rho = 0.11$ mm where $\hat{z} = 3.7$ is the number of power gain lengths in the undulator. Since the bunch length L = 2.2mm (FWHM) we have $L/L_c = 11.3$, and M=27. If we subtract quadratically the standard deviation of the background from the IR signal we obtain a standard deviation for the IR distribution of 18%, corresponding to M~30, in qualitative agreement with our estimate. A more complete analysis of the data, using a convolution of the X-ray background and IR intensity distributions, will be presented in a future publication.

To summarize, we have observed amplification of the spontaneous radiation, with an increase of the first harmonic intensity by 600% over the spontaneous intensity. We have also observed for the first time the intensity fluctuations of the output amplified radiation. Analysis of the data shows a good agreement with the analytic theory of SASE, the time dependent Ginger simulations, and the experimental results.

ACKNOWLEDGMENTS

We wish to express our gratitude to the many people who, with their help, have made this experiment possible, in particular Herman Winick, Massimo Cornacchia, Rich Sheffield, Todd Smith, Glen Westenskow,and Ilan Ben-Zvi.

REFERENCES

1. S.C. Hartman et al., *Nuclear Instruments and Methods* **A340**, 219 (1994)
2. R. Zhang, C. Pellegrini, and R. Cooper, "Study of a Novel Compact Standing Wave RF Linac", in course of publication in Nuclear Instruments and Method.
3. A.A. Varfolomeev, S.N. Ivanchenkov, and A.S. Khlebnikov, *Nuclear Instruments and Methods* **A318**, 813 (1992).
4. A.A. Varfolomeev, Yu. P. Bouzouloukov, S.N. Ivanchenkov, and A.S. Khlebnikov, N.S. Osmanov, S.V. Tolmachev, *Nuclear Instruments and Methods* **A359**, 85 (1995).
5. J. Rosenzweig et al., *Nuclear Instruments and Methods* **A 341**, 379 (1994).
6. R. Bonifacio, C. Pellegrini, and L. Narducci, *Optics Comm.*. **50** 373 (1984).
7. K.-J. Kim, *Nuclear Instruments and Methods* **A250**, 396 (1986).
8. W. M. Fawley, Private Communication.
9. M.C. Teich, T. Tanabe, T.C. Marshall and J. Galayda, *Phys. Rev. Lett.* **65**, 3393 (1990).
10. R. Bonifacio, et al, *Phys. Rev. Lett.* **73**, 70 (1994)
11. E.L. Saldin, E.A. Schneidmiller, and M.V. Yurkov, "Statistical properties of Radiation from VUV and X-ray Free Electron Laser", *DESY rep. TESLA-fel* **97-02**, (1997).

Superradiance in a Short Pulse FEL Oscillator and its Relevance to the X-ray FEL

D.A. Jaroszynski

The University of Strathclyde, Department of Physics and Applied Physics
John Anderson Building, 107 Rottenrow, Glasgow, G4 0NG, Scotland, UK

Abstract. Superradiance (SR) in the short pulse free-electron laser (FEL oscillator is the optimal way of extracting optical radiation from an electron beam. SR is characterised by a peak intracavity optical power, $\mathcal{P}$, scaling as the square of the electron charge, Q, ($\mathcal{P} \propto Q^2$), an optical pulse duration, σ_z, scaling inversly with the square root of the charge,($\sigma_z \propto Q^{-1/2}$) and an efficiency, η, scaling with the inverse of optical pulse length ($\eta \propto 1/\sigma_z \propto Q^{1/2}$). The latter scaling also implies that the relative spectral brightness, $\eta/(\sigma_\lambda/\lambda)$, is constant and close to $\sqrt{3}/2$. Similar scaling of the peak power, temporal width, energy and efficiency with the cavity quality factor also exist. The efficiency in SR emission is enhanced above the usual natural efficiency, $\eta = 1/2N_u$, for the weakly saturated CW FEL, where N_u is the number of undulator periods. We show that SR in the oscillator has analogous properties to that in a high gain amplifier and discuss the relevance of these analogies to the X-ray FEL starting from self-amplified spontaneous emission (SASE).

INTRODUCTION

The most reputed properties of free-electron lasers (FELs) are a wide tunability range, scalability and high power. However, another important characteristic of considerable scientific and practical significance is the possibility of controlling the radiation waveform in time and frequency, and thus obtaining exceptional optical characteristics. Recent evidence suggests that pulses as short as one optical cycle can be obtained by superradiance (SR) in FEL-like devices [1]. Optical pulses less than 6 cycles long in the infrared and far infrared are at present routinely produced [2]. Superradiance has been a fascinating area of research since the pioneering work by Dicke in 1954 [3] and was shown theoretically to occur in free-electron lasers, initially in high-gain single-pass amplifiers, by R. Bonifacio et. al. in 1985 [4,5]. Recently SR has also been demonstrated theoretically [6–10] and confirmed experimentally [1]

CP413, *Towards X-Ray Free Electron Lasers*
edited by R. Bonifacio and W. A. Barletta
© 1997 The American Institute of Physics 1-56396-744-8/97/$10.00

in the FEL oscillator and has also been linked to spiking due to synchrotron instabilities [6,1,11,12] in the post-sideband regime in CW FEL amplifiers and oscillators [13,14]. SR has been observed, in numerical simulations, in the SASE regime, arising from shot-noise, in the form of uncorrelated spikes.

The isolated SR spikes observed experimentally in the low-gain high-slippage oscillator driven by short electron bunches [1] occur as a result of the optical pulse shortening due to the increase in the intracavity optical power. The ultimate limit of one cycle pulses (or even sub-cycle), should be achievable experimentally by reducing the optical cavity losses. Moreover, the efficiency of the SR process depends inversely on the optical pulse duration so that short optical pulses are always associated with high efficiency operation. It is clear therefore that SR is a promising way to produce ultra-short high power pulses. Similar phenomena have also been observed in the cyclotron resonance maser (CRM) [15] and, theoretically, in the collective atom recoil laser (CARL) [16].

BACKGROUND

In 1954, Dicke [3] showed that cooperative emission of radiation can lead to an enhancement of the radiation rate of emitters through a collective radiation damping. Each radiator does work against its own field plus the fields of all of its neighbours. Initially no correlations exist between the radiators and correlations develop by interaction through the common field. Dicke showed that the enhancement rate or superradiant rate is proportional to the number of emitters, n_e, leading to a peak intensity scaling with n_e^2 and a temporal width scaling with n_e^{-1}. He also showed that SR was the most efficient way of extracting energy from an inverted two-level atomic system.

Superradiance in the high-gain FEL amplifier [4] is characterised by an intensity proprtional to Q^2, where Q is the total charge. However, the pulse temporal width in FEL SR scales as $Q^{-1/2}$. Cooperation between the emitters leads to a build up of correlations between electrons on a wavelength scale. The mutual coupling between radiators also occurs through the electromagnetic field causing bunching of the electrons. In this paper we review the first direct experimental observation of superradiance from an FEL oscillator [1] which was carried out on the FELIX [17] FEL, situated in The Netherlands. We also show that for small cavity detuning $\delta\mathcal{L}$, the FEL oscillator behaves in an analogous way to the high-gain FEL amplifier and we demostrate that spikes due to synchrotron instabilities have very similar properties to the SR spike produced in the short pulse FEL oscillator.

SUPERRADIANCE

Superradiance can be defined as *coherent emission of optical radiation from an ensemble of phase-correlated infinitesimal radiators* [3] (relativistic free electrons in our case). We shall distinguish between two forms of superradiance which we shal call linear and nonlinear superradiance respectively.

i) *Linear superradiance* results from electrons that are already correlated in phase, or "pre-bunched", on entering the interaction region. This form of superradiance is often called *coherent spontaneous emission* (CSE) [18–20] and occurs when the electron pulses are short compared to the radiation wavelength, or at least have a strong density modulation on a wavelength scale [18,19]. In CSE, shaping of the electron pulses, prebunching and bunching due to energy/phase correlations are important experimental issues to be considered which have important implications in the design of electron transport systems for linear colliders and X-ray SASE FELs such as at DESY [21]. In these devices, the electron bunch is so short that it always radiates coherently in compressors and transport systems with deleterious effects on the electron bunch length and peak current.

In FEL-like devices, the radiation field grows by amplification of the spontaneous emission, which may be coherent (superradiant) emission or incoherent shot noise. The study of the characteristics of SASE will contribute to the understanding of the physics of X-ray lasers currently being developed around the world.

Linear SR is a promising method for achieving the ultimate limit in pulse duration because the optical pulse length is determined mainly by the interaction time. However, in this case the efficiency is usually low.

ii) *Nonlinear superradiance*, or *superfluorescence* as it is often called in the atomic case [22], arises when an initially uncorrelated pulse of electrons develop correlations and becomes bunched due to nonlinear interaction through the common radiated field. This is a spontaneous self-organisation process. When FEL-like amplifiers and oscillators, operating in the large slippage regime, enter the nonlinear stage of power evolution, nonlinear SR may occur and pulses of the order of one cycle are theoretically possible [2]. This provides an important means of achieving ultra-short ultra-high power optical pulse with good efficiencies using free-electron lasers.

In the FEL, saturation is due to electron trapping in the ponderomotive potential associated with the beat wave of the radiated and the undulator waves, and occurs when the electrons perform half a synchrotron oscillation during their interaction with radiation along the undulator. The synchrotron period decreases as the optical power increases leading to higher optical powers. When the synchrotron period approaches the interaction time the very favourable situation occurs : the shortening of the optical pulse is accompanied by an increase in both peak optical power and the overall efficiency, ie. superradiance [1]. The production of short pulses by nonlinear SR is achievable by either operating with low cavity losses (preferably less than 2% for the highest efficiencies) or increasing the electron charge. If the electron pulses are long, secondary instabilities lead to spiking of the radiated pulse, which also leads to an increase in efficiency [13,14].

Theory and experiments have shown that, under certain circumstances, shorter pulses and higher efficiencies are unstable and limit cycles appear [23,9], followed by more complicated dynamics that lead to chaos [9]. Stable operation and smooth pulses in these high efficiency regimes can be achieved by either cavity dumping, varying the synchronism between optical and electron pulses (dynamic cavity desynchronisation) [24,25] and injection locking a regenerative FEL amplifier [26]. Although numerical evaluation using 3D codes under realistic conditions show that SR is not critically influenced by the quality of the electron beam, and is a robust phenomenon, further systematic numerical and analytical studies will need to reconsider the validity of the standard "slowly varying envelope approximation" (SVEA), and possibly go beyond when the optical pulses approach one period in duration.

Both types of superradiance are closely related: short electron bunches will radiate both by linear superradiance and due to shot noise, and both of these contributions will be amplified (*self amplified spontaneous emission* (SASE)) [27,28], possibly leading to nonlinear superradiance. This is why these phenomena should be studied together.

THEORY OF SUPERRADIANCE

We begin by considering an RF driven short electron pulse FEL with an electron bunch length, l_b, less than or of the order of the slippage length, $\Delta = N_u\lambda$, where $l_b = Q/\mathcal{I}_b$ and $\mathcal{I}_b$ is the peak current, λ the resonance wavelength and N_u the number of undulator periods [29]. The optical field amplitude $A(\xi,\tau)$ and electron bunching $B(\xi,\tau) = \left\langle \exp(-i\tilde{\theta}) \right\rangle$ of the low loss FEL oscillator can be described by scaled equations in terms of the slippage distance, $\Delta = N_u\lambda$, scaled gain parameter,

$$\mathcal{G} = g_0(l_b/\Delta), \tag{1}$$

where g_0 is the usual cw small signal gain coefficient [30,29],

$$g_0 = N_u^3\gamma_0(\mathcal{I}_b/\mathcal{I}_A)(\lambda/z_R)F(a_u)^2 \tag{2}$$

where $\mathcal{I}_A = 4\pi\epsilon_0 mc^3/e \sim 17kA$ is the Alfven limit current, ϵ_0 is the permittivity of free space, z_R the Rayleigh length for the optical cavity, $F(a_u) = 4\pi[a_u/(1 + a_u^2)][J_0(\chi) - J_1(\chi)]$, with $\chi = a_u^2/2(1 + a_u^2)$ and a_u is the rms undulator parameter. The beam radius, r_b, has been assumed to match the beam waist radius, $(\lambda z_R/\pi)^{1/2}$. The scaled longitudinal coordinate, roundtrip number, cavity detuning and cavity loss are, respectively, defined as

$$\xi = z/\Delta, 0 < \xi < 1$$
$$\tau = \mathcal{G}n$$
$$\nu = 2\delta\mathcal{L}/\Delta\mathcal{G}$$
$$\alpha = \alpha_0/\mathcal{G}, \tag{3}$$

where α_0 is the total cavity loss per passage. The cooperation length is defined in the usual way by

$$l_c = \lambda/4\pi\rho \tag{4}$$

where

$$\rho = g_0^{1/3}/(4\pi N_u) \tag{5}$$

is the FEL parameter [31], proportional to the third root of the beam current, $I_b^{1/3}$.

The phase space evolution of the electrons is governed only by parameters ν and α and can be described by the first few moments of the electron distribution : Bunching, $B = \langle \exp(-i\tilde{\theta}) \rangle$, mean momentum, $R = \langle p \rangle$, momentum bunching, $P = \langle p \exp(-i\tilde{\theta}) \rangle$, and $S = \langle p^2 \rangle$ and $T = \langle p^2 \exp(-i\tilde{\theta}) \rangle$ where $p = \partial\tilde{\theta}/\partial\xi = 4\pi N_u(\gamma - \gamma_0)/\gamma_0$. The electrons are characterised by their phase, $\theta = (k + k_u)z - \omega t$ and their momentum, $p = d\theta/dz$, in the ponderomotive potential.

The complex nonlinear phenomena such as nonlinear superradiance, limit-cycles, chaos etc. are well described, assuming the truncation

$$\left\langle (p - \langle p \rangle)^2 \exp(-i\tilde{\theta}) \right\rangle = \left\langle (p - \langle p \rangle)^2 \right\rangle \left\langle \exp(-i\tilde{\theta}) \right\rangle$$

which gives

$$T = SB - 2R^2B + 2RP,$$

by the evolution of the moments of the electron distribution, B, P, R and S, and the field amplitude, A. The evolution of the field and moments can thus be described by a closed system of equations:

$$\frac{\partial p}{\partial \xi} = -\left\{ A \exp[i\tilde{\theta}] + c.c. \right\}$$
$$\Pi B = \frac{\partial A}{\partial \tau} - \nu \frac{\partial A}{\partial \xi} + \frac{\alpha}{2} A$$
$$\frac{\partial B}{\partial \xi} = -i \left\langle p \exp(-i\tilde{\theta}) \right\rangle = -iP$$

$$\frac{\partial^2 B}{\partial \xi^2} = -\left\langle p^2 \exp(-i\tilde{\theta}) \right\rangle - i \left\langle \frac{\partial p}{\partial \xi} \exp(-i\tilde{\theta}) \right\rangle$$

$$= -T + iA + iA^* \left\langle \exp(-i2\tilde{\theta}) \right\rangle \tag{6}$$

where $\Pi = 1$ for $0 < \xi < 1$ and $\Pi = 0$ elsewhere. The last term in Equation(6) is usually dropped as it represents coupling between the fundamental and the second harmonic. The momentum spread of the electrons is governed by $\sigma_e = \sqrt{S - R^2}$. These equations are similar in form to those decribing the backward wave oscillator [32].

The evolution of P, R and S can be numerically calculated and a number of important analytical equations can also be deduced [9]. From the evolution of the electron energy, R and the field intensity $|A|^2$ we obtain the energy balance :

$$\alpha |A|^2 = \left(\nu \frac{\partial}{\partial \xi} - \frac{\partial}{\partial \tau} \right) |A|^2 - \Pi \frac{\partial R}{\partial \xi}. \tag{7}$$

The efficiency is obtained from the electron energy loss by integrating R over the interaction region:

$$\eta(\tau) = \left(\frac{\Delta\gamma}{\gamma} \right) = \frac{\gamma_0}{\gamma} \left\{ \frac{\Delta p}{4\pi N_u} \right\} \approx - \left[\frac{R(\tau)}{4\pi N_u} \right]_{\xi=1} \tag{8}$$

which, from the energy balance equation, we can calculate η in terms of the optical field energy:

$$\eta(\tau) = \frac{1}{4\pi N_u} \left(\frac{d}{d\tau} + \alpha \right) \mathcal{E}(\tau) \tag{9}$$

where the total stored radiation energy is $\mathcal{E}(\tau) = \int_{-\nu\tau}^{1} d\xi |A|^2$. We can thus calculate the efficiency from the electron energy loss or the outcoupled optical field assuming knowledge of the cavity loss. Experimentally, the measured electron efficiency can be used to calibrate the measured optical power.

GENERAL CONSIDERATIONS

Following reference [1] we show that the SR scaling can be deduced from the condition that at saturation the optical pulse narrows to equal half a synchrotron length and the efficiency increases until the gain at saturation is equal to the loss. The electric field strength, $|E|$, for a planar undulator is related to the normalised vector potential $|a|$ by [29]

$$|a| = \pi N_u^2 F(a_u)(e\lambda |E| / \sqrt{2} mc^2). \tag{10}$$

60

Saturation of the oscillator occurs when the electrons excecute half a synchrotron oscillation in the ponderomotive potential, when the peak intracavity power gives a normalised field potential, $|a| \approx \pi^2$. The synchrotron length L_{syn}, is inversly proportional to the fourth root of the field intensity: $L_{syn} = 2\pi N_u \lambda / |a|^{1/2}$. The peak intracavity power can be calculated from the electric field strength $|E|$:

$$P_{int} = c\epsilon_0 \pi^2 z_R \lambda |E|^2 / 8. \tag{11}$$

The SR scaling is obtained if we assume that the optical pulse shortens to an optical pulse shape with a rms dispersion of σ_z a fraction of a synchrotron length, consistent with analytical and numerical studies of the evolution of short pulses in FEL amplifiers and oscillators [9,10]. At saturation the electrons drift through the optical pulse only interacting with the optical field for a time equal to the duration of the optical pulse. This has the effect of limiting the effective number of undulator periods to

$$N_s = \frac{\mathcal{F}\sigma_z}{\lambda} \tag{12}$$

where $\mathcal{F}(\approx 2)$ is a factor depending on the optical pulse shape. The electrons drift freely after interaction and may rebunch due to *inertial bunching* and re-radiate small secondary pulses in an analagous way to "Burnham Chaio" ringing observed in Dicke SR systems [33]. The ponderomotive potential will exist only for a time duration of $\tau_s \approx \sigma_z \mathcal{F}/c$, as is shown schematically in Figure(1). At saturation the oscillator will be in equilibrium with the gain equal to the loss. The power lost by the electrons, $\mathcal{E}_{out}$, on each passage will be $\mathcal{E}_{out} = \alpha_0 \mathcal{E}_{int}$ giving an efficiency $\eta = \mathcal{E}_{out}/\mathcal{E}_e$ where $\mathcal{E}_e = mc^2 \gamma_0 (Q/e)$ is the total energy of the electron bunch. The optical energy at equilibrium can then be related to the optical pulse length:

$$\mathcal{E}_{out} = \frac{\mathcal{E}_e \lambda}{2\mathcal{F}\sigma_z}. \tag{13}$$

The energy stored inside the optical cavity,

$$\mathcal{E}_{int} = \frac{\mathcal{E}_e \lambda}{2\mathcal{F}\sigma_z \alpha_0}, \tag{14}$$

is deduced from the normalised potential at saturation:

$$\mathcal{E}_{int} = \frac{\sqrt{2\pi}\sigma_z P_{int}}{c} = \frac{\sqrt{2\pi}\sigma_z \epsilon_0 \pi^2 z_R \lambda |E|^2}{8} \tag{15}$$

to obtain a steady state value for σ_z :

$$\sigma_z = \left(\frac{\mathcal{E}_e \lambda c}{2\sqrt{2\pi}\mathcal{F}\alpha_0 P_{int}} \right)^{1/2}. \tag{16}$$

61

Assuming an interaction length of N_s undulator periods, from equation(12) [1], σ_z is then obtained self-consistently from the intracavity power:

$$\sigma_z \approx \left(\frac{a_0 z_R}{r_c}\right)^{1/2} \left(\frac{mc^2}{\mathcal{E}_e}\right)^{1/2} \left(\frac{\lambda}{\mathcal{F}^{3/2}F(a_u)}\right) \tag{17}$$

where r_c is the classical radius of an electron. The SR efficiency, determined by the effective number of undulator periods $\mathcal{F}\sigma_z$, is given by,

$$\eta = \frac{0.22\lambda}{\mathcal{F}\sigma_z}, \tag{18}$$

which is much larger than the usual low gain CW FEL oscillator natural efficiency, $\eta_0 \approx 1/2N_u$.

All the SR scaling laws can thus be calculated from σ_z as shown in Table 1.

It should be noted that when the electron bunch is longer than the slippage length ($l_b > \Delta$) a single spike can still form, depending on the shape of the electron bunch. However, in this case only electrons in the slippage region Δ surrounding the peak of the electron bunch are involved in the interaction. The effective total electron energy

$$\mathcal{E}'_e = \mathcal{E}_e \Delta / l_b$$

with

$$l'_b = Q/\mathcal{I}_b$$

should be used in the above calculations, for comparing theory with measurements. The measurable efficiency, η', taking into account all the electrons in the bunch, is then

$$\eta' = \eta\Delta/l_b. \tag{19}$$

For ($l_b \gg l_c$) many independent spikes, separated by an average distance Δ, may form as in the spiking regime. However, the above analysis can still be used to calculate the peak power and width of the largest spikes in the post-sideband regime [13,14], provided that the total energy and charge of the electrons in a slippage length is substituted for the respective bunch values.

CAVITY DETUNING CONDITION

In this section we review the cavity detuning conditions necessary for superradiance following reference [9]. The optical pulse shape of the FEL oscillator will evolve at first to a steady state shape characterised by supermodes of the linearised equations of motion (ie. the last of equations(6) with $\partial^2 B/\partial\xi^2 = iA$). The optical width, σ_z, depends only the normalised cavity detuning, ν and cavity loss α [9,10,34]. The maximum gain, at the peak of the small signal detuning curve, is at [9]

$$\delta\mathcal{L} \approx 0.22\Delta\mathcal{G} \approx 0.02 l_b g_0. \tag{20}$$

However, for smaller values of $\delta\mathcal{L}$, the gain per pass, G, drops to zero as

$$G = \frac{3^{3/2} l_b^{1/3} \delta\mathcal{L}^{2/3}}{l_c} \tag{21}$$

and becomes independent of the undulator length but proportional to $\mathcal{I}_b^{1/3}$ through ρ, as in the linear stage of the SR regime in amplifiers [35–37]. A finite cavity detuning is necessary for synchronism between electron and optical pulses in the cavity because of swept gain lethargy, which effectively slows down the optical pulse [38]. Typical measured gain detuning curves for the FELIX FEL are shown in Figure(2) exhibit these features. The experimental conditions for the detuning curves will be discussed below.

To solve the nonlinear superradiant evolution and describe saturation, of the optical field and the electron phase-space distribution we substitute the fundamental eigenfunction (supermode) of the linearised Equations(6), $\Phi_1(\xi,\tau)$, corresponding to the first supermode, λ_1, into the fully nonlinear Equations(6), projecting the field on the fundamental supermode, with a complex gain coefficient, $A(\xi,\tau) = \Lambda_1(\tau)\Phi_1(\xi)$ to obtain, to third order in Λ_1, a Landau Ginzburg equation [9]:

$$\frac{d\Lambda_1}{d\tau} = (\lambda_1 - \frac{\alpha}{2})\Lambda_1 - \beta_1\Lambda_1|\Lambda_1|^2. \tag{22}$$

β_1 can be calculated from the first supermode, $\Phi_1(\xi)$ [9]. At saturation the effect of lethargy is suppressed and the optimum cavity detuning is close to zero desynchronism. By solving Equation(22) it can be shown [9] that the maximum efficiency,

$$\eta_{max} = \rho \left(\frac{l_b}{\alpha_0 l_c} \right)^{1/2}, \tag{23}$$

occurs at

$$\delta\mathcal{L}_{max} \approx 0.18 l_b \left(\frac{\alpha_0 l_c}{l_b} \right)^{3/2}. \tag{24}$$

At $\mathcal{L}_{max}$ the peak intracavity power is

$$\mathcal{P}_{int} = \rho P_b \left(\frac{l_b}{\alpha_0 l_c} \right)^2, \tag{25}$$

where P_b is the electron beam power. We can thus deduce the following scaling laws:

$$\mathcal{P}_{int} \propto \left(\frac{Q}{\alpha_0}\right)^2$$

$$\eta \propto \left(\frac{Q}{\alpha_0}\right)^{1/2}$$

$$\mathcal{E}_{int} \propto \left(\frac{Q}{\alpha_0}\right)^{3/2}$$

$$\sigma_z \propto \left(\frac{Q}{\alpha_0}\right)^{1/2}, \tag{26}$$

showing that the SR scaling [9,10] is obeyed in the oscillator. The experimental measurements shown in Figure(2), clearly showing the enhancement of the power and efficiency at small desynchronism and the enhancement of the small signal gain at larger desynchronism.

ANALOGY WITH PULSE PROPAGATION IN THE HIGH GAIN AMPLIFIER

The similarity between superradiance in the oscillator and in the high gain amplifier suggests analagous behaviour of short pulses propagating in the two systems. Moreover, both the high gain amplifier and the short pulse oscillator are characterised by an exponential growth period followed by self-similar pulse evolution that obey the SR scaling laws. To explore the analogy further we begin by showing that a short pulse in the amplifier develops in the same way an optical pulse developing in the oscillator when the pulses are shorter than a slippage length, $\sigma_z < \Delta$, and the optical cavity detuning is set close to the synchronous condition, $\delta\mathcal{L} \approx 0$. We can consider the optical pulse as effectively propagating in an amplifier with a continuous electron beam and an infinitely long undulator, by unfolding the oscillator cavity [1], as shown schematically in Figure(3). The interaction region of length $(\Delta - 2\delta\mathcal{L})$ can be considered as co-propagating with the optical pulse at a velocity $v_g = c(1 - \delta\mathcal{L}/\mathcal{L}_u)$, where $\mathcal{L}_u$ is the undulator length of the oscillator. In the steady state, the optical pulse that evolves is given by a sum of supermodes of Equations(6), and has a duration fixed by the scaled cavity detuning, ν, and cavity loss α or conversly, by the velocity, v_g, which exactly matches the reduced velocity due to the lethargy of the pulse. When the FEL reaches saturation, lethargy is reduced and the optical pulse velocity increases, causing the optical pulse to slip out of the window unless the window is moved by dynamically varying the detuning of the oscillator [24,25]. This process may enter a limit cycle as new pulses grow up [23,9]. For large detuning the above analogy then only describes the evolution of the last pulse of the train propagating in the amplifier. However, for a single spike, or distantly separated spikes in an amplifier, it can still be used to evaluate the behaviour of the amplifier at the onset of saturation. Not only does the model allow us to examine the evolution of short pulse in time in the linear and nonlinear stages of evolution but it also helps us decide whether we can consider SR in the oscillator as true SR. Furthermore, evaluation of pulse propagation in the amplifier could, conversely, be considered in terms of the evolution of coupled supermodes, as in the oscillator. Moreover, the startup of the amplifier and low gain oscillator could be considered on equal footing. This may be useful for studying (SASE) [35,36] in infrared [28], VUV and X-ray lasers [21] and in particular when trying to enhance the startup power by enhancing the coherent spontaneous emission component of the initial spontaneous emission [27,19,18].

In a SR FEL amplifier, the optical power of the SR pulse increases as it propagates over the electron beam with a peak power

$$\mathcal{P} \approx \rho P_b (l_b'/l_c)^2$$

and an efficiency

$$\eta \approx \rho \sqrt{l_b'/l_c},$$

where $l_b' = l_s$ is the electron beam section swept by the SR pulse in N_u undulator periods [37,35,36]. The SR oscillator, driven by electron pulses with length l_b shorter than the slippage length, l_s, behaves as a SR amplifier in which the effective section of the beam swept by the optical pulse after n passes through the cavity is $l_b' = n l_b$. This is only possible when the optical pulse length is less than the slippage length, ensuring that the electrons interact with the radiation for a number of undulator periods that is determined only by the width of the optical pulse. For small cavity detuning and loss, the energy of the SR pulse reaches a stationary state when the number of roundtrip passes, $n \approx 1/\alpha_0$. The effective beam length in the amplifier analogy is, therefore, $l_b' = l_b/\alpha_0$. These conclusions are in broad agreement with the stationary solution of the Landau Ginzburg model for the fundamental supermode described above, and of ref. [10], where the transient evolution of SR pulses in FEL oscillators is analysed for zero cavity detuning.

We can apply all the usual criteria for strong and weak SR in the high gain FEL amplifier [35,36,39]. The generalised "slippage parameter" $\mathcal{S}$ is calculated in this context with an effective slippage length $l_s' = n l_b$ and effective bunch length $l_b' = l_s'$, giving $\mathcal{S} = l_s'/l_b' = 1$. The evolution of the oscillator can be tracked by dividing the propagation distance in the amplifier by l_b. To establish the condition for strong SR, the effective gain parameter, $G' = 4\pi\rho N_i$, is calculated, where $N_i = n l_b/\lambda$ is the effective number of undulator periods in an interaction length. $G' = 4\pi\rho n l_b/\lambda$ and the high gain region is entered when $G' > 1$ [39] which is equivalent to a few roundtrip passages in the case of the oscillator for typical FEL parameters. The slippage in a gain length is given by $K = l_c/l_b' = 1/G'$ and strong SR occurs when $K \ll S$ or when $n \gg (\lambda/4\pi\rho l_b)$ [35,36,39]. This places an upper limit on α_0 when n is replaced by $1/\alpha_0$ in the expression for K :

$$\alpha_0 < \frac{4\pi\rho l_b}{\lambda}. \tag{27}$$

In the nonlinear stage, or SR stage of evolution a self-similar optical pulse evolves [39] with an approximately hyperbolic secant in shape [7], a peak power of $\mathcal{P} \propto \mathcal{I}\rho(l_b'/l_c)^2 \propto n^2\mathcal{I}_b^2$, a pulse width σ_z diminishing as $\sigma_z \propto (G')^{-1/2} \propto (n\mathcal{I}_b)^{-1/2}$ and an efficiency increasing as $\eta = \rho\sqrt{l_b'/l_c} \propto \mathcal{I}_b^{1/2}n^{1/2}$. In the limit of $n = 1/\alpha_0$ the steady state oscillator SR scaling with loss and intensity is obtained. In the linear stage of evolution, the optical pulse has a constant shape and group velocity, at small signal levels, defined by stationary super-modes [9,10,34] whereas in the nonlinear stage the optical pulse shape evolves. Complicated pulse evolution such as a limit cycle [23] may also occur in the amplifier under certain conditions. In both the linear and nonlinear stages of evolution, the pulses have properties of self-similar pulses [40,41].

From this discussion, we conclude that SR optical pulses in the saturated oscillator evolve in a similar way to that in the high gain amplifier.

EXPERIMENTAL OBSERVATION OF SUPERRADIANCE

Experimental results verifying that superradiance has been observed in the the FEL2 oscillator of the FELIX experiment situated in The Netherlands are presented [1]. The FELIX FELs operate in the large slippage regime [17] producing short optical pulses in the IR to FIR, with durations as short as 6 optical cycles [2], which are used by a large and varied user community [17]. In Table(2) we give experimental parameters of the FELIX FEL relevant to the experimental results presented here. The hole output coupler accounts for a dependence of the cavity loss, α_0, on wavelength. In the experiments the cavity losses were determined from the measured "ring-down" time, of optical power stored in the cavity, at the end of the optical macropulse.

To measure optical spectra an evacuated grating spectrometer has been utilised. The optical pulse shapes and durations are inferred from measurements of the second-order autocorrelation spectra using a zero-background technique based on a non-linear CdTe doubling crystal [2] with an overall resolution of between $100fs$ at $5\mu m$, and $500fs$ at $35\mu m$. The efficiency of the FEL interaction is determined from measured electron spectra using a 0.22% resolution, transient multichannel electron spectrometer, placed after the FEL undulator [42]. The larger efficiencies are inferred from extrapolate the efficiencies using calibrated optical power measurements. Beyond a threshold of about $\eta' \approx 1\%$ the electron spectra have very wide tails and the widely dispersed electrons in the spectrometer generated x-rays after striking components outside the target OTR region. This results in spurious signals which swamp the detectors used for monitoring the spectra [42].

The electron spectra were much wider than expected from the efficiencies measured. This is due, in part, to the fact that the efficiency in a slippage length η is usually much higher than the measurable efficiency η', sometimes as much as 5%, as discussed above. In the FELIX FEL2 oscillator, for small cavity detunings, the largest total efficiencies, η', are more than 3.5% at the longer wavelengths where slippage is strongest, as shown in Figure(4). In Figure (4) the SR efficiency have been measured as a function of wavelength by varying a_u from 0.73 to 1.58 while keeping the electron energy and charge constant, at 30 MeV and 200pC, respectively. These high efficiencies are much larger than the predicted natural efficiency of $\eta = 1/2N_u \approx 1\%$, and indicate the presence of superradiance. Moreover, there is excellent agreement of measured and predicted SR efficiencies, η', as shown in Fig. 4 where the predicted efficiencies have been calculated using the measured losses. The actual efficiency in a slippage length, η, is shown by the dashed line in the figure. The largest typical value of more than 5% is a record for operation of a low gain oscillator with no additional system for enhancing the efficiency, such as tapering [43] or reverse-step tapering [44]. The high values of η account for the larger than expected electron energy spreads observed in electron spectra [42].

The higher efficiencies were determined either by extrapolating the efficiencies, measured from electron spectra at low efficiencies, using simultaneously measured optical micropulse pulse energy, $\mathcal{E}_{out}$, evolution. The optical macropulse evolution signal, from the detector, is proportional to the average out-coupled power or the outcoupled micropulse energy, $\mathcal{E}_{out}$. Measured efficiency η', is related to the optical energy $\mathcal{E}_{out}$, by conservation of energy,

$$\eta' = \frac{\eta \Delta}{l_b} = \frac{\Delta}{l_b \mathcal{E}_e} \left[\frac{d\mathcal{E}_{out}(n)}{dn} \frac{1}{\alpha_0} + \mathcal{E}_{out}(n) \right] \qquad (28)$$

where n is the number of roundtrips. In Fig. 5 we show the η' derived from the optical macropulse for a typical set of data. Using these extrapolation techniques we are able to observed efficiencies much larger than 1.0% although the signal to noise ratio suffers for the largest efficiencies extrapolated. In all cases the electron data is also used to calibrate the measured optical power, $\mathcal{E}_{out}$, at low efficiencies. The calibrated $\mathcal{E}_{out}$ data are then used to determine the efficiencies from the optical power in an experiment. However, it should be noted that calibration becomes invalid if α_0 varies with wavelength.

Using the cavity loss, determined from the macropulse shape, in Equation(28), the efficiency determined from the optical power is fitted to the electron data efficiencies at low efficiencies for each measurement carried out as shown in the example of Figure(5).

Superradiance scaling laws in the FEL oscillator have been confirmed in the FELIX FEL2 oscillator by measuring the efficiency and the optical pulse duration as a function of the charge Q, using the electron spectrometer and the autocorrelator respectively. The electron bunch charge has been varied by adjusting the electron beam current to the linac while compensating for energy decreases by increasing the RF power driving the accelerator. We have thus varied Q from 100pC to 200pC while keeping the electron bunch length constant with the linac set at 39 MeV and $a_u = 1.53$ giving $\lambda = 18.5$. The oscillator's cavity detuning curves have been measured for each value of Q by varying $\delta\mathcal{L}$ to determine the maximum efficiency operating point as shown in Fig. 2. At low currents the electron macropulse was extended to compensate for the diminished gain which delays the onset of saturation. From each measurement of η', $\mathcal{E}_e$ and σ_z, the values of $\mathcal{P}$ and $\mathcal{E}_o$ have been calculated. Measured optical pulse widths determined from the second order autocorrelation traces varied between $\sigma_z = 50\mu m$ (6 optical cycles) and $\sigma_z = 110\mu m$. The autocorrelation curves were a close fit to both an hyperbolic pulse shape [2], and the shape of the first supermode. At the highest measured powers there was evidence of a tail in the second order autocorrelation traces. The shape indicated low level sub-pulse structure consistent with the excitation of higher order supermodes [9,10] which we considered as evidence of "Burnham Chaio" ringing [33] due to inertial bunching. The main SR scaling laws were confirmed by measurements of the peak intracavity power as a function of the electron charge. Fig. 6 shows $\mathcal{P}_{int}$ vs Q^2 for two different wavelengths clearly showing a square law dependence and confirming SR evolution in the FEL oscillator. The absolute values of the experimental data compare extremely well with those calculated : e.g for $Q = 200pC$ we expect i) $\mathcal{P} = 264MW$ at 18.5μm for $\alpha_0 = 5.2\%$ and $l_b = 1.5mm$; and ii) $\mathcal{P} = 202MW$ at 8μm for $\alpha_0 = 7\%$ and $l_b = 1.3mm$. The corresponding largest intracavity power, $\mathcal{P}_{int}$, for $\lambda = 18.5\mu m$ is estimated to be 3.2GW.

In Figure(7) the optical pulse energy stored in the cavity, $\mathcal{E}_{int}$, is shown as a function of $Q^{3/2}$. The largest measured value of $\mathcal{E}_{out}$ at 200 pC compares very well the predicted value of $180\mu J$. The enhancement and scaling of the efficiency with charge has also been observed. The efficiency scaling with charge, $\propto \sqrt{Q}$, is shown in Figure(8) However, the actual efficiency, η, for electrons in a single slippage length is much larger than η' for the whole bunch. We calculate $\eta \approx 5.2\%$ for $Q = 200pC$. This value is very much higher than the "natural" efficiency of $\approx 1\%$ for the CW FEL oscillator usually predicted in the literature.

To support the oscillator-amplifier analogy the dependence of the various parameters on cavity loss were determined. α_0 was varied between 5% and 15.5% by inserting very thin wires into the cavity. Fig. 9 shows $\mathcal{P}_{int}$, derived from measurements of η' and σ_z as a function of $1/\alpha_0^2$, again giving the SR scaling. Unfortunately, because of difficulties in carrying out the experiment, $\mathcal{P}_{int}$ has only been evaluated for three different losses. However, more con-

68

clusive measurement of the efficiency as a function of cavity loss have been carried out. Figure(10) shows the measurable efficiency, η', as a function of $\alpha_0^{-1/2}$ for a wavelength of $15.2\mu m$ [45]. Here again, we have observed the SR scaling.

Finally, to provide a link between SR in the short pulse FEL and the spiking regime we have measured η' and the relative spectral bandwidth, σ_λ/λ, of the optical pulses to determine the relative spectral brightness defined by $\mathcal{B} = \eta/(\sigma_\lambda/\lambda)$. We have found it to be constant and close to $\sqrt{3}/2$ over a wide range of electron bunch charges, as shown in Fig. 11. In reference [13,14] it has been shown both theoretically and experimentally that the spiking regime in the long pulse FEL can be characterised by $\mathcal{B} = \sqrt{3}/2$. Our measurements of $\mathcal{B}$ also consistently obtain the same numerical value. We can conclude from our measurements that this is quantitative evidence that the efficiency enhancement in the short pulse SR oscillator arises through the same process as that observed in the long pulse FEL operating in the spiking regime. We can draw a general conclusion from these observations: The FEL pondero-motive potential will continue to grow (ie. the efficiency will increase) at the expense of the optical pulse duration (ie. the interaction time or the lifetime of the ponderomotive potential) to optimise the efficiency until the process is stopped when the gain becomes equal to the losses. If we allow the optical pulse to continue evolving by reducing the losses then the optical pulses will continue to narrow and the efficiency will continue to increase. The fact that $\mathcal{B}$ is constant is reflected by the fact that the Fourier transform product is approximately constant, $2\pi\sigma_\nu\sigma_\tau \approx 1.5$, as shown in Fig. 12. The slight tendency for $\sigma_\nu\sigma_\tau$ to increase with current would indicate a frequency chirp in the optical pulse increasing every slightly with intracavity power [9,8]. The above measurements indicate that it may be fruitful to reduce the cavity loss to both shorten the optical pulses further, possibly to the single cycle limit, while at the same time increasing the efficiency still further.

CONCLUSION

We have demonstrated experimentally that the superradiant regime in an FEL oscillator can be reached, studied, and used for optimal generation of radiation. We have also derived the main scale laws for the superradiant regime of short pulse FEL's from the scaled nonlinear equations and furthermore, from the assumption that the steady state optical pulse duration is a fraction of the synchrotron length. We have also shown that there exists an analogy between the high gain amplifier and the oscillator. Our work shows that superradiance is not exclusive to high gain devices but also occurs in short pulse low gain systems. SR emission in the oscillator can also be considered consistent with the definitions of strong superradiance in the amplifier. Superradiance is, in fact, a rather general phenemenon occuring in a wide variety of devices and for very different choices of operating parameters. It is for instance now well established that high current long pulse FEL's produce spikes of radiation characterised by a constant spectral brightness $\mathcal{B} \approx 0.86$, where the typical length of spikes and the efficiency also scale as $Q^{-1/2}$ and $Q^{1/2}$ respectively. By measuring $\mathcal{B}$ we have shown that spiking and SR have the same general characteristics. As superradiance is the optimum way of extracting radiation, methods could be devised to : i) cavity dump as an efficient way of obtaining a high power FEL pulse; ii) dynamic desynchronism to combine high gain and high efficiency at saturation [24,25]. We can also learn about the operation of the amplifier and apply the theoretical techniques developed to describe the limit cycle [23] and chaos in amplifiers in deep saturation, using supermodes, by analogy with the oscillator. We should also examine conditions necessary to the ultimate limit to the optical pulse duration of a single optical cycle or less.

The author D.A.J. would like to acknowledge the indispensible support of the FELIX experimental team, Dr. D. Oepts and Dr. L. van der Meer, at the FOM Instituut voor Plasmafysica, Nieuwegein, The Netherlands, in the experimental work, and the support of P. Chaix at the CEA, Bruyeres le Chatel in France, Dr. N. Piovella at the University of Milan, Italy, for their valuable theoretical input.

REFERENCES

1. D. Jaroszynski et al., Phys. Rev. Lett. **78**, 1699 (1997).
2. G. M. H. Knippels et al., Phys. Rev. Lett. **75**, 1755 (1995).
3. R. Dicke, Phys. Rev. **93**, 99 (1954).
4. R. Bonifacio and F. Casagrande, J. Opt. Soc. Am. **B2**, 250 (1985).
5. R. Bonifacio, B.W.J.McNeil and P.Pierini, Phys. Rev.A. **40**, 4467 (1989).
6. G. Moore and N. Piovella, IEEE J. Quantum Electron **QE- 27**, 2522 (1991).
7. N. Piovella, Opt. Comm. **83**, 92 (1991).
8. N. Piovella. et. al., Phys. Rev. Lett. **72**, 88 (1994).
9. N. Piovella, P. Chaix, G. Shvets, and D. Jaroszynski, Phys. Rev. E. **52**, 5470 (1995).
10. N. Piovella, Phys. Rev. E. **51**, 5147 (1995).
11. B. Richman, J. Madey, and E. Szarmes, Phys. Rev. Lett. **63**, 1682 (1989).
12. F. Glotin et al., Phys. Rev. Lett. **71**, 2587 (1993).
13. D. Iracane, P. Chaix, and J. Ferrer, Phys. Rev. E **49**, 800 (1994).
14. P. Chaix and D. Iracane, Phys. Rev. E **48**, R3257 (1993).
15. N. Ginzburg et al., Phys. Rev. Lett. **78**, 2365 (1997).
16. R. Bonifacio, L. D. Salvo, L. Narducci, and E. D'Angelo, Phys. Rev. A **50**, 1716 (1994).
17. D.Oepts, A. van der Meer, and P. van Amersfoort, Infrared Phys. Technol. **36**, 297 (1995).
18. D. Jaroszynski et al., Phys. Rev. Lett. **71**, 3798 (1993).

19. D. Jaroszynski *et al.*, Nucl. Instr. Meth. A **A341**, ABS24 (1994).

20. A. Gover *et al.*, Phys. Rev. Lett. **72**, 11921195 (1994).

21. J. Rossbach, Nuc. Inst. and Meth. A **A375**, 269 (1996).

22. M. Gross and S. Haroche, Phys. Reports **93**, 301 (1982).

23. D. Jaroszynski *et al.*, Phys. Rev. Letters **70**, 3412 (1993).

24. D. Jaroszynski *et al.*, Nucl. Instr. Meth. **A 296**, 480 (1990).

25. R. J. Bakker *et al.*, Phys. Rev. E **48**, R3256 (1993).

26. G. Shvets and J. Wurtele, Phys. Rev. E **accepted for publication**, .

27. D. Jaroszynski *et al.*, Nucl. Instr. Meth. A **A358**, 228 (1995).

28. R. Prazeres, F. Glotin, D. Jaroszynski, and J. Ortega, Phys. Rev. Lett, accepted for publication (1997).

29. W. Colson, in *Laser Handbook Vol. 6*, edited by W. Colson, C. Pellegrini, and A. Renieri (North Holland, Amsterdam, 1990), p. 115.

30. J. Madey, Nuovo Cimento **50**, 1 (1979).

31. R. Bonifacio, C. Pellegrini, and L. Narducci, Opt. Commun. **50**, 373 (1984).

32. N. Ginzburg, S. Kuznetsov, and T. Fedosseva, Radio Phys. Quant. Electron **21**, 728 (1978).

33. D. Burnham and R. Chiao, Phys. Rev. **188**, 667 (1969).

34. G. Dattoli and A. Renieri, in *Laser Handbook Vol. 4*, edited by M. Stitch and M. Bass (North Holland, Amsterdam, 1985).

35. R. Bonifacio, L. D. Salvo, and B. W. J. McNeil, Opt. Comm. **93**, 179 (1992).

36. R. Bonifacio *et al.*, Rivista del Nuova Cimento **15**, 1 (1992).

37. R. Bonifacio, L. D. S. Souza, P. Pierini, and N. Piovella, Nucl. Instr. Meth. A **A296**, 358 (1990).

38. F. Hopf, P. Metstre, and M. Scully, Phys. Rev. Lett. 35 **35**, 511 (1975).

39. R. Bonifacio *et al.*, Phys. Rev. Lett. **73**, 70 (1994).

40. T. Zhang and T. Marshall, Phys. Rev. Lett. **74**, 1916 (1995).

41. S. Cai and A. Bhattacharjee, Phys. Rev. A **43**, 6934 (1991).

42. W. A. Gillespie *et al.*, Nucl. Instr. Meth. A **A358**, 232 (1995).

43. N. Kroll, P. Morton, and M. Rosenbluth, IEEE J. Quantum Electron **QE-17**, 1436 (1981).

44. D. Jaroszynski *et al.*, Phys. Rev. Lett. **74**, 2224 (1995).

45. A. MacLeod, W. Gillespie, D. Jaroszynski, and A. van der Meer, Nucl. Inst. and Meth. **to appear**, (1997).

71

	Q and α_0 dependence	
Power	$\mathcal{P}_{out} = \mathcal{E}_{out}/(\sqrt{2\pi}\sigma_z)$	$\propto Q^2/\alpha_0$
Energy	$\mathcal{E}_{out} = \eta\mathcal{E}_e \propto \mathcal{I}_b^{3/2}/\sqrt{\alpha_0}$	$\propto Q^{3/2}\alpha_0^{-1/2}$
Duration	$\sigma_z \approx (\alpha_0 z_R \lambda^2 mc^2/\mathcal{E}_e \mathcal{F}^3 F(a_u)^2 r_c)^{1/2}$	$\propto Q^{-1/2}\alpha_0^{1/2}$
Efficiency	$\eta = 0.22\lambda/\mathcal{F}\sigma_z$	$\propto Q^{1/2}\alpha_0^{-1/2}$

TABLE 1. Superradiant scaling laws

Electron energy	30MeV and 39MeV
Maximum bunch charge (Q)	200 pC
Bunch length (l_b)	1.5 mm
Electron energy spread	0.5%
Normalized rms emittance	100 $\pi mmmrad$
Undulator period	65 mm
Number of undulator periods	38
Maximum rms undulator strength	1.9
Optical cavity length	6 m
Roundtrip loss	5-15 %
Rayleigh length	1.2 m
Output coupler	2mm diam. on-axis hole

TABLE 2. FELIX FEL2 parameters.

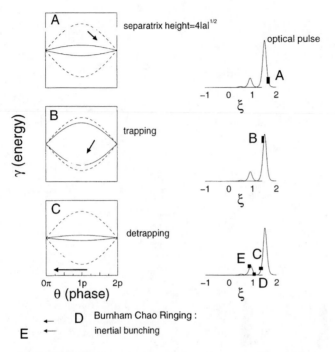

FIGURE 1. Phase space evolution of a superradiant pulse

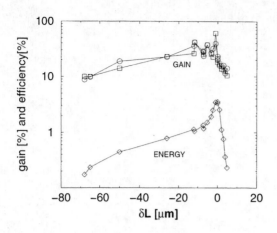

FIGURE 2. Cavity detuning curves on FELIX of small signal gain and efficiency, at $\lambda = 19\mu m$, measured as a function of $\delta\mathcal{L}$.

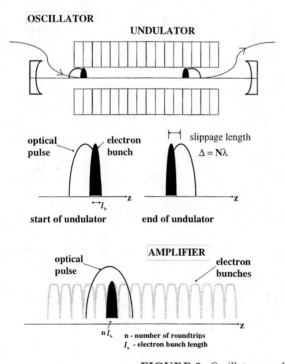

FIGURE 3. Oscillator amplifier analogy

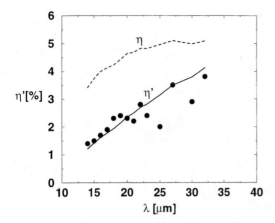

FIGURE 4. Measured η' vs λ by varying a_u between 0.73 and 1.58. Solid line represents η' and the dotted line η calculated using measured α_0 values and assuming $l_b = 1.5mm$.

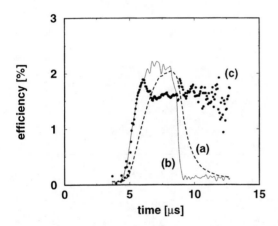

FIGURE 5. Typical extrapolation of efficiency from optical data. (a) Raw electron data (b) Raw optical data (c) Efficiency calculated from optical data and fit to electron data for times less than $4.7\mu s$.

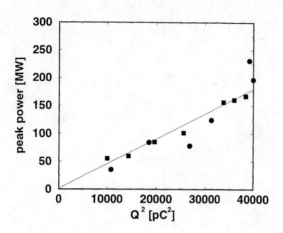

FIGURE 6. $\mathcal{P}_{out}$ vs Q^2. Squares for $\lambda = 18.5\mu m$ (39 MeV, $\alpha_0 = 5.2\%$) and dots $\lambda = 8\mu m$ (30 MeV, $\alpha_0 = 7\%$).

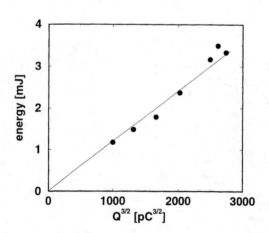

FIGURE 7. Intracavity energy, $\mathcal{E}_{int}$ vs $Q^{3/2}$ for $\lambda = 18.5\mu m$.

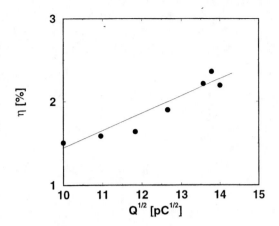

FIGURE 8. η' vs $Q^{1/2}$ for $\lambda = 18.5\mu m$.

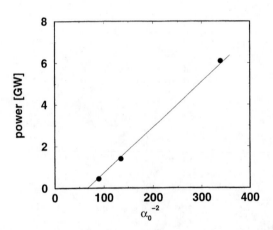

FIGURE 9. Peak intracavity power vs α_0^{-2}

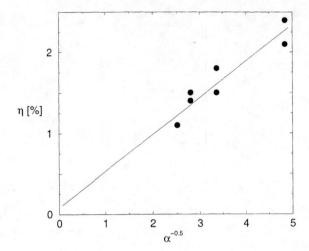

FIGURE 10. Efficiency, η', measured as a function of $\alpha_0^{-1/2}$

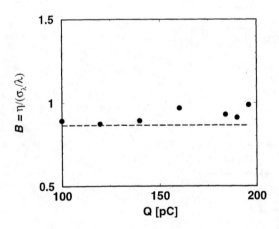

FIGURE 11. $\mathcal{B} = \eta/(\sigma_\lambda/\lambda)$ showing a constant value of $\mathcal{B} = 0.86$ where $l_b = 1.5mm$ in the estimate of η

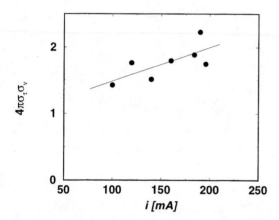

FIGURE 12. Measured Fourier transform product $4\pi\sigma_\nu\sigma_t$ as a function of current.

Observation of Coherent and Self-Amplified Spontaneous Emission in the mid-infrared at the CLIO Free-electron laser

J.M.Ortega, R.Prazeres, F.Glotin, D.A.Jaroszynski*

*LURE, bât. 209d, Université de Paris-Sud, 91405 Orsay cedex, FRANCE and
*University of Strathclyde, Dept. of Physics, John Anderson Building, 107
Rottenrow, Glasgow G4 0NG, UK*

Abstract : We have produced and analysed the "Coherent Spontaneous Emission" (CSE) and the "Self-Amplified Spontaneous Emission" (SASE) emitted by a relativistic electron beam passing through an undulator, for the first time in the mid-infrared. In this spectral range, the two phenomena are both very noisy and can be distinguished only by detailed spectral measurements. The spectral behaviour of the SASE line is different from CSE and exhibits an unexpected growth at the start-up of the process.

I - Introduction :

There are many free-electron lasers (FEL) in the world, operating in various wavelength regions from milimetre waves to Ultra-Violet[1,2]. In principle, the FEL will operate in the X-ray spectral range. However, at present, the quality of the optical cavity mirrors and of the electron beam have not been sufficient to produce the FEL oscillation at wavelengths shorter than 240 nm[3]. In the infrared range however, the gain can be larger than 100% much larger than the few percents achieved in the UV. Other related techniques, noticeably the VUV harmonic generation in an undulator[4] and X-ray generation by FEL intracavity Compton backscattering[5] have been demonstrated, but are producing small power. Nevertheless, a solution has been proposed about ten years ago to reach the X-ray range[6] by operating with a very high gain in single pass configuration, thus avoiding mirrors : the "spontaneous emission" produced by the electrons is amplified in a very long undulator (20 to 40 m), and reaches saturation in one pass. This so-called "Self-Amplified Spontaneous Emission" (SASE) requires a very high quality electron beam (high peak current, low energy spread, small emittance). The SASE has so-far only been observed in cm waves[7] and in the millimetre[8] spectral ranges. Indeed, the electron beam requirements needed for SASE are more and more demanding as the wavelength decreases. Thus study of SASE in the mid-infrared region is an important step in understanding the process and in extrapolating to the possible developement of SASE sources in X-rays. We have been able to produce it in the spectral region of 5 - 10 μm[9].

CP413, *Towards X-Ray Free Electron Lasers*
edited by R. Bonifacio and W. A. Barletta
© 1997 The American Institute of Physics 1-56396-744-8/97/$10.00

The gain medium of the FEL is a relativistic electron beam crossing a magnetic device (called "undulator"). This system produces synchrotron radiation (called "spontaneous emission") that is amplified by transfer of energy from the electrons to the optical wave. In an oscillator configuration, the undulator is installed in an optical cavity which stores the spontaneous emission and allows laser operation. The wavelength is easily tunable over a large range, by adjusting the magnetic field of the undulator and the electron beam energy. In a superradiant configuration using SASE, the undulator has to be much longer if one wants to reach saturation intensity . However, SASE can be observed below saturation in order to study the start-up of the process.

The SASE process is only one aspect of the radiation which occurs in the FEL, and which involves either "Spontaneous Emission", "Coherent Spontaneous Emission", "FEL Gain" or "SASE" :

The "Spontaneous Emission" (SE) is the radiation produced by a single electron, travelling along an undulator of N magnetic periods. It is peaking at the so-called resonance wavelength $\lambda_R = \lambda_u (1+K^2/2)/2\gamma^2$ where λ_u and K are respectively the period and the "deflection parameter" of the undulator, and $\gamma\, mc^2$ is the electron beam energy. The single electron radiation is a wave-train of N periods, corresponding to a length of $N\lambda_R$. The specral linewidth of the radiation is :

$$\Delta\omega/\omega \cong 0.9/N$$

Considering a bunch of N_e electrons, and assuming that the electrons density is uniformly distributed in the bunch, the electrons are incoherent sources, and the energy produced by SE scales as $N.N_e$ (on-axis and in a bandpass < 1/N, this energy is proportionnal to N^2, due to the undulator interference effect)

The "Coherent Spontaneous Emission" (CSE) is observed if the dimension of the electron bunch is smaller than the emitted wavelength. In this case, the individual sources (electrons) are in phase and the total emitted energy scales as $N.N_e^2$, i.e. linearly with the undulator length, and quadratically with the electron current. In most FELs, including CLIO, the electron bunch length is larger than the emitted wavelength, and the CSE should be negligible. Nevertheless, a partial coherence still occurs if the electron bunch longitudinal density exhibits structures like sharp edges, or more generally, strong components at high frequency in its Fourier spectrum. When CSE is strong, the SE, which is the remaining phase noise, becomes negligible. When CSE is weak, compared to its maximum level all electrons contributing), one can consider that, at first order, a CSE component of intensity $I_{coh}(\omega)$ adds to the SE intensity $I_{se}(\omega) = I_0(\omega).N_e$, where I_0 is the intensity emitted by one electron.

82

The CSE component is described by the diffraction theory which yields a coherent intensity :

$$I_{coh}(\omega) = I_0(\omega).N_e^2.f^2(\omega)$$

where $f(\omega)$ is the Fourier transform of the longitudinal electron density.

The total intensity is :

$$I_T(\omega) = I_{coh} + I_{se} \approx N.(i + f^2.i^2),$$

where i is the electron beam average current. f is also a function of i, when the electron longitudinal shape varies with the average current.
Both SE and the CSE are "spontaneous emission" and have, in first approximation, the same spectral distribution.

Another method of increasing the emission of light is through the FEL gain process, induced by the interaction between the electron bunch and an optical wave, which creates a periodical modulation (micro-bunching) on the electron beam distribution, and produces a strong Fourier component at the resonnant wavelength (and possibly its harmonics). This component adds to the initial optical wave and constitutes the "gain" : with an adequate optical cavity, it produces the FEL oscillation[2]. Also, due to the gain, the spontaneous emission which is produced along the undulator is somewhat amplified in a single pass. This kind of radiation is called "Self-Amplified Spontaneous Emission" (SASE). The analysis of this process has been carried out by several authors[6]. The power of SASE grows exponentialy along the undulator axis z, with :

$$P_{sase} \propto \exp(z/L_g) \qquad \text{where } L_g = \lambda_u/(4\pi\rho\sqrt{3})$$

L_g is the "gain length", which characterizes the exponential growth of SASE and depends on a dimensionless "Pierce parameter", ρ, which is proportionnal to $(\hat{i}/\sigma_e\gamma^3)$. $\hat{i}$ is the peak electron current, σ_e, its transverse size and γmc^2, its energy. Therefore SASE requires higher intensities and smaller emittance (transverse size) as the energy increases, i.e. as the desired wavelength is smaller. The saturation occurs for $\rho N \cong 1$. Far from saturation ($\rho N \ll 1$), the radiation is the spontaneous emission of an incoherent electron beam, the spectral width being $\sim 1/N$. In the exponential growth regime, one expects a reduction of the spectral linewidth to $\Delta\omega/\omega \cong (\rho N)^{1/2}/N$.

II - The CLIO infrared FEL facility :

We present here the successful production and observation of SASE in the mid-infrared region at $\lambda \cong 5\ \mu m$ and $10\ \mu m$. These observations have been carried out with the "CLIO" FEL. CLIO is an infrared free-electron laser and a user facility since 1993[10]. CLIO is offering more than 2000 hours/year of beam time to users working mostly in the fields of Surface and Solid state physics, Electrochemical interfaces, Molecular dynamics, Near-field microscopy and Medicine. Although devoted to users, CLIO is using some of the beam time to study FEL basic processes and improvements[4,5,9-12]. For example, the 2-color FEL is an original property of CLIO[12], which enhances its applications capabilities in the field of time resolved pump-probe experiments.

CLIO is based on a linear accelerator. The machine characteristics are the followings :

Accelerator :	type	linear, radio-frequency, 3 GHz
	gun	thermo-ionic, 1 ns pulse
	pre-buncher	500 MHz reentrant cavity
	buncher	3 GHz, 5 MeV, SW
	accelerator	3 GHz, 50 MeV, TW
	maximum energy	50 MeV
	minimum energy	20 MeV
	magnetic bend	doubly achromatic
		nearly isochronous ($\cong 1$ ps/%)
	peak current	100 A
	90% emittance	150 π.mm.mrad (normalized)
	energy spread (FWHM)	1 %
Time structure :	macropulse length	11 μs
	repetition rate	1 - 50 Hz
	micropulse length	8 ps (measured)
	micropulse separation	4 - 32 ns
FEL :	undulator period	50.4 mm
	nb. of periods : N	19 (for each undulator section)
	measured gain per pass	up to 500 %
	(laser rise time at start-up)	
	Pierce parameter : ρ	1.9×10^{-3}
	Laser range	3 - 53 μm

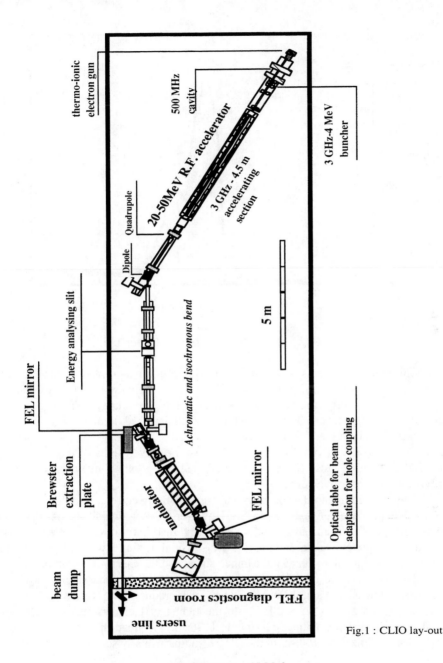

Fig.1 : CLIO lay-out

FIGURE 1 : CLIO lay-out

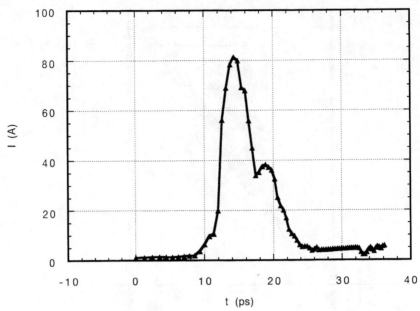

FIGURE 2 : Measured shape of the electron pulse

The pulse length has been measured with a collimator placed in an energy dispersive section of the magnetic bend : the beam is dephased in the accelerating section so as to transform its time dependance into an energy dependance. The resolution of the method can be very good if it is well designed. In our case it is limited by the fact that some bunch evolution may still occur in the long accelerating section and we estimated it to 1 ps (300 μm), well above the wavelengths of interest (5 - 10 μm). However, the pulse is very asymetric[11] so that it may have Fourier components at short wavelengths.

The same collimator has been used to vary the beam current, in order to look at the current dependance of SASE and CSE. However, moving one slit across the beam selects parts of the beam having different energy distribution which make it more difficult to interprete the data. In fact, it is very difficult to find a way to vary the peak current while maintaining constant the bunch shape and the energy distribution.

The thermo-ionic gun makes the machine very reliable and stable. However, there are small instabilities in the bunch shape which produce large fluctuations on the observed CSE as well as on SASE. Also, the tuning of the machine is much more critical for CSE and SASE than for the FEL itself, the quadrupoles acting on the achromatism and the isochronism of the bend being particularly sensitive.

The CLIO spectral range spans 3 to 50 μm at electron energies ranging from 50 to 20 MeV. The present experiment have been performed at 50 MeV because, on one hand, we always observed a better FEL at this energy and, on the other hand, in order to avoid too much CSE, since the observed SASE is far from saturation.

The undulator of CLIO has 38 magnetic periods, divided in two half-undulator sections of N=19 periods, for which each gap is independantly adjustable. This feature is made to run the FEL in a two colour mode[12], but it also should allow discrimination between SASE and CSE effects, because CSE scales linearly with the undulator length and quadratically with the current, whereas SASE scales exponentialy.

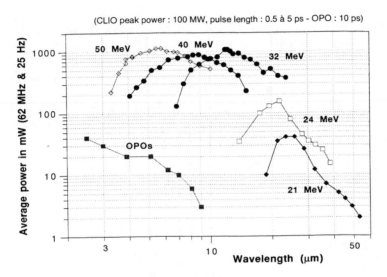

FIGURE 3 : CLIO spectral range

III - Observation of CSE :

The spontaneous emission enhancement is monitored through the signal delivered by an infrared detector (InSb or HCT) placed on-axis in front of the undulator. The radiation is observed in an angular aperture sufficiently small to avoid spectral broadening when the spectrum is recorded. Production of either CSE or SASE requires different settings of the accelerator :

- Quadrupoles settings in the bend : they act on the beam focusing in the undulator. Normally, they do not influence the CSE, which is sensitive only to the longitudinal shape of the bunch. However, this shape is sensitive to the magnetic elements acting on the bend isochronicity. Indeed, we found that CSE was less critically depending on these tunings, since SASE is strongly influenced by the beam focusing in the undulator, as we will see below.
- RF phases between the pre-buncher, buncher and accelerating section. These tunings act strongly on both the electron peak current and longitudinal shape, in a manner that

cannot be predicted in practice : the CSE and SASE levels obtained being far from saturation, either one can be produced by a small sub-structure of the electron bunch such as a spike (CSE or SASE) or a sharp edge (CSE).

When one maximises the spontaneous emission level to its maximum, one never knows whether the path taken will lead ultimately to CSE rather than SASE. Apparently, the optimisation seems to favor only one type, rather than make CSE and SASE to coexist. Therefore, they have to be distinguished by intensity and spectral measurements. Let us begin with CSE.

The intensity dependence of CSE has been measured, using the collimator placed in the bend. The curve exhibits a step, far from the quadratic dependence expected for CSE. It can be understood, if we assume that the bunch has a substructure exhibited only by the particles possessing nearly the nominal energy (center of the curve). It is reasonable to think that the particles on the wings of the energy distribution have a smoother longitudinal distribution : therefore, they cannot contribute to CSE although they can still contribute to the electron peak current and, hence, to SASE, as we will see below.

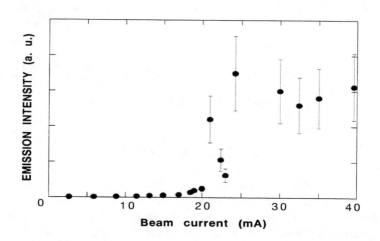

FIGURE 4 : Intensity dependence of CSE with the collimator slit position

The spectral behavior of the emitted radiation is the more convincing evidence : at any level, adjusted by the phases, the spectral distribution is identical and possesses a constant spectral width of 3.3% (fig.5). The simple theory yields 2.3%, but the spectral width has always been found to be larger, due to inhomogeneous effects (emittance, in particular).

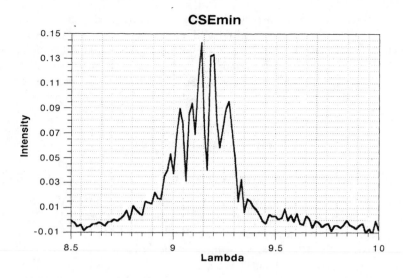

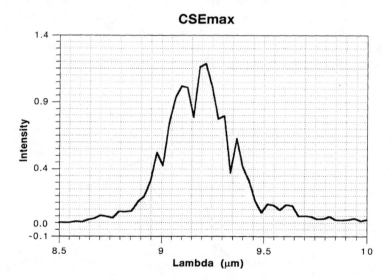

FIGURE 5: Spectra of CSE at various (minimum and maximum) levels

89

For clarity the signals displayed on fig. 5 are averaged : in fact, the noise affecting the CSE is of the order of 100%. This is due to the fact that very small changes in the bunch longitudinal shape produce very large changes in the high ranking Fourier coefficients : indeed, this would not happen at wavelength longer than the electron bunch length. In our case, it is worth to notice that we produce coherent emission at wavelengths (5-10 μm), of the order of 1/1000 of the electron bunch length.

IV - Observation of SASE :

With the above parameters, the gain length L_g is about 1m, and the saturation parameter is $\rho.2N=0.07 \ll 1$ in our case. This indicates that the SASE radiation is necessarily far from saturation, which would occur for 500 periods. For proper settings of the machine parameters, an appreciable level of SASE can be observed. Its is very sensitive to the RF phase adjustments. The influence of such phase tuning, leaving the average current unchanged, is displayed on fig. 6 : the spontaneous emission intensity is strongly affected. One set of curves ("phase ON") has been obtained with optimum RF phase adjustement, corresponding to a strong maximum of emission intensity, and the other set of curves ("phase OFF") has been obtained with a detuned RF phase. While the average current remains constant, the peak current diminishes by approximately 50%[11]. This is an evident proof that SASE or CSE process occurs, since the SE is strictly proportionnal to the average current. The very strong observed effect pleads for SASE rather than CSE. This we discuss below.

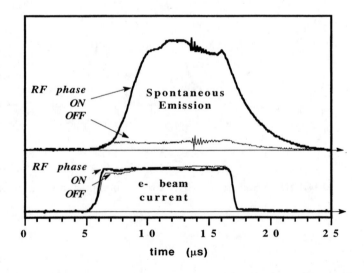

FIGURE 6 : Influence of the RF phase tuning, at nearly constant average beam current, on the spontaneous intensity (top curves) during the electron macropulse.

90

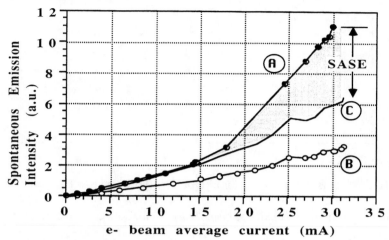

FIGURE 7 : Intensity of SASE versus electron beam current. Curves A and B correspond respectively to 2N and N (= 19) periods undulators. Curve C is 2 times curve B.

Fig. 7 shows the spontaneous emission intensity as a function of the electron beam current with one or two undulators of equal number of periods. Beam current has been varied by controling the aperture of a beam slit of the collimator. The curve A (2 undulators) exhibits clearly a nonlinear behaviour (the curve B, for 1 undulator, also, though less obvious) which implies a coherence effect such as CSE and/or SASE. The curve C is the curve B multiplied by a factor of 2. Since the CSE (and the SE) scales linearly with the undulator length, the difference between curves A and C is necessarily due to the presence of SASE. Therefore SASE is present, which is also shown by its nonlinear behavior, although some CSE may also exist. The SASE amplification acts on the total spontaneous emission : SE+CSE. It may occur also in the first undulator, which may be responsible for the small nonlinearity of curve B, but the accuracy is not sufficient to determine whether this behaviour is exponential (SASE) or quadratic (CSE).

These curves have been taken with an electron beam transverse size adjusted for the FEL oscillator. When one adjusts the size to maximise SASE, the intensity is only slightly increased by the presence of the second undulator : in this case the ρ parameter is maximized by a very small electron beam size in the center of the first undulator. Then ρ becomes almost negligible in the second undulator, due to the divergence of the beam following a very small focus.

91

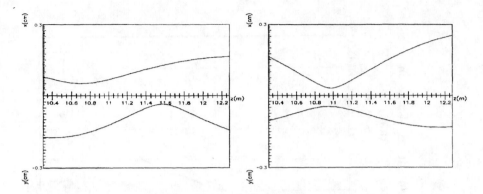

FIGURE 8 : Beam envelopes for FEL or CSE (left), and SASE experiments (right)

The spectrum of the SASE has been measured for various intensities of SASE :

- A spectrum is displayed on fig. 9 for the case corresponding to the FEL beam adjustment of fig.8. It is taken at 5 μm, so that we can use a sensitive InSb detector and measure both the SE and SASE : clearly a moderate amplification appears which is located at a slightly longer wavelength $\Delta\lambda/\lambda = 1.4\%$, close to the theoretical value of 1/2N expected for FEL gain.

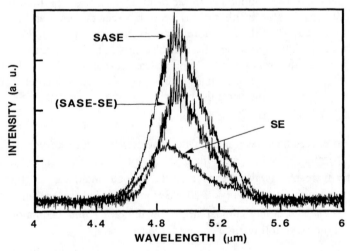

FIGURE 9 : Spectra of the emission with and without SASE and their difference at $\lambda = 5$ μm, with the "FEL" adjustement.

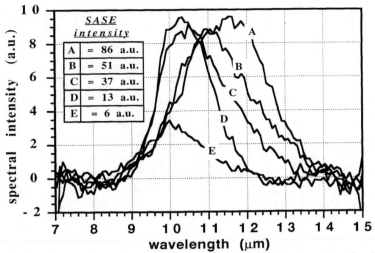

FIGURE 10 : Spectra of SASE for various SASE intensities (varying the linac peak current) for the best beam tuning at λ = 10 μm.

- In fig. 10, we display the spectra obtained with the best beam adjustment, at 10 μm, where the detector (HgCdTe) is not sensitive enough to measure the SE : When SASE increases, the spectrum linewidth increases and the central wavelength shifts toward large values. The larger spectrum, corresponding to the higher SASE intensity, is displaced by 15% (at 11.5 μm), and has a linewidth of 23%. The resonance wavelength shift cannot be explained either by an angular error of 8 mrad or by a relative energy variation of 7%, which would cause large beam losses.

The experimental increase of the linewidth up to 23% is not in agreement with the SASE theory, which predicts a narrowing of the spectrum as compared to the SE. However, this theory considers the exponential growth of intensity rather than the start-up regime, as it is the case here. Here, if we assume this width to be Fourier limited, a linewidth of $\Delta\lambda/\lambda = 23\%$ corresponds to a wave train of about $\Delta z \cong 40$ μm. This value is 10 times shorter than the length, $N\lambda = 400$ μm, of the SE wave train. Such a short pulse regime is likely to be due to the fact that the electron bunch profile has a sharp maximum, that produces SASE, of the order of 1 ps, i.e. 300 μm or even shorter. This effective electron pulse length is of the order of the slippage length : 200 μm for one undulator (since only one undulator is efficient in the SASE best adjustment). Therefore, only a fraction of the SE wave train can be amplified, leading to the observed linewidth increase. The best phase tuning for SASE corresponds necessarily to the shortest spike in the electron bunch structure since the average current remains constant.

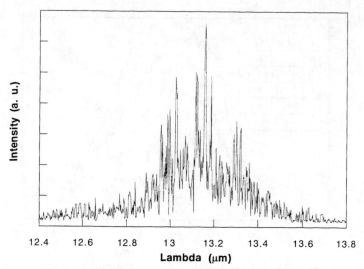

FIGURE 11 : SASE spectrum without averaging

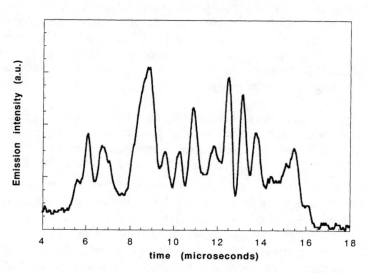

FIGURE 12 : SASE recorded with a time resolving detector

Finally, let us point out that CSE, like SASE, is very noisy. Fig. 11 shows a SASE specrum recorded without averaging : It consists in a series of lines. However, it is not clear whether this behaviour comes from the expected spectral fluctuations of SASE or from its intensity fluctuations. The time dependance of SASE during the macropulse is shown on fig. 12, similar to fig. 5 except it is recorded with a fast detector : It is extremely noisy along the macropulse. Beside the intrinsic unstable nature of below saturation SASE, this may be due to small RF phase variations (of the order of one degree[10]) which modify the bunch shape during the bunching process. Similar temporal behaviours have been recorded with CSE. Therefore a discrimination between CSE and SASE based on an analysis of the level of noise is very difficult.

In conclusion, the high gain, which can be obtained with the CLIO infrared Free Electron Laser, has allowed us to observe "Self-Amplification Spontaneous Emission" (SASE) and Coherent spontaneous emission (CSE) in the mid-infrared region, around 5 and 10 μm. Careful spectral measurements have allowed us to discriminate between them. Although SASE is far from saturation, since the amplification is only one of order of magnitude, we have been able to measure the spectral behaviour at start-up and to observe an unexpected growth of the linewidth. In certain conditions, SASE is absent and CSE can be observed at wavelength as short as 1/1000 of the electron bunch length. Therefore CSE is likely to be observed on experiments aimed at producing SASE, since they plan to use very short electron bunches. These will be produced, as in the case of CLIO, by velocity modulation imprinted in an accelerating element followed by a drift or a dispersive magnetic bend (magnetic compression). Indeed, CSE could be an easier way to produce coherent light at short wavelength, since it is sensitive only to the longitudinal structure of the bunch but much less than SASE to its emittance and energy spread.

REFERENCES :

1. Proceedings of FEL'95 conference, *Nucl.Instr.and Meth.*, **A375** (1996)
2. *"Laser Handbook"*, vol.6 (W.B.Colson, C.Pellegrini, A.Renieri. ed., North-Holland -1990)
3. G. Kulipanov et al., *Nucl.Instr.and Meth.*, **A296**, 1 (1990)
4. R. Prazeres et al., *Europhys. Lett.*, **4**, 817 (1987)
5. F. Glotin et al., *Phys. Rev. Lett.*, **77**, 3130 (1996)
6. B.Bonifacio et al. , *Opt. Commun,.* **50**, 373 (1984)
    J. Murphy, C. Pellegrini, *Nucl.Instr.and Meth.*, **A237**, 159 (1985)
    K-J. Kim, *Phys. Rev. Let.*, **57**, 1871 (1986)
7. T. Orzechowski, *Nucl.Instr.and Meth.*, **A250**, 144 (1986)
8. D. Kirkpatrick., *Nucl.Instr.and Me.*, **A285**, 43 (1989)
9. R.Prazeres, J.M.Ortega, F.Glotin, D.A.Jaroszynski, O.Marcouillé
    Phys. Rev. Lett., **78**, 2124 (1997)
10. J.M. Ortega et al., *Nucl.Instr.and Met.*, **A375**, 618 (1996)
    R. Chaput et al., *Nucl.Instr.and Met..* **A331**, 267 (1993)
11. F. Glotin et al., *Nucl.Instr.and Met.*, **A341**, 49 (1994)
12. D.A.Jaroszynski et al., *Phys. Rev. Lett.*, **72**, 2387 (1994)

The TESLA Test Facility FEL: Specifications, Status and Time Schedule

B. Faatz

Deutsches Elektronen-Synchrotron (DESY),
Notkestrasse 85, 22603 Hamburg, Germany

Abstract

The TESLA Test Facility (TTF) FEL is a user facility under construction at DESY in Hamburg. The radiation wavelength will reach 6 nm, not including the planned higher harmonic radiation generated in a 1 to 1.5 meter long radiator, at a power level of several GW. Specific to the FEL are the low emittance photo injector gun, the bunch compressors, and the undulator with integrated FODO lattice.

Introduction

The TTF-FEL project will be built in two stages [1]. The first stage, which is already under construction, has as main purpose to test the different (FEL specific) components, such as low-emittance injector, bunch compressors and undulator. In addition, it will be the first proof of principle of Self Amplified Spontaneous Emission (SASE) in this wavelength range. For phase two of the project, both accelerator and undulator will be extended to go down in wavelength to 6 nm. In this configuration the machine will be employed as user facility. The peak brilliance of the radiation delivered to the user is shown in Fig. 1. Note its increase compared to existing synchrotron sources by several orders of magnitude.

Major research efforts for the FEL

A large number of subjects is under investigation in order to improve and characterize the performance of the TTF FEL. They can be subdivided in to several areas.

- Make/preserve low emittance. For the FEL process, the beam emittance and energy spread have to be as small as possible along the entire beam line. Because the influence of the electron gun and the several bunch

CP413, *Towards X-Ray Free Electron Lasers*
edited by R. Bonifacio and W. A. Barletta
© 1997 The American Institute of Physics 1-56396-744-8/97/$10.00

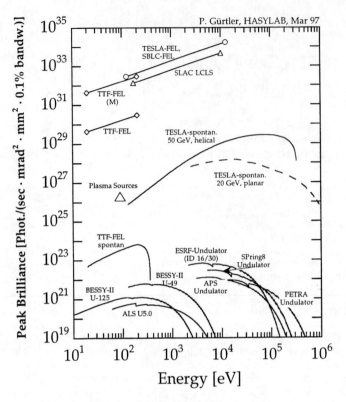

Figure 1: *Peak brilliance of different FELs and synchrotron sources.*

compressors in the entire system are the main source of emittance growth, they have been studied in detail [2, 3].

- Coherence/monochromaticity properties of the SASE FEL radiation. For certain types of experiments, the spectral characteristics of the FEL output radiation, e.g. the bandwidth and/or spikes, needs to be improved. Studies have shown that this can be done by using a two stage undulator. The first undulator is used to generate radiation at low power, which is then monochromatized. The electron beam is debunched and amplifies the small bandwidth radiation in the second undulator. The spectral peak brilliance increases approximately by two orders of magnitude (see in Fig. 1 TTF-FEL (M)) [4].

Table 1: *Parameters of the TTF VUV-FEL*

	Phase I	Phase II
Electron beam		
Energy	390 MeV	1 GeV
Peak current	500 A	2500 A
Normalized rms emittance	2π mm mrad	2π mm mrad
rms energy spread	0.17 %	0.1 %
External β-function	1 m	3 m
rms electron beam size	57 μm	55 μm
Undulator		
Type	Planar	Planar
Period	27.3 mm	27.3 mm
Peak magnetic field	0.497 T	0.497 T
Magnetic gap	12 mm	12 mm
Undulator section length	4.5 m	4.5 m
Total undulator length	15 m	30 m
Drift section length	327.6 mm	327.6 mm
FODO-lattice		
Period	955.5 mm	955.5 mm
Quadrupole length	136.5 mm	136.5 mm
Quadrupole strength	18.3 T/m	18.3 T/m
Radiation		
Period	42 nm	6 nm
Saturation length	13.5 m	21 m

- Electron beam diagnostics. A specific problem for a single pass, high-gain FEL, is the tight tolerance on the electron beam alignment. Several methods have been proposed to measure and correct the beam trajectory. One of them is beam based alignment, in which several beam position monitors along the undulator are used to measure the dispersion of the electron beam trajectory which is then corrected, resulting in a straight trajectory [5, 6]. Another method uses the undulator radiation emitted by the electron beam under a small angle. A pinhole camera placed off-axis measures the radial dependence of this radiation and determines the complete electron trajectory, from which corrector settings are calculated [7].

- Photon beam diagnostics. Although some aspects, such as the energy loss spectrum of the electron beam, give information about the interaction process that has taken place, most information is obtained from measurements of the photon beam properties. In order to determine what kind of amplification process has taken place, one has to measure both power and spectral properties of the radiation field. For users it is important to know what the properties of the radiation are that they use for their experiments, such as spectrum, peak power and its fluctuations, etc.

Time schedule of the Project

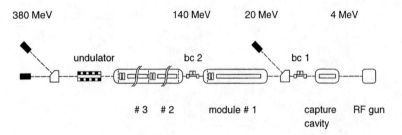

Figure 2: *Layout of phase I of the TESLA Test Facility FEL at DESY*

Phase I of the project, used for the proof of principle of the SASE process and for testing of the different components and procedures, will take the next two to three years. The year 1997 is planned for testing the first accelerator module with a thermionic gun and first magnetic measurements of a prototype of the undulator. The next year is planned for the entire accelerator with a high charge photocathode gun. At the end of that year and the beginning of 1999, the FEL gun will be installed, as well as the undulator. In this configuration, given in Fig. 2, the first FEL tests will be performed. For the final user facility, the accelerator will be extended with five additional accelerator modules, one additional bunch compressor and the undulator length will be doubled.

References

[1] *A VUV Free Electron Laser at the TESLA Test Facility: Conceptual Design Report*, DESY TESLA-FEL 95-03, Hamburg, 1995.

[2] B. Dwersteg, K. Flöttmann, J. Sekutowicz, Ch. Stolzenberg, *RF Gun for the TESLA VUV Free Electron Laser*, Presented at the FEL96 Conference, August 26-31 in Rome, Italy.

[3] M. Dohlus, T. Limberg, *Emittance growth due to Wakefields on Curved Bunch Trajectories*, Presented at the FEL96 Conference, August 26-31 in Rome, Italy.

[4] J. Feldhaus, E.L. Saldin, J. Schneider, E.A. Schneidmiller, M.V. Yurkov, *Possible Application of X-ray Optical Elements for Reducing the Spectral Bandwisth of an X-ray SASE FEL*, to appear in *Optics comm.*, August first 1997.

[5] P. Castro, *TTF Beam Based Alignment by Dispersion Correction using Micado algorithm*, DESY TESLA-FEL 97-04, Hamburg, 1995.

[6] K. Flöttmann, B. Faatz, E. Czuchry and J. Roßbach, *Beam Based alignment Procedure for an Undulator with Superimposed FODO Lattice*, DESY TESLA-FEL 97-05, Hamburg, 1995.

[7] J.S.T. Ng, *A Beam Trajectory Monitor using Spontaneous Undulator Radiation*, DESY TESLA-FEL 96-16, Hamburg, 1995.

[8] J. Feldhaus, Private communications.

LCLS: Status and Simulation Results*

Heinz-Dieter Nuhn

*Stanford Linear Accelerator Center,
Stanford University, Stanford, CA 94309-0210, USA*

Abstract. The collaboration for the Linac Coherent Light Source (LCLS) under the leadership of the Stanford Linear Accelerator Center (SLAC) has been preparing the design report since the middle of 1996 to be completed before the end of calendar year 1997. The work on the report greatly improved the understanding of problems and helped to develop solutions for many of them. The present article will focus on one particular aspect, the simulation of the FEL process, itself.

INTRODUCTION

The proposed x-ray free-electron-laser (FEL) project, LCLS [1], is based on ideas first formulated during the Workshop of Fourth Generation Light Sources held at SSRL in 1992 [2, 3]. It was recognized then that recent developments in ultra-low emittance guns, the advanced understanding of beam transport in high energy linacs from SLC experiments and NLC studies, as well as advances in precision undulator design make it now conceivable to push the high energy end of FEL technology from the UV into the x-ray regime. The workshop on scientific applications of coherent x-rays recommended moving the proposed range of operations down to wavelengths of 1.5 Å [4].

Such a device has to be built as a single pass high gain FEL based on the physics of Self Amplified Spontaneous Emission (SASE) since there are no practical solutions for mirrors to build oscillator resonators in the x-ray regime. For the x-ray FEL to saturate within a reasonable length, demanding requirements have to be met by the electron beam, such as multi-GeV energy, diffraction limited emittance, and several thousand amperes of peak current. It was also recognized that the parameters can not be achieved in a storage ring: a high energy linac is needed. The LCLS group was established to take advantage of the SLAC linac, the only device that is presently capable of delivering the required electron parameters. The cost of a new linac greatly exceeds the costs of the other components, needed.

The DESY laboratory at Hamburg in Germany adopted the LCLS idea and is including x-ray FELs in their future NLC proposal. They are presently building a test FEL at longer wavelengths at the TESLA Test Facility [5].

*Work supported by Department of Energy contract DE-AC03-76SF0015 and Office of Basic Energy Sciences, Division of Chemical Sciences.

CP413, *Towards X-Ray Free Electron Lasers*
edited by R. Bonifacio and W. A. Barletta
© 1997 The American Institute of Physics 1-56396-744-8/97/$10.00

The central part of the LCLS is the interaction of electromagnetic radiation and electrons in the FEL undulator that will enhance the radiation power and eventually saturate at multi GW power levels. The understanding of this process, the properties of the radiation and the tolerances that have to be put on the electron beam parameters as well as the design, construction and alignment of the FEL undulator are of central importance for the successful operation of such a device.

While the theoretical understanding of the SASE process has been greatly advanced during the last decade [6, 7, 8, 9, 10], extensive computer simulations are still required, especially to establish parameter tolerances. The present report will discuss some of the LCLS issues that are studied using computer simulations.

LCLS Contributors

LCLS is a multi-laboratory effort with contributions from the

Stanford Linear Accelerator Center
J. Arthur, K. Bane, V. Bharadway, G. Bowden, R. Boyce, R. Carr, J. Clendenin, W. Corbett, M. Cornacchia, T. Cremer, P. Emma, A. Fasso, C. Field, A. Fisher, R. Gould, R. Hettel, J. Humphrey, K. Ko, T. Kotseroglou, Z. Li, D. Martin, B. McSwain, R. Miller, C. Ng, H.-D. Nuhn, D. Palmer, M. Pietryka, S. Rokni, R. Ruland, J. Sheppard, R. Tatchyn, V. Vylet, D. Walz, H. Winick, M. Woodley, A. Yeremian, R. Yotam

University of California, Los Angeles
C. Pellegrini

Lawrence Berkeley National Laboratory
W. Fawley, K. Halbach, K.-J. Kim, S. Lidia, R. Schlueter, M. Xie

Lawrence Livermore National Laboratory
L. Bertolini, K. van Bibber, L. Griffith, M. Libkind, R. Moore, E.T. Scharlemann

European Synchrotron Radiation Facility
A. Freund

University of Rochester
D. Meyerhofer

INFN, University of Milan
L. Serafini

Los Alamos National Laboratory

LCLS Layout

After the SLAC B-Factory starts operating in 1998, the last kilometer of the SLAC linac, which is not used for B-Factory injection, will be available to accelerate electron bunches to final energies between about 5 and 15 GeV. The electron bunches will come from a photocathode rf gun capable of producing several picosecond long bunches with a charge of about 1 nC at normalized emittance values down to about 1 mm mrad.

Bunch compression, necessary to increase the peak current from 100 to 3400 A, as required for the FEL process, will be done in two stages in the linac, at the 285 MeV and 6.5 GeV energy points. During compression the bunchlength becomes extremely short, which will lead to very strong coherent synchrotron radiation at wavelengths equal to or larger than the bunch length. The radiation fields are expected to interact with the electrons, thus producing longitudinal misalignments of transverse bunch slices (integrated emittance increase). The design of the second bunch compressor stage was recently changed to suppress the effects of coherent synchrotron radiation [11].

A 100 m long FEL undulator will be housed in the existing FFTB tunnel. Experimental stations are being designed to use the 280 fsec (fwhm) long x-ray pulses produced by the undulator.

X-RAY FEL SIMULATIONS ISSUES

Many aspects of the FEL process are being studied using computer modelling, either based on a semi-analytical approach [12] or on numerical simulation codes [13, 14, 15]. These x-ray FEL modeling issues include:

- General FEL Parameter Optimization
- Magnet Error Tolerances (Performance degradation as function of random walk amplitudes and phase error)
- Beam Position Control

    - Steering Algorithm
    - Optimum Steering Station Separation
    - Steering Error Tolerance

- Focusing

    - Focusing Method (Lattice)
    - Optimum Beam Size
    - Tolerable Beta-Function Modulation-Amplitude
    - z Dependence of Optimum Beam Size

- Spontaneous Radiation

    - Average Energy Loss (Effect and its Compensation)
    - Energy Spread Increase
    - Emittance Increase

- Output Power Control
    - Peak Current
    - Phase Shifter at Saturation
- Pulse-to-Pulse Output Power Stability
- Initial Phase Space Distribution (Coupling to LINAC Simulation Results)
    - Transverse Halo
    - Momentum Distribution
    - Electron Beam Pointing Tolerances
    - Beam Size Matching
- Optical Klystron
- Wiggle Plane vs. Out of Plane Tolerances
- z Dependence of Error Tolerances
- Effect of External Dispersion Function
- Wakefields
- Coherent Synchrotron Radiation

FEL Simulation Codes

TABLE 1. List of FEL simulation codes.

Name	Avail.	3D	Noise Startup	Error Treatment	Time Domain	Contact Person
FRED	LLNL	No				E.T. Scharlemann
NUTMEG	LLNL	No				E.T. Scharlemann
GINGER	Public	No	Yes		Yes	W.F. Fawley
FRED3D	LLNL	Yes		Yes		E.T. Scharlemann
TDA3D	Public	Yes		Yes		G. Travish
FELEX	LANL	Yes		Yes		J. Goldstein
MEDUSA	(Public)[1]	Yes	Yes	Yes		H.F. Freund
FEL3D	(Public)[1]	Yes	Yes	Yes	Yes	V.N. Litvinenko
	APS[1]	Yes	Yes	Yes		N. Vinokurov / R. Dejus

[1] Under development

Simulation codes are important tools to characterize the physics of x-ray FELs. The higher dimensional simulation codes used for the LCLS design are GINGER (Noise Startup, Time dependent), FRED3D (Magnet Error Analysis, Beam Position Control), TDA3D (Effects of Focusing), and NUTMEG (Harmonic Generation). The codes NUTMEG, GINGER and FRED3D were developed about 10 years ago at LLNL for the ELF experiment. They have been extensively cross-checked with experimental results from ELF and are under full control of the authors. TDA3D was developed during the last 10 years starting at MIT. It is now widely available but has very little version control.

As is apparent from table 1, there is no code available, yet that can handle all aspects related to the design of an x-ray FEL. New codes are presently under development aiming towards the goal to achieve such a completeness.

FEL Undulator Length History

During the development of the LCLS the proposed total undulator length was increased several times as studies indicated that some of the selected parameters needed to be moved to a more conservative regime, thus increasing the saturation length L_{sat}. Papers published over the course of this development, therefore discuss LCLS performance at very different levels. Table 2 summarizes that development, hopefully taking some of the confusion away.

TABLE 2. Development of saturation length during the LCLS design process.

	Undulator Type	Peak Current	normalized Emittance	Saturation Length
- 1994/95	helical	5000 A	1 mm mrad	30 m
- 1995	planar	5000 A	1 mm mrad	55 m
1996	planar	5000 A	1.5 mm mrad	78 m
1997	planar	3400 A	1.5 mm mrad	94 m

LCLS Parameters for the 1.5 to 15 Å range

Table 3 lists parameters of the LCLS for the boundaries of the proposed operational range from 15 Å down to 1.5 Å.

FOCUSING

As the electron beam is transported through the undulator, its radius needs to be controlled using appropriate focusing. A planar undulator focuses in the plane perpendicular to the wiggle motion, only (natural focusing). The focusing strength can be expressed by specifying the beta-function that is kept constant by that amount of focusing:

$$\beta_x^{nat} = \sqrt{2}\gamma/k_w K$$

By appropriately shaping the pole faces one can direct half of the focusing into the wiggle plane. This type of constant focusing in both planes is called 'Ted-Pole' focusing

$$\beta_{x,y}^{TP} = \beta_x^{nat}/\sqrt{2}$$

107

TABLE 3. LCLS parameter list.

Parameter	Equation	$\lambda_r = 1.5\,\text{Å}$	$\lambda_r = 15\,\text{Å}$	
Period Length	λ_w	3.0	3.0	cm
Undulator Wavenumber	$k_w = 2\pi/\lambda_w$	209.4	209.4	m
Magnetic Field	$\hat{B}_w$	1.32	1.32	T
Undulator gap	g	6	6	mm
K (undulator parameter)	$K = \hat{B}_w c / k_w m_e c^2$	3.71	3.71	
Quadrupole Focussing	$\beta_{x,y}^{ext}$	18.0	6.3	m / rad
Normalized Emittance	ϵ_n	1.5	2.0	mm mrad
Electron Beam Size	$\sigma_{x,y}$	31	37	μm
Electron Energy	E	14.35	4.54	GeV
Peak Current	$\hat{I}$	3400	3400	A
Bunch Length (FWHM)	L_b/c	280	280	fs
Cooperation Length	$2\pi\, L_c/c = \lambda_r/\rho c$	1.1	3.8	fs
Rayleigh Length	$Z_r = \pi 2\sigma_{x,y}^2/\lambda_r$	39	5.7	m
Pierce Parameter	ρ	0.00047	0.00131	
Gain Length (1D)	$L_G = \lambda_w/4\pi\sqrt{3}\rho$	2.93	1.05	m
Saturation Length (3D)	L_{sat}	94	32	m
Periods	N_w	3312	3312	
Saturation Power (3D)	P_{sat}	9.3	10.6	GW
Peak Spont. Power	$\hat{P}^{spont}$	81	8.1	GW
Peak Brilliance	$\hat{B}$	6.1	0.8	10^{32}*

* photons/s/mm^2/mr^2/.1%

The amount of focusing that can be obtained this way is often smaller than required for optimum FEL performance, especially for high energy and short wavelength applications.

Optimum Beam Size

The smaller the beam size, the larger the electron density for equal pulse length, which in turn results in a larger value for the FEL parameter ρ. 3D effects, especially diffraction, will eventually lead to a decrease in FEL performance when the beam size gets too small, in spite of the continuous increase of the ρ parameter. Figure 1 shows the effect as calculated with the fit formula developed by M. Xie [12]. The beta-function values that are required for minimum saturation length and maximum output power are different at a given energy and also change as the operational energy is varied between 15 and 5 GeV. To reduce costs, it is important to optimize the FEL for minimized undulator length, which means minimized saturation length at the shortest wavelength.

'Ted-Pole' focusing would provide about 70 m/rad beta function at 15 GeV and 22 m/rad at 5 GeV. The optimum beta-function values that result in minimum saturation length are 18 m/rad at 15 GeV and 6 m/rad at 5 GeV. As discussed later, it turns out that about the same amount of additional magnetic field gradient could be used in both cases. Using 'Ted-Pole' focusing, alone, would result in a 22 % increase of saturation length at 15 GeV (about 30-40% increase at 5 GeV).

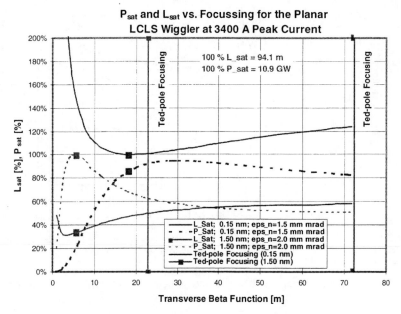

FIGURE 1. Effect of Electron Beam Focusing on LCLS Performance

Focusing Method (Lattice)

Focusing beyond natural or 'Ted-Pole' focusing can only be done with additional, external quadrupoles, which are focusing in one plane while defocusing in the perpendicular plane. A combination of two (FODO), three (Triplet), or more quadrupole magnets is required to provide focusing in both planes. Quadrupole focusing causes the longitudinal velocity of the electrons to change as they execute betatron oscillations away from the axis.

With quadrupole focusing the beta-function is no longer independent of z. We call the ratio $\Delta\beta/\beta = (\beta_{max} - \beta_{min})/\langle\beta\rangle$ the beta-function modulation. Beta-function modulation causes the beam envelope to change as a function of the longitudinal z axis. Beta-function modulation also causes amplitude dependent path length changes to the modulated electron trajectory. For the LCLS, a FODO lattice was selected. The amount of beta-function modulation is given by the length of the FODO cell.

Tolerable Beta-Function Modulation Amplitude

In storage ring and beam transport applications, the quadrupole strength of FODO cells are often optimized for a reduced maximum beta-function at a given cell spacing. This leads, for round beams, to a criterion that the optimum phase advance per cell is 90 degrees. For an FEL undulator, instead, the cell spacing should be optimized to reduce the maximum beta-function at a *constant value of the average beta-function*. The optimum cell spacing under this consideration would be zero, the continuous gradient

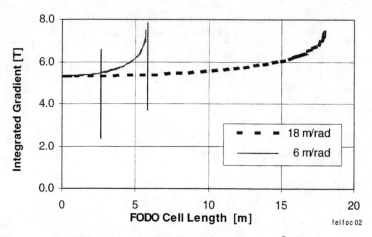

FIGURE 2. Integrated field gradient required to maintain constant average beta function values for a "lumped" FODO cell as function of cell length. The average beta function values are 18 m/rad at 14.35 GeV and 6 m/rad at 4.54 GeV.

lattice. This, of course, can not be achieved with a FODO lattice; a compromise is needed.

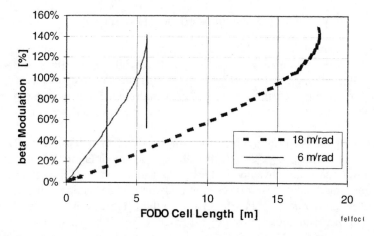

FIGURE 3. Beta-function modulation as function of quadrupole distance for LCLS boundary values of the average beta-function

The required strength of the quadrupoles does not strongly depend on the cell spacing as long as the cell spacing is less than about half the targeted average beta-function. For LCLS parameters, if a FODO cell length of less than about 3 m is chosen, the required integrated gradient would be the same over the 5 to 15 GeV energy range (figure 2). Here, permanent magnet quadrupoles can be used. For cell spacings close to the smallest average beta-function, quadrupole strength has to be adjusted to keep the optimum beta-function over the entire operational energy range. The beta-function modulation amplitude increases slowly with the FODO cell length (figure 3). The largest cell spacing possible for a given average beta-function is $L_{FODO}^{\max} = \langle \beta \rangle$, which corresponds to a beta-modulation amplitude of about 141% and a phase advance of 90 degrees per FODO cell. Larger beta-modulation amplitudes can be achieved with smaller FODO cell lengths but larger quadrupole gradients.

The question of the maximum allowable FODO cell length can only be answered by computer simulations. For the LCLS, the effect of beta-modulation amplitude on FEL performance has been studied with a version of the TDA3D [15] code as presently maintained by B. Faatz at DESY[16]. That version contains an implementation of lumped focusing by G. Travish [17].

TDA3D analysis for alternate gradient focusing done to date, which does not yet include magnet error and steering analysis, predicts the effects of beta-function modulation to be rather small. At the largest cell length that still would allow to produce the optimum beta-functions, the increase in saturation length is a few percent at 15 GeV (beta modulation < 30 %) and slightly above 10 % at 5 GeV (beta modulation ˜140 %). The increase of saturation length at the lower energy is not as critical since the saturation length is about one third of the length of the undulator.

A FODO cell length of 4.32 m is presently proposed for the LCLS, for which no significant performance degradation compared to constant gradient focusing is expected.

UNDULATOR SECTIONS

The LCLS undulator design uses a modular layout. Identical short sections are combined to form the 100 m device. Originally, separations between the sections were avoided due to potential detrimental effects on FEL performance. An uninterrupted long undulator becomes very complicated and costly. Focusing, diagnostics, orbit correctors and pumping have to fit into the very narrow undulator gap.

Recent studies, done both analytically and with computer simulations, found the effect of separations on FEL performance to be small[18, 19].

TABLE 4. Results from two "error-free" FRED3D runs simulating 4.32 m long undulator section separated by drift sections, 1 mm and 23.5 cm of length, respectively.

Separation Length	Total Inserted Drift to Saturation	L_{sat}	P_{sat}
0.001 m	0.022 m	94.5 m	14 GW
0.235 m	5.170 m	101 m	13.2 GW

Table 4 summarizes FRED-3D simulations for the LCLS: L_{sat} increases by 6.5 m when a total of 5.17 m of drifts is inserted each 23.5 cm long. The increase in actual undulator length needed to saturate is only 1.33 m.

The separations have to be designed to allow the electron bunches to slip an integral number of wavelengths behind the electromagnetic wave. This can be accomplished by choosing the correct length for the drift space (an integral multiple of $\lambda_w(1 + K^2/2)$, which is 23.5 cm for the LCLS) or by adding a phase shifter that corrects the slippage of a separation of different length.

The LCLS design now includes 23.5 cm long separations between undulator sections. A section length was changed to about 1.92 m (64 periods) so that every quadrupole of the 4.32 m FODO cell lattice, together with beam position monitors, orbit correctors and other components, will fit into the separations.

MAGNET ERROR TOLERANCES

One of the most significant sources of performance reductions comes from errors in the on-axis magnetic field [20]. The effect of magnetic field errors on LCLS FEL performance has been studied with the FRED-3D [13] simulation code on the NERSC computer systems at LLNL and LBNL. FRED-3D simulates the interaction between the electron beam and the optical field in the undulator of an FEL amplifier. The effects of random pole-to-pole errors in the undulator magnetic field on the centroid motion of the electron beam and on relative electron-to-radiation phase are included. In each half-period, a transverse momentum increment corresponding to the magnetic field error at that magnetic pole is added to the motion of each particle. The field errors are chosen from a truncated Gaussian distribution. The RMS fractional field error and the truncation level are specified as input parameters.

The random walk of the electron beam, generated by these errors, reduces the overlap between the electron and photon beams and also causes de-phasing of the electrons with respect to the FEL ponderomotive potential wells.

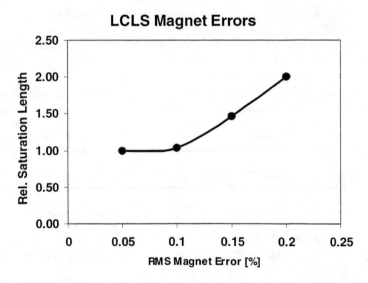

FIGURE 4. Saturation length as function of RMS Magnet Errors for 1.5 A LCLS FEL. The random walk orbit has been corrected using point-to-point steering with 5 m station separation. The connecting line is for visualization purposes, only.

The result shown in figure 4 have been obtained with FRED-3D for the 15 GeV LCLS using error free trajectory corrections spaced 5 m apart (see description below). They indicate a threshold in the sensitivity of the saturation length to rms magnet errors. If appropriate orbit correction is used, magnet errors have little or no influence on saturation length below an rms error distribution of about 0.1 %. Above that threshold, saturation length increases with error amplitude.

Steering Algorithm

The random walk can be partially corrected in FRED-3D by introducing "steering stations", at which the position of the electron beam is measured and a transverse momentum kick is applied to steer the electron beam onto the axis of the next steering station. The position measurement is assumed to be imperfect, with specifiable errors in the accuracy with which the beam position monitors are aligned and the accuracy with which they can measure the beam position. The positions of steering stations along the undulator axis and the magnitude of the steering errors are inputs to the code.

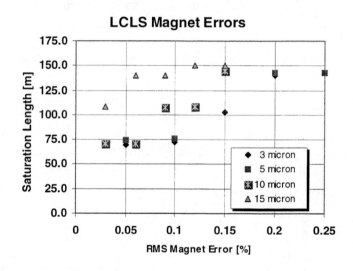

FIGURE 5. Saturation length as function of rms on-axis magnet error for 15 GeV operation. Includes orbit correction with steering stations spaced 5 m apart.

Steering Error Tolerance

When steering correction is imprecise due to misalignment errors in the position monitors, or noise in the processing electronics, maximum achievable FEL performance can be reduced. Figure 5 shows the effect of four different rms steering errors on LCLS output power as result of the simulations. For the 15 GeV LCLS with 5 m steering station separation, steering errors need to be kept below 5 μm rms.

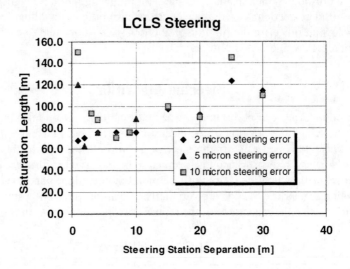

FIGURE 6. LCLS Error Sensitivity for 0.1 % rms per pole field error.

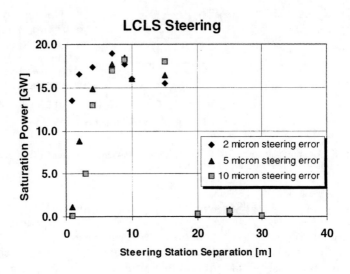

FIGURE 7. Saturation power optimum for steering station separation between 5 and 15 m for 0.1 % rms per pole field error.

The point-to-point correction scheme used in the FRED-3D code could work for the LCLS if errors in beam position measurement are small. When steering errors are larger, point-to-point steering can degrade as well as improve performance. Figure 6 shows FEL saturation length while figure 7 shows FEL saturation power as a function of steering station separation for the 14.35 GeV LCLS as result of FRED-3D simulations. Performance close to the optimum can be expected if steering station separation is kept between roughly 5 and 15 m and the steering error is kept below 5 μm. Steering becomes insufficient for larger spacings. For shorter spacing the effect of large steering errors severely reduces performance.

Recent studies of beam based alignment yield that beam position monitor calibration for the LCLS can be achieved to rms values significantly below 5 μm [21].

Steering algorithms that limit the maximum kick angle are expected to do better at small steering station separations.

SUPERRADIENT SPIKES

Figure 8 shows the simulation results obtained with the time dependent computer code GINGER [14], at a wavelength of 1.5 Å, an emittance of 1.5 mm mrad, and a peak current of 3400 A. Self-Amplified-Spontaneous Emission relies on longitudinal electron density fluctuations (shot-noise bunching). Regions in which initial bunching is larger produce more radiation, thus accelerating the lasing process. Due to slippage during the transport through the undulator, those centers will expand to form spikes on the scale of $2\pi L_c = \lambda_r/\rho \approx L_{slip} = \lambda_r N_{sat}$[22], at saturation. Where N_{sat} is the number of undulator periods to saturation and L_{slip} is the slippage length at saturation. For the LCLS, the time scale of the spike structure is about 1 fs at 1.5 Å and about 3 fs at 15 Å.

OUTPUT POWER CONTROL

For applications that use the x-rays that are produced by the LCLS FEL, it is important that output power levels can be controlled. This can be achieved by using a gas absorption cell after the FEL undulator. Several mechanisms that aim at controlling (reducing) the FEL output power production, including peak current variations and a phase shifter at the saturation point, have been analyzed.

Peak Current

Changing peak current, either by reducing the amount of charge per pulse or by increasing the pulse length, can be used to control (reduce) FEL power production over many orders of magnitude. Doing this, not only reduces saturation power, but also increases saturation length. The LCLS undulator will be designed to be not much longer than the expected saturation length at the highest achievable peak current. Reduction

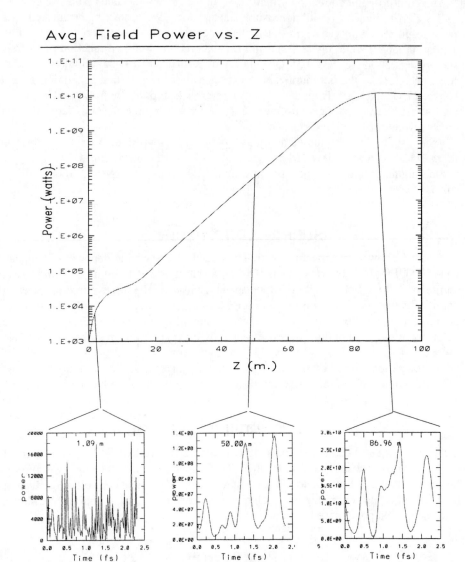

FIGURE 8. Development of the temporal structure of the LCLS x-ray pulse during the FEL process as predicted by GINGER simulations.

of peak current at the highest energy (and shortest wavelength) will move the saturation point beyond the end of the undulator. The FEL output will then be in the exponential gain regime rather then in the saturation regime, resulting in much stronger pulse to pulse variation.

Pulse-to-Pulse Output Power Stability

Sensitivity to electron parameter jitter at a fixed undulator position are expected to be much larger in the exponential gain regime than they are at saturation. According to 1D FEL theory, the derivative of peak power with respect to peak current is $dP_{sat}/d\hat{I} = 4/3\ P_{sat}/\hat{I}$ at saturation and $dP_{sat}/d\hat{I} = (1 + 1/3\ z/L_G)\ P_{sat}/\hat{I}$ in the exponential gain regime. The sensitivities can be defined as

$$\frac{\Delta P_{sat}/P_{sat}}{\Delta \hat{I}/\hat{I}} = \frac{4}{3}, \quad \frac{\Delta P_z/P_z}{\Delta \hat{I}/\hat{I}} = 1 + \frac{1}{3}\frac{z}{L_G}.$$

In the exponential gain regime, the sensitivity to fluctuations in peak current increases as do the sensitivities to fluctuations in normalized emittance, and energy spread. The relative sensitivities as obtained from GINGER simulations for the 1.5 Å LCLS are:

TABLE 5. Sensitivities of LCLS performance to electron beam parameters at the end of the 100 m long undulator in the exponential gain regime ($@z/L_G \approx 11.3$ for $\hat{I} = 1500$ A) and at saturation ($\hat{I} = 3400$ A).

Sensitivity at $\hat{I} = 3400$ A Saturation		Sensitivity at $\hat{I} = 1500$ A Exp. Gain Regime	
$\dfrac{\Delta P/P}{\Delta \hat{I}/\hat{I}} =$	1.8	$\dfrac{\Delta P/P}{\Delta \hat{I}/\hat{I}} =$	6.4
$\dfrac{\Delta P/P}{\Delta \epsilon_n/\epsilon_n} =$	−1.6	$\dfrac{\Delta P/P}{\Delta \epsilon_n/\epsilon_n} =$	−9.3
$\dfrac{\Delta P/P}{\Delta \delta\gamma/\delta\gamma} =$	0.2	$\dfrac{\Delta P/P}{\Delta \delta\gamma/\delta\gamma} =$	−1.8

The 3D sensitivities are larger than the 1D formulae predict, 1.3 compared to 1.8 at saturation and 4.8 compared to 6.4 at $z/L_G \approx 11.3$. With the expected pulse-to-pulse jitter of the LCLS electron beam as provided by the linac ($\Delta \hat{I}/\hat{I} \approx 10\ \%, \Delta\epsilon_n/\epsilon_n \approx 2\ \%$, $\Delta\delta\gamma/\delta\gamma \approx 3\ \%$), x-ray power jitter will be large when operating at saturation but will be unacceptable when operating in the exponential gain regime. The peak current is therefore not a suitable variable for controlling FEL output power.

Phase Shifter at Saturation

Another method of controlling output power was suggested, which makes use of the relative stability at saturation by changing the relative position between the electron beam and the electromagnetic wave. This could be implemented by installing a chicane at the saturation point of the undulator. A second short undulator after the chicane would accelerate the electrons by an amount determined by the phase shift, thus reducing the power in the electron beam. The estimated effect is relatively small, though. Even 1D simulations predict about one order of magnitude of output power reduction. This would come at a cost of a 20 % increase in undulator length.

OPTICAL KLYSTRON

The great length of the LCLS undulator makes it very desirable to find a mechanism that would shorten the undulator. The suggestion to use an optical klystron configuration has been brought up again and again when the topic of undulator shorting was discussed. The LCLS undulator could be converted into an optical klystron by splitting it in two parts and inserting a dispersive section. In the first part the interaction between the electron beam and the emerging electromagnetic wave imprints an energy modulation on the electron beam that is transformed into a spacial modulation in the dispersive section. The bunching created by the dispersive section increases the coherent emission of radiation in the second part of the undulator.

The theory of the high-gain optical klystron has been presented by Bonifacio et al. [23]. They find that the effect is strongly dependent on initial energy spread and that the condition $\Delta\gamma/\gamma \ll \rho$ must be fulfilled. For the 1.5 Å LCLS the energy spread condition is badly violated: $\Delta\gamma/\gamma \approx \rho/2$, and performance improvement from the optical klystron configuration can not be expected. This has been confirmed with LCLS computer simulations, which indicate that the reduction of saturation length can be significant for very small values of energy spread but that the effect vanishes as energy spread approaches the expected LCLS value of $\Delta\gamma/\gamma = 2 \cdot 10^{-4}$.

ACKNOWLEGEMENTS

The author would like to thank the members of the LCLS design group for their support during this study, especially Roger Carr, Max Cornacchia, Paul Emma, Bill Fawley, Ted Scharlemann and John Sheppard.

REFERENCES

1. H.-D. Nuhn et. al., "Short Wavelength FELs Using the SLAC Linac," in *Proceedings of the EPAC94*, pp. 855–857, 1994. SLAC-PUB-6541, SLAC/SSRL-0082, LBL-35905.

2. C. Pellegrini, "A 4 to 0.1 nm FEL Based on the SLAC Linac," in *Workshop on 4th Generation Light Sources* (M. Cornacchia and H. Winick, eds.), pp. 364–375, Feb. 24-27 1992. SSRL-Report-92/02.

3. W. Barletta, A. Sessler, and L.-H. Yu, "Using The SLAC Two-Mile Accelerator For Powering An FEL," in *Workshop on 4th Generation Light Sources* (M. Cornacchia and H. Winick, eds.), pp. 376–384, Feb. 24-27 1992. SSRL-Report-92/02.

4. J.Arthur, G. Materlik, H. Winick edts., "Workshop on Scientific Applications of Coherent X-Rays, Stanford, CA," February 12, 1994. SLAC-Report-437.

5. "A VUV Free Electron Laser at the TESLA Test Facility at DESY. Conceptual Design Report," June 1995. DESY Print, TESLA-FEL 95-03.

6. R. Bonifacio, C. Pellegrini, and L. Narducci, "Collective instabilities and high gain regime in a free electron laser," *Opt. Commun.*, vol. 50, no. 6, 1985.

7. J. Murphy and C. Pellegrini, "Generation of high intensity coherent radiation in the soft x-ray and VUV regions," *J. Opt. Soc. Am. B2*, no. 259, 1985.

8. K.-J. Kim et al., "Issues in storage ring design for operation of high gain FELs," *Nucl. Instr. and Meth. in Phys. Res.*, vol. A239, no. 54, 1985.

9. K.-J. Kim, "Three-dimensional analysis of coherent amplification and self-amplified spontaneous emission in free electron lasers," *Phys. Rev. Letters*, vol. 57, p. 1871, 1986.

10. C. Pellegrini, "Progress towards a soft x-ray FEL," *Nucl. Instr. and Meth. in Phys. Res.*, vol. A272, no. 364, 1988.

11. P. Emma and R. Brinkmann, "Emittance Dilution Through Coherent Energy Spread Generation in Bending Systems," in *Proceedings of the 1997 Particle Accelerator Conference (PAC97), Vancouver, B.C., Canada*, May 12-16, 1997. SLAC-PUB-7554.

12. M. Xie, "Design Optimization for an X-ray Free Electron Laser Driven by SLAC Linac," *LBL Preprint*, vol. No-36038, p. 3, 1995.

13. E.T. Scharlemann, "Wiggle plane focusing in linear wigglers," *J. Appl. Phys.*, vol. 58(6), pp. 2154–2161, 1985.

14. W.M. Fawley, "An Informal Manual for GINGER and its post-processor XPLOTGIN," December 1995. LBID-2141, CBP Tech Note-104, UC-414.

15. T. Trans and J. Wurtele, "TDA3D," in *Computer Phys. Commun.*, vol. 54, pp. 263–272, 1989.

16. B. Faatz, "Private Communication," 1997.

17. G. Travish and J. Rosenzweig, "Numerical Studies of Strong Focusing in Planar Undulators," in *Proceedings of the 1993 Particle Accelerator Conference, Washington D.C.*, vol. 2, pp. 1548–1550, 1993.

18. K.-J. Kim, "Private Communication," 1997.

19. V. Vinokurov, "Private Communication," 1997.

20. L.H. Yu et. al., "Effect of wiggler errors on free-electron-laser gain," *Phys. Rev. A*, vol. 45, pp. 1163–1176, 1992.

21. P. Emma, "Private Communication," 1997.

22. R. Bonifacio et. al., "Spectrum, Temporal Structure, and Fluctuations in a High-Gain Free-Electron Laser Starting from Noise," *Phys. Rev. Let.*, vol. 73(1), p. 70, 1994.

23. R. Bonifacio, R. Corsini, and P. Pierini, "Theory of the high-gain optical klystron," *Phys. Rev. A*, vol. 45(6), pp. 4091–4096, 1992.

119

Overview of the SPring-8 Linac Future Plan

Shinsuke Suzuki, Kenichi Yanagida, Tsutomu Taniuchi,
Yasuaki Kishimoto, Akihiko Mizuno, Hirosi Abe, Hiroshi
Yoshikawa and Hideaki Yokomizo

SPring-8, Kamigori, Ako-gun, Hyogo 678-12, Japan

Abstract. We carry out incidence operation in proportion to the commissioning to the storage ring at present, after SPring-8 linac started beam commissioning from August, 1996, and after it confirmed the beam performance early. In the future, the improvement in beam performance and equipment performance is tried, while we advise the improvement of the beam monitor. We develop the new applied research with it. We consider the application using various beams as part of the effective utilization of SPring-8 linac. In this paper, we introduce the plan of the free electron laser with the aim of the lasing at 4nm wavelength as one of the inside. And, we also introduce simulation result of the RF photocathode electron gun for it.

INTRODUCTION

The linac is the 140m overall length from the electron gun, and 1GeV output line through 10m matching section, and the beam is bend 15 degrees in the right. There is the synchrotron in end, and electron is accelerated to 8GeV by the synchrotron. It is possible to take out the electron beam outside, by bending 30 degrees in the left. In the end of new beam transport line, there is 1.5GeV storage ring called "New SUBARU" under this construction at present [1]. Another beam transport line (L3 beam transport line) is branched from the middle point of the "New SUBARU" transport line further and leads to the assembly hall. The appearance figure of final phase beam line is shown in FIGURE 1.

For improvement of the linac technological development of beam performance and application research, we have planned low emittance and short bunch of the electron beam. In present, the electron gun is used the cathode of the traditional thermal type with the grid. When in the trouble of filament or so, we are taken in the about 1 week in order to restore in operating

CP413, *Towards X-Ray Free Electron Lasers*
edited by R. Bonifacio and W. A. Barletta
© 1997 The American Institute of Physics 1-56396-744-8/97/$10.00

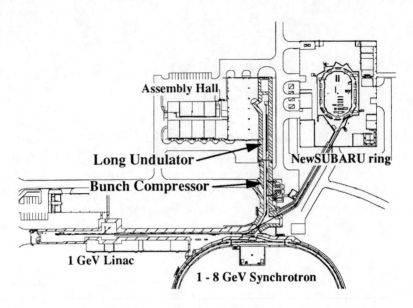

FIGURE 1. Layout of the beam line design

states such as cathode exchange, evacuation, aging. Future, we switch two electrons guns by the on-line. We also consider the dualization update. One is the small-diameter heat cathode electron gun, and another is the technological development of the photocathode RF gun. In the research of the free electron laser, densifying by short bunch of the electron beam is important. Considering the Coulomb force, we carry out examination and development of the beam transport system for doing the short bunch without the degradation of the beam performance. The upgrading of accelerating structures and upgrading of magnet alignment of the transport system, phase control of the acceleration high frequency, etc. is also indispensable.

I FEL CALCULATION

We chose the water window region at a wavelength of the goal of FEL. "Water window" is a region located in K absorption edge (4.3nm) of the carbon and K absorption edge (2.3nm) of the oxygen. In this region, it is possible of selectively observing in the situation with small contribution of the oxygen, in respect of the carbon which is an important component of the biological material. The observation in this region seems to be effective in order to observe the organic substance by techniques such as biological imaging. However, it is very difficult to aim at such small wavelength region for technically difficult like the long undulator, and present linac (1.2GeV) need energy up to 1.8GeV. Therefore, we consider the improvement in the linac with the aim of the lasing in the wavelength range of the 20nm as a first step. We confirm the principle by the lasing in this region.

TABLE 1. Results of the single pass FEL calculation

Wavelength	20 nm	4 nm
Energy (GeV)	0.69	1.55
Und. period (mm)	32	32
Und. Gap (mm)	13	13
K	>1.62	>1.62
Peak Current (kA)	10	10
Norm. Emittance (πmm•mrad)	1	1
Und. Length (m)	10	10
Peak Brilliance (photon/sec/mm^2/mrad^2/0.1%)	5.40×10^{28}	2.60×10^{29}
Repetition (Hz)	10	10
No. of Bunch	1	1
Ave. Brilliance (photon/sec/mm^2/mrad^2/0.1%)	5.40×10^{17}	2.60×10^{18}

It is shown the parameter of the calculation by the original three-dimensional code and the calculation result in TABLE 1. The lasing at 20nm which is a first phase occurs in 0.69GeV at 1πmm•mrad normalize emittance and peak currents 1kA. The length of undulator is 10m. In this case, the photon flux is 5.4×10^{28}photon/sec/mm^2/mrad^2/0.1% as the peak brilliance. The lasing of second phase in the water window is possible using the 1.55GeV electron beam and the beam current equal to the superscription. However, the length of the undulator becomes 30m. We get the flux of 2.6×10^{29}photon/sec/mm^2/mrad^2/0.1% as the peak brilliance. The flux of the X-ray is 7 or 8 order higher than the flux of the synchrotron radiation from the third generation storage ring. And the X-ray got, because the linac is the pulse operation, becomes a very short pulse, and it is coherent. The experiment of the beam diagnostic using L3BT line will be start in 1998 autumn.

II PHOTOCATHODE RF GUN

We carry out the research and development of the RF photocathode RF gun in order to obtain 1πmm•mrad normalized emittance necessary for SASE. The cavity under processing is a single cell at present, as to compare with the simulation, is easy. It is made to be the structure for keeping axial symmetry of the electric field with two couplers. We input RF from unilateral, and it is output from the opposite side. Output RF will be absorbed in the dummy load. The cathode material was selected the copper for the easy handling. By this, the processing of the cavity became easy in the sake equal to the cavity material. We use the 262nm result which the laser is the 4th harmonic of YLF

(the 1047nm fundamental wavelength) laser. The external form of the cavity is shown in FIGURE 2.

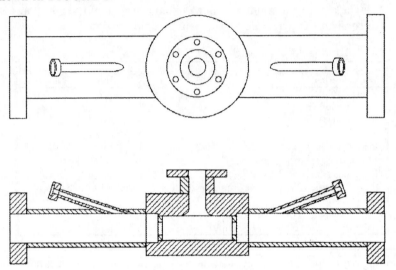

FIGURE 2. The external form of the cavity

The electric field on the cathode surface has made 150MV/m to be a goal. The behavior of the emittance as the electric field changes is shown in FIG-URE 3. The calculation code is MAFIA. Accelerated charge is 1nC/bunch. The normalized emittance in giving 150MV/m electric field to the cathode surface improves about 20% further than the normalized emittance of 100MV/m electric field. However, there is not the improvement in the emittance in the electric field over 150MV/m. In 300MV/m electric field, normalized emittance reversely deteriorates on the contrary.

It is considered that transverse component of the electric field in the disk vicinity in the cavity causes the increase of the emittance. Therefore, we set the goal of the electric field on the cathode surface at 150MV/m. Normalized emittance by this cavity is obtained 1.5πmm$\bullet$mrad and beam energy is 3.4MeV. However, the focusing effect by the solenoid magnetic field does not come in this calculation. We carry out the calculation which let the solenoid magnetic field in at present. We will prepare the cavity in 4 types. We do all HIP treatment of the material oxygen-free copper. We processed the first cavity by usual processing. The second cavity carried out the high-pressure ultrapure water cleaning. Using the copper single crystal material in the cathode surface, the third cavity carried out the high-pressure ultrapure water cleaning. It will be processed the fourth cavity in order to we conduct the thin coating of the titanium on the copper surface, and in order to the dark current decrease. It will be experimented in 1998.

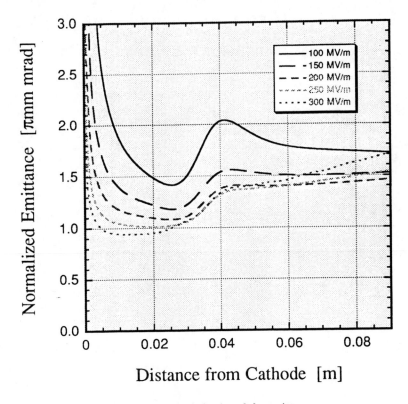

FIGURE 3. The behavior of the emittance

III CONCLUSION

We formally and not yet decided future utilization research program of the linac. It is the stage in which the volunteer examines the possibility. The improvement in reliability and beam performance of the equipment of the linac is tried, while we perform original purpose as an injector of synchrotron radiation facilities and we expect that it also utilizes it to these future utilization research and develops. For the SASE project, it is considered that we should approach the last goal by the procedure of the three phase. In the first phase, we construct the transport line, and it is supposed to be sure in respect of the technology of the beam diagnostic. And, we develop the photocathode RF gun in the off line. This stage is progressing at present, and the experiment will begin in 1998. Next step installs the photocathode RF gun in the linac, and we develop the technology of the bunch compression. And, we will insert comparatively short (about 10m) undulator in second phase. We aim at the lasing at 20nm. We insert the long undulator to the beam line as a last phase, and we aim at the lasing of the "water window" region.

REFERENCES

1. A. Ando et al., Proc. Particl Accl. Conf. (1997) to be publish.

UV-FIR FEL Facilities and Application Research at the FELI

T.Tomimasu, T.Takii, S.Nishihara, E.Nishimura and K.Awazu.

Free Electron Laser Research Institute, Inc. (FELI),
2-9-5, Tsuda-Yamate, Hirakata, Osaka 573-01, Japan

Abstract. Four FEL facilities FEL-1, FEL-2, FEL-3 and FEL-4 achieved lasing at the FELI in Oct. 1994, in Feb. 1995, in Dec. 1995 and in Oct. 1996, respectively. They are open for users to supply several MW-level FELs covering the wavelength range of $0.3\text{-}40\,\mu$ m. The present maximum average power of the IR-FEL is about 1 W. Total operation time in 1996 was about 2400 hours. The visible- and UV-facility FEL-3 has broken the world record for the shortest wavelength oscillation of linac-based FELs with a thermionic gun up to $0.278\,\mu$ m. The FIR facility FEL-4 ($20\text{-}80\,\mu$ m) was installed at the 33-MeV beam line, downstream of FEL-1. The FEL beams are delivered from the optical cavities to the diagnostics room and four user`s rooms (eighteen stations) through the pipe lines. The beam can be shared by using fan-shaped mirrors to several stations including two manipulators simultaneously for isotope separation, bio-medical and semiconductor applications.

FELI FACILITY

The Free Electron Laser Research Institute, Inc.(FELI) was established in March 1991. The building used for the development of FEL facilities and application studies was designed and constructed in Tsuda Science Hills, Hirakata City in Nov. 1993.

The main facility consists of an S-band, 165-MeV electron linac with a thermionic gun,beam transport(BT) lines and four FEL facilities(FEL-1,FEL-2, FEL-3 and FEL-4) as shown in Fig. 1. The compact facility, composed of a 20- MeV linac with an RF-gun and FEL-5, is also shown in Fig. 1. FEL-1, FEL-2 and FEL-3 were designed in 1993 and 1994 to cover the wavelengths from $0.3\,\mu$ m to $22\,\mu$ m of our appointed research target. However, FEL-4 and FEL-5 were designed in 1995 to extend our FEL applications to semiconductor devices at far-infrared wavelengths from 20 to $100\,\mu$ m.

The 165-MeV linac consists of the 6-MeV injector (1) and seven Electrotechnical Laboratory(ETL)-type waveguides(2). The seven accelerating waveguides with a

CP413, *Towards X-Ray Free Electron Lasers*
edited by R. Bonifacio and W. A. Barletta
© 1997 The American Institute of Physics 1-56396-744-8/97/$10.00

length of 2.93m are of linearly narrowed iris type to prevent beam blow up (BBU) effects at high peak current acceleration.

The injector consists of a 120-kV thermionic triode gun, a 714-MHz prebuncher and a 2856-MHz standing wave type buncher. The gun with a dispenser cathode (EIMAC Y646B) usually emits 500-ps pulses of 2.3A at 22.3125MHz. Recently (Sept. 1996) this was increased to 89.25 MHz. The grid pulser was manufactured by Kentech Instruments,Ltd., U.K. These pulses are compressed to 60A × 10ps by the prebuncher and the buncher.

The electron beam of the FELI linac consists of a train of several picosecond micropulse repeating at 22.3125MHz or at 89.25MHz. The train of the micropulse continues for 24μ s(macropulse) and the maximum number of the micropulse in a macropulse is 2140 with a micropulse separation of 11.2ns. The repetition rate of the macropulse is 10Hz or 20Hz.

The rf system for linac-based FELs requires rf sources with long pulse duration and high stability. Our rf sources are a 714-MHz klystron(1VA88R, 15kW for 20Hz, 24-μs flat top pulses) for the prebuncher and two 2856-MHz klystrons (E3729, 25MW for 20-Hz, 24-μ s flat top pulses per each) for the buncher and seven accelerating waveguides. The latter klystrons are based on a E3712 klystron, used for 80-MW, 4-μ s pulse operation, modified for 25-MW, 24-μ s pulse operation (3,4). The modulator for the klystron 1VA88R uses MOS-FET modules (5). However, the modulator for the E3729 klystron consists of 4 parallel networks of 24 capacitors and 24 variable reactors, and it has a line-switch of an optical thyristor stack. The flatness of our klystron modulator for E3729 is 0.067% at 24-μ s pulse operation (6).

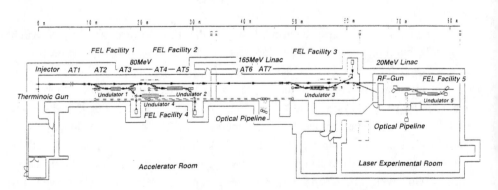

FIGURE 1 Layout of main FEL facilities(165-MeV linac and four FEL facilities) and compact FEL facility(20-MeV linac and FEL facility)

OPERATION OF FEL FACILITIES AND LASINGS
AT UV-FIR RANGE

The electron beam size and position are always monitored and controlled to pass through the center of accelerating waveguides using screen monitors installed at the inlet and outlet of every accelerating waveguide and quadrupole magnet. Further, using five screen monitors installed in an S-type BT line (7) for each FEL facility, the beam size and position are adjusted along the axis so as to pass through the center of a narrow vacuum chamber inside each undulator. The first 22.5° -bending magnet of each S-type BT line and the water absorber are used as an energy spectrometer.

The beam emittance was measured using three Al-foil OTR beam profile monitors installed in the undulator vacuum chamber. The two profile monitor emittance measurement method is used for its simplicity and short time for data acquisition at each FEL facility. The normalized emittance ε n of a 155-MeV, 60-A electron beam is estimated to be 26π mm mrad. Each Al foil has a 1-mm ϕ aperture. The S-type BT line can focus about 80-90% of the electron beam to pass through the aperture.

Each optical cavity is a Fabry-Perot cavity which consists of two mirror vacuum chambers (8). The spontaneous radiation and FEL beam from the main facility are delivered from the optical cavity on the first floor to the diagnostic room on the third floor with six Al-coated mirrors. Their macropulse shape, macropulse power and spectrum are simultaneously observed in the diagnostics room and the linac control room.

The FEL oscillation experiment was begun at FEL-1(5-22 μ m) in August and succeeded in lasing at 5-7 μ m in October 1994 (9). FEL-2 (1-6 μ m) and FEL-3 (0.23-1.2 μ m) achieved lasing at 1.88-3.1 μ m in February (10) and at 0.339-0.353 μ m and in December 1995, respectively (11).

It has been said that no lasing at ultraviolet wavelengths can be achieved by using an electron linac with a thermionic gun from the wavelength limit due to optical diffraction, because of beam emittance growth in the bunching process from the thermionic gun to the linac. The FELI linac, therefore, was carefully designed to reduce the emittance growth in the bunching process and FEL-3 has broken the world record for the shortest wavelength oscillation up to 0.278 μ m on June 6, 1996, using a 155-MeV, 60-A electron beam at FEL-3(0.23-1.2 μ m). Fig. 2 shows a 0.278- μ m FEL spectrum. A radiation damage on Ta_5/SiO_2 mirrors was found for the first time after about fifty hours FEL operation at visible range. We are challenging to lase at 0.23 μ m.

Experimental verifications of gain estimates given by Dattoli,et al.(12,13) and of the criteria on electron beam quality and wavelength limit due to photon diffraction given by Sprangle, et al.(14) are tested with our data of UV-FEL oscillation experiments (15).

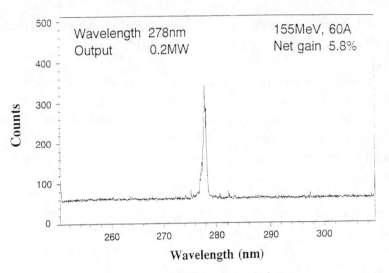

FIGURE 2. 0.278-μm FEL spectrum.

Despite having taken great care with beam handling in order to maximize beam duct throughput, a low energy tail of the beam is lost near the bending magnet and mishandling of the quadrupole magnets increases irradiated radiation dosage to the undulator permanent magnets and the mirrors in the optical cavity. Wakisaka et al measured irradiated radiation dosages for multilayer mirrors in optical cavity 3 and NdFeB permanent magnets with thermoluminescence dosimeters (TLDs) (16).

FEL-4 (20-80μm) including a 2.7-m long hybrid type undulator with electro- and permanent magnets (λ_u=9cm, N=30) and a 6.72-m long optical cavity was installed at the 33-MeV beam line, downstream of FEL-1 in March 1996. The FEL beam is delivered from the downstream optical cavity to the user's rooms through the pipeline of FEL-2. Since the middle of June, we tried to pass a well collimated beam through undulator 4 (17) and succeeded in lasing at 18μm on October 24, 1996.

However, six pairs of steering coil had to be attached on the beam duct to reduce the vertical magnetic field deviation induced by the built-in electromagnets, and then the minimum gap of the hybrid type increased up to 52mm. Therefore, the hybrid type undulator was replaced by a new Halbach type one (λ_u=8μm, N=30) after the first lasing. The new FEL-4 was installed in the middle of Dec. 1996 and first lasing at 18.6μm was achieved on Dec.26, within ten hours operation. The new FEL-4 achieved lasing at 18.6~40μm in Jan.1997 (18). Fig.3 shows a 34.5-μm FEL macropulse shape measured with an HgCdTe detector and electron current macropulse shape measured with a button monitor(19).

FEL-5(50-100μm) of the compact facility, including a 2.24-m long undulator (λ_u=8cm, N=28) and a 5.04-m long optical cavity, was installed at the S-type beam line of the 20-MeV linac with an RF-gun in the Laser Instrument Room in March 1996. The FEL beam is delivered from the upstream optical cavity to the Laser Experimental

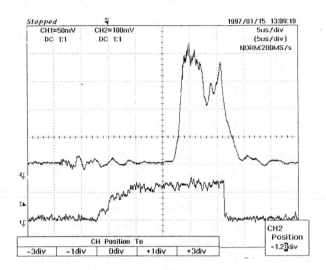

FIGURE 3. (a) 34.5-µm FEL macropulse shape and (b) electron current macropulse shape

Room through the pipeline. The operation of FEL-5,including final RF aging, was begun in May and the beam acceleration was started in the middle of June. Spontaneous emission was observed in October 1996. It is an RF linac based facility designed to extend our FEL applications to semiconductor devices at the far-infrared wavelength from 50 to 100 μ m. The features of the FEL facility 5 are the use of a thermionic-cathode RF gun, a PV 3020 klystron driven by a 10-μ s modulator with a PFN of two parallel by 24 sectors, and a 2.24 m long, 28 periods NdFeB Halbach type undulator.

The schematic layout of FEL-5 is shown in Fig.1. As an injector, the RF gun consists of an S-band half cavity(2856 MHz) and an alpha magnet. The electrons are emitted from a 3 mm diameter LaB_6 cathode. They are accelerated by an RF field in the gun cavity. The electrons emerge from the cavity through steering and triple quadrupole magnets and then pass through the alpha magnet.

Since the electron energy is of the order of 1 MeV at most, the emerging bunches will shorten as they pass through the drift spaces before and after the alpha magnet. This indicates that electron bunches are compressed to a few picoseconds at the five screen monitors installed in the S-type BT line for the undulator 5, the beam size and position are adjusted along axis so as to pass through the center of a narrow entrance of an ETL type 2.2-m long accelerating waveguide(2).

It is well known that the back-bombardment effect limits the electron macropulse length due to the S-band RF field. This is the main reason we can not use the full 10-μ s pulse length of the klystron. However, the photo-cathode RF gun can use the full 10-μ s pulse length.

The Rf system consists of a 20 MW high power klystron driven by a 10 kW klystron,TH2436. The modulator for the PV3020 klystron consists of 2 parallel networks of 24 capacitors and 24 variable reactors, and it has a line-switch of the F-241 thyratron. The flatness of our klystron modulator for PV3020 is 0.3% at 10-μ s pulse operation.

Along the electron beam line, seven screen monitors are prepared. The electron beam size and position are always monitored and controlled to pass through the center of into the undulator the 2.2-m long accelerating waveguide using the screen monitors. Further, after passing through a vacuum chamber inserted into the undulator, the beam is focused and sent to a beam dump using a 45°–bending magnet.

The Ge-Ga detector cooled by liquid helium is used for detecting the spontanueous emission. The detector spectral response is provided by micron cut-on filter set at the cone entrance aperture. The signal at preamplifier output at 1000 Hz of this detcctor is 0.6 V. The chopped power falling on the detector is 0.455 μ W.

Fig.4 shows both the electron macropulses(lower 3 parts) and the spontaneous emission (upper part) at 80 μ m (20). The electron macropulses 1-3 are of before and after the accelerating waveguide and before the undulator, respectively. The macropulse peak current is about 130 mA. The beam current of 200 mA was not observed because of RF breaking in the RF gun cavity and of short life time of the LaB$_6$ cathode(about two months). It will need more RF aging for the RF gun cathode.

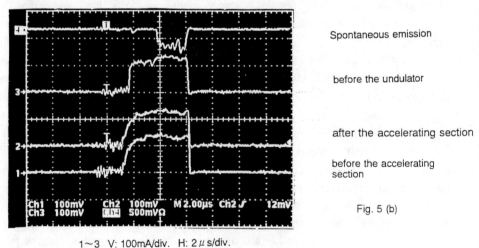

Spontaneous emission

before the undulator

after the accelerating section

before the accelerating section

Ch1 100mV Ch2 100mV M 2.00µs Ch2 ∫ 12mV
Ch3 100mV Ch4 500mVΩ

Fig. 5 (b)

1~3 V: 100mA/div. H: 2 μ s/div.

FIGURE 4. Electron macropulses and spontaneous emission at 80 μ m.

FEL BEAM QUALITIES

The FEL micropulse structure depends on the electron micropulse and macropulse structures, the detuning effect of the optical cavity, and the small signal gain of the undulator.

The electron beam of the FEL consists of a train of several picosecond pulses (micropulses) repeating at 22.3125 MHz. Recently(Sept. 1996) this was increased to 89.25 MHz, although the results presented in this paper were all taken at the lower frequency. To estimate the electron micropulse duration, the optical transition radiation induced by the electron micropulse was measured with a streak camera.

The usual FEL micropulse was measured with the streak camera. The train of micropulses continues for $24\,\mu$ s(macropulse) and repeats at 20Hz. The number of micropulses of in a macropulse is about 2140 with a micropulse separation of 11.2 ns and the FEL macropulse duration of FEL-1 is about $18\,\mu$ s, at the 22.3125 MHz operation, since the small signal gain is near 20% (13). Therefore, the number of FEL micropulses is about 400. The usual FEL micropulse duration is determined by the detuning effect of the optical cavity and is shorter than half of the electron micropulse duration (bunch length) of several picosecond. The detuning effect on visible- and ultraviolet-FEL micropulse duration was measured with the streak camera by Wakita et al.(21).. A typical micropulse duration is 1.7ps at 0.6-0.35 μ m. Fig. 4 shows a detuning curve of the FEL-1 for 9.2-μ m FEL (9). The micropulse duration of MIR-FEL near 10µm was estimated to be 3.2 ps from a pump-probe profile measured with GaAs/AlGaAs multi quantum wells by Suzuki et al.(22). The FEL peak and average powers of FEL-1 are 6MW and 0.2W at 7.5 μ m,

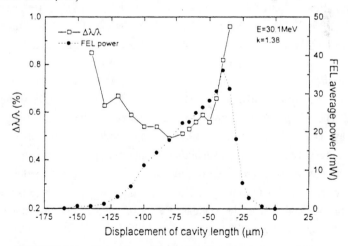

FIGURE 5. Detuning curve of FEL-1 for 9.2-μ m FEL.

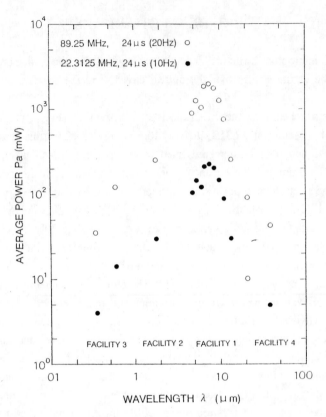

FIGURE 6. Average FEL powers as a function of wavelength.

respectively, at the outlet of Au-coated front mirror with 1-mm aperture. The FEL peak power of FEL-2 is 2MW at $1.88\,\mu$ m at the 0.5-mm aperture of Au-coated mirror and those of FEL-3 are 1.8MW at $0.35\,\mu$ m and 4.7MW at $0.6\,\mu$ m at the multilayer mirror.

Fig.6 shows average FEL powers as a function of wavelength measured at each front mirrors. In order to increase the average FEL power, we have tried to increase the repetition rate of the micropulse up to 89.25MHz and the repetition rate of the macropulse up to 20Hz. These improvements are expected to increase average powers up to a few W.

The wavelength dependence of FEL spectral width $\Delta\lambda$ is also an important parameter, especially in Si isotope separation by the multiphoton decomposition process. Experimental data on the wavelength dependence of $\Delta\lambda/\lambda$ are plotted in Fig.7(23). Solid lines were calculated from $\Delta\lambda/\lambda=(1/\pi)(\lambda/N\sigma z)^{1/2}$, where N is the number of undulator periods and σz is the standard deviation of the electron bunch length. Our experimental data agrees with those of CLIO and FELIX.

134

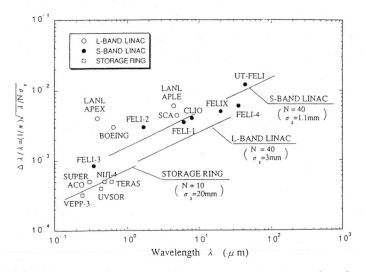

FIGURE 7. Relation between spectral width and wavelength.

FEL BEAM SHARING FOR USER`S ROOMS.

The FEL beams are delivered from the optical cavities to the diagnostics room and four user's rooms (eighteen stations) on the third floor through the optical pipes as shown in Fig.8 (23,24). Since October 1995, FEL-1 was open for FEL users for two days a week. Three days a week were used to upgrade FEL-2 and FEL-3 and to achieve lasing at Facilities 4 and 5 in the far-infrared range. Since the end of October 1996, FEL-1, FEL-2 and FEL-3 are open for users every other week. Total operation time in 1996 was about 2,400 hours.

The FEL beam extracted from the narrow aperture becomes thick and round due to hole coupling and is delivered to the diagnostics room and four user's rooms. The FEL power and spectrum are always monitored at the linac control room with a fan-shaped mirror at the diagnostic room.

The thick and round FEL beam is suitable for simultaneous FEL beam sharing to the diagnostics room and user's stations, using fan-shaped mirrors as shown in Fig. 9. The opening angle of the fan-shaped mirror can change a sharing ratio of delivering FEL average power to the diagnostics room and the user's rooms. For the present FEL diagnosis, a quarter of the round FEL beam is shared with a fan-shaped mirror with a 90° -opening angle. A quarter of the FEL power is enough to do some experiments and to diagnose the FEL spectrum, power level and macropulse duration. Three quarters can be shared equally with a fan-shaped mirror with a 180° -opening

135

angle at the following user's rooms. Two manipulators were installed in Lab.2. Each can focus the FEL beam to a 0.1-mm ϕ MIR-FEL spot size inside of 2-m diameter working area as shown in Fig.10.

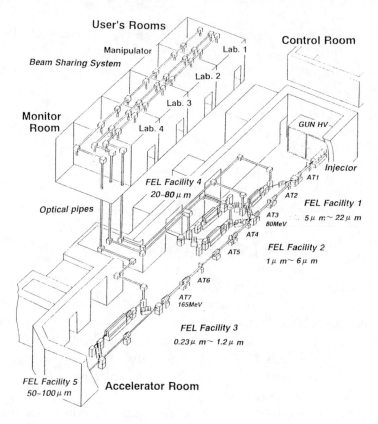

FIGURE 8. Bird's eye view of the FELI facilities and user's rooms.

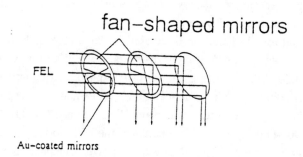

FIGURE 9. FEL beam sharing with fan-shape mirrors.

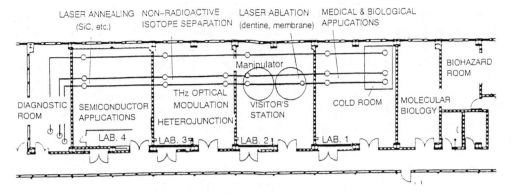

FIGURE 10   FEL user's facilities at the FELI

APPLICATION RESEARCH

Basic studies carried out in FEL user's facilities are as follows.

(1) Introduction of genes into living cells (with Kyoto Univ.in Lab.1) (25)

(2) FEL microscopic irradiation device (in Lab.1) (26)

(3) FEL ablation and annealing of dental hard materials (in Lab. 2) (27)

(4) FEL ablation of hard materials (with Toyohashi Univ.of Technology in Lab.2)

(5) Multiphoton ionization time-of-flight mass spectrometry(in Lab.2) (27)

(6) FEL-induced molecular operation (in Lab.2) (28)

(7) Resonant excitation of molecular vibrations and structural relaxation (with Univ. of Osaka Prefucture in Lab.2) (29)

(8) Feasibility study on application of FEL to material analysis, device physics and processing (in Lab.3)(30).

(9) Band discontinuity measurement of semiconductor heterojunctions (in Lab.3) (31)

(10) Ultrafast all-optical modulation in semiconductor quantum wells (with   Kyoto Univ.and Osaka Univ. in Lab.3)(22)

(11) Isotope separation of carbon and silicon by multiphoton decomposition process (with Kyoto Institute of Technology in Lab.3) (32)

(12) Semiconductor processing (in Lab.4) (33)

(13) Photodynamic Therapy (with Fukui Medical Univ. in Diagnostic room)

Introduction of genes into living cells (25) and FEL-induced molecular operation (25) are useful bio-medical applications of FEL-induced stress transient.   We established model systems for these studies.   We could introduced Green Fluorescent Protein genes into Cos-1 cells by irradiation of 6.1μ m FEL with an average power density of $50mW/mm^2$ using an FEL microscopic irradiation device (26).

The effect of FEL irradiation on human dentine is also studied to develop new possibility for laser dentistry. Ogino et al. found that the 9.4μm FEL irradiation causes selective ablation of phosphoric acid and annealing (27).

Multiphoton ionization TOF mass spectrometry (27) and FEL isotope separation (32) are also useful applications of FEL-induced multiphoton absorption.

Tunable,high peak power FELs enable resonant and selective excitation of molecules or lattice vibrations. Semiconductor processing (33),structural relaxation of hydrogenated amorphous silicon(29) and annealing of dentine are due to these FEL irradiation effects.

CONCLUSIONS

Four FEL facilities (FEL-1, FEL-2, FEL-3, FEL-4) can supply several MW-level FELs covering the wavelength range of 0.3-80μm.   FEL-5 is in the commissioning stage.

Although people have said that no lasing at ultraviolet wavelengths can be achieved by using an electron linac with a thermionic gun from the wavelength limit due to optical diffraction, the FELI linac was carefully designed to reduce the emittance growth in the bunching process and FEL-3 has broken the world record for the shortest wavelength oscillation up to 0.278μm. We are challenging to lase at 0.23μm. FEL-4 (20-80μm) also achieved lasing at 18.6-40μm.

Experimental verifications of gain estimates given by Dattoli, et al. and of the criteria on the electron beam quality and wavelength limit due to photon diffraction given by Sprangle, et al. are tested with our data of ultraviolet-FEL oscillations.

FEL-1, FEL-2, FEL-3 and FEL-4 are open for users to supply several MW-level FIR-UV FELs covering the wavelength range of 0.28-40μm.   Total operation time in 1996 was about 2400 hours. The maximum average power of IR-FELs will be a few W as shown in Fig.6. The FEL beams are delivered from the optical cavities to the diagnostics room and four user's rooms through pipe lines. The beams can be shared by using fan-shaped mirrors to several stations including two manipulators simultaneously for isotope separation, bio-medical and semiconductor applications.

ACKNOWLEDGEMENTS

The FELI project was financially supported by the Japan Key Technology Center (70 %) and thirteen corporations(30%) for six years and four days. We greatly appreciate their support.

The successful development of our FEL facilities and application researches are due to the collaboration of many scientists and engineers. All systems of these facilities were designed, installed and operated by FELI group. However, the 89.25-

138

MHz operation of the 500-ps grid pulser was performed in cooperation with Dr. J>Hares of Kentech Instruments, Ltd., UK. The first author would like to express his sincere thanks to all members concerned with this project for their collaboration.

REFERENCES

1. T. Tomimasu,et al.,Nucl.Instr.Meth.,**A358,ABS11-12**(1995).

2. T. Tomimasu,IEEE Trans.,**NS-28(3),3523-3525**(1981).

3. Y. Morii,et al.,"THE RF SYSTEM OF FELI" in *Proceedings of the. 9ᵗʰ Symp. on Accelerator Science and Technology* (KEK,Aug. 25-27, 1993) pp.225-227

4. Y. Ohkubo,et al.,"S-band Long Pulsed Klystron for the FELI Linac" in *Proceedings of the 20ᵗʰ Linear Accelerator Meeting in Japan* (FELI,Osaka,Sept.6-8,1995) pp.72-74

5. S. Abe,et al.,"DEVELOPMENT OF KLYSTRON DRIVER AMPLIFIER USING TRANSISTOR SWITCH" in *Proceedings of the 19ᵗʰ Linear Accelerator Meeting*(JAERI,July 20-22,1994) pp.225-227

6. E.Oshita, et al.,"24-MW,24-μ s Pulse RF Power Supply for Linac Based FELs" in *IEEE Proceedings of PAC'95* (Dallas, May 1-5, 1995) pp.1608-1610.

7. Y. Miyauchi, et al.,"LATTICE DESIGN OF BEAM TRANSPORT SYSTEM OF FELI" in Ref. 3, pp.416-418.

8. K. Saeki,et al., Nucl.Instr.Meth.,**A358,ABS56-58**(1995).
   K. Saeki,et al., Nucl.Instr.Meth.,**A375,pp.10-12**(1996).

9. T. Tomimasu,et al.,"FIRST LASINGS AT VISIBLE AND IR RANGE OF LINAC-BASED FELs" in IEEE Proceedings of PAC '95 (Dallas, May 1-5, 1995) pp.257-259.

10. A. Kobayashi, et al., Nucl.Instr.Meth.,**A375,pp.317-321**(1996).

11. T. Tomimasu,et al., Nucl.Instr.Meth.,**A383,337-341**(1996).

12. G. Dattoli,et al., IEEE J.Quantum Electron.,**QE-20**,637 (1984).

13. Y. Miyauchi,et al., Nucl.Instr.Meth.,**A375**,ABS42-45(1995).

14. P. Sprangle,et al., Nucl.Instr.Meth., **A331**,6-11(1993).

15. T. Tomimasu,to be published in Nucl.Instr.Meth.A

16. K. Wakisaka,et al.,"Irradiated radiation dose measurements of multilayer mirrors and permanet magnets used at FELI facilities"in *Proceedings of AFEL`97*(Hirakata, Jan.21-24,1997) pp.91-94.

17. A. Zako,at al., Nucl.Instr.Meth.**A375**,ABS80-82 1996).

18. T. Takii,et al.,to be published in Nucl.Instr.Meth.A.

19. A. Zako,et al.,"Diagnostics of Picosecond Pulsed Electron beam Position and Current using Button-type Beam Position Monitors at the FELI" in *Proceedings of*

2nd Asian Symp. On FEL(Novosibirsk,June 13-16,1995) pp.57-60.

20. S.Nisihara,et al.,"Committioning of FELI facility 5 covering the wavelength from 50-100 m" in Proceedings of AFEL`97(Hirakata, Jan.21-24, 1997) pp.81-86.

21. K.Wakita,etal.,"Picosecond-pulse duration measurements of UV-and visible-FELs and bunched electrons at FELI facility 3" in *Proceedings of*

  AFEL`97(Hirakata,Jan.21-24) pp.87-90.

22. T.Mitsuyu,et al.,"Observation of ultrafast interband-light modulation by intersubband-light in n-doped quantum wells by using free electron laser" in *Proceedings of IEEE Lasers and Electro-Optics Society, 1996 Annual Meeting, Boston, MN4 pp.133-135.*

  T.Suzuki,et al., Appl. Phys. Lett.,**69**,4136-4138(1996).

  T. Suzuki,et al.,"Ultrafast All-Optical Modulation in Semiconductor Quantum Wells" in *Proceedings of AFEL`97*(Hirakata,Jan.21-24, 1997)pp.331-338.

23. T. Tomimasu,et al., Nucl.Instr.Meth.,**A375**,626-631(1996).

24. S. Okuma,et al., Nucl.Instr.Meth.,**A375**,654-656(1996).

25. E. Nishimura,et al.,"Introduction of genes into living cells" in

  Proceedings of AFEL '97 (Hirakata, Jan. 21-24, 1997) pp.271-278

26. M. Yasumoto,et al.,Nucl.Instr.Meth.,**A387**,459-462(1997)

  M. Yasumoto,et al.,"FEL microscopic irradiation device for gene introduction" in Proceedings of AFEL`97(Hirakata,Jan.21-24,1997)pp.279-286.

27. S. Ogino,et al., SPIE,**2922**,184-192(1996).

  S. Ogino,et al., SPIE,**2973**,29-38(l997).

  S. Ogino,et al.,"The FEL ablation from tooth dentine" in *Proceedings of AFEL`97*(Hirakata,Jan.21-24,1997) pp.287-296.

28. K. Awazu,et al.,SPIE,**2975**,302-309(1997).

  K. Awazu,et al.,"FEL induced molecular operation on cultured fibroblast and cholesterol" in *Proceedings of AFEL`97*(Hirakata,Jan.21-24,1997) pp.297-305.

29. Y. Maeda, et al.,"Structural relaxation of hydrogenated amorphous silicon using resonant vibrational excitation with free electron lasers"in *Proceedins of AFEL'97* (Hirakata, Jan. 21-24, 1997) pp.307-314

30. T. Mitsuyu et al.,"Application of FEL to semiconductor resarech-an overview" in *Proceedings of AFEL'97* (Hirakata, Jan. 21-24,1997)pp.315-320

31. K. Nishi,et al., Jpn. J. Appl. Phys.,**35**,L760-L761(1996)

  K. Nishi,et al., MRS-J Trans.,**20,** 759-761(1996).

  K. Nishi,et al., Appl. Phys. Lett.**70,** 2171-2173(1997).

  K. Nishii,et al., Appl. Phys. Lett.**70**, 3585-3587(1997).

32. S. Kuribayashi,et al., submitted to Appl. Phys.

S. Kuribayashi,et al.,"Experimental studies on laser isotope separation of carbon and silicon by a free electron laser" in *Proceedings of AFEL `97*(Hirakata,Jan.21-24.1997) pp.263-270.

33. H. Ohyama,et al., Jpn. J. Appl. Phys., **35**,L683-684(1996)

H. Ohyama,et al., J. Electronic Materials, **26(3)**,183-186(l997).

H. Ohyama,et al.,"Semiconductors processing with free electron laser" in *Proceedings of FEL `97*(Hirakata, Jan.21-24,1997) pp.339-346.

Design of an X-ray FEL Undulator

Roger Carr

Stanford Synchrotron Radiation Laboratory
Stanford Linear Accelerator Center
Stanford, CA 94309 USA

Abstract: An undulator designed to be used for an x-ray free electron laser has to meet a set of stringent requirements. With no optical cavity, an x-ray FEL operates in the single pass Self Amplfied Spontaneous Emission (SASE) mode; an electron macropulse is microbunched by an undulator and the radiation it creates. The microbunched pulse emits spontaneous radiation and coherent FEL radiation, whose power may reach saturation in a sufficiently long and perfect undulator. The pulse must have low emittance and high current, and its trajectory in the undulator must keep the radiation and the pulse together with a very high degree of overlap. In this paper we use the case of the Linear Coherent Light Source (LCLS) x-ray free electron laser to illustrate design concepts for long free electron laser undulators. [1,2] The LCLS is intended to create 1.5 Å x-rays using an electron beam with 15 GeV energy, 1.5π mm-mrad normalized emittance, 3400 A peak current, and 280 fsec FWHM bunch duration. According to our simulations, this $2\sigma_r = 65$ mm rms diameter beam must overlap its radiation with a walkoff of no more than 5 μm RMS per 11.7 meter field gain length for efficient gain. The energy spread of the beam is $\sigma_E = 0.0002\ E_e$. This places severe limitations on the magnetic field errors and on mechanical tolerances. In this paper we shall discuss how to meet these requirements.

INTRODUCTION

In a single pass free electron laser operating in the Self Amplified Spontaneous Emission (SASE) mode, exponential gain of coherent radiation intensity is predicted by theory. [3] Power saturation is achieved in about 10 amplitude gain lengths, assuming no tapering of the undulator. It is desirable to build an FEL undulator to a full saturation length, so that the output is more stable than it would be on the exponential part of the gain growth curve. Given the desired output radiation wavelength and the energy of the available electron beam, saturation implies a certain output power, desired or not. If one reduces the beam current or enlarges the emittance in order to lower the output power, saturation will not be achieved.

The task of the undulator design is to achieve saturation with an undulator that is not excessively longer than the saturation length based on ideal undulator parameters. In the case of the LCLS, a perfect undulator is predicted to saturate in 94 m, based on both GINGER simulations [4] and semi-analytic models [5]. We hope that building a 100 m (or slightly longer) device will allow for imperfections and reach saturation. The undulator length refers just to the length of magnets; separations between undulator segments will make the total device about 12% longer.

MAGNET DESIGN

The undulator magnetic design involves the design of the undulator proper, the termination of the undulator at separations between segments, and the design of quadrupole focusing magnets, which are also used as correctors and alignment devices. It also requires the measurement of the fields of fabricated components, and the design of shimming strategies to achieve desired tolerances.

CP413, *Towards X-Ray Free Electron Lasers*
edited by R. Bonifacio and W. A. Barletta
© 1997 The American Institute of Physics 1-56396-744-8/97/$10.00

We studied several undulator options, but chose a plane polarizing design that uses hybrid permanent magnet technology. We examined harmonic generation strategies where longer period bunching would feed radiation to successive shorter period devices, but no overall length savings were predicted [6] We also looked at putting optical klystron style dispersive sections in the undulator to allow the energy modulated bunch to become spatially modulated, but found that energy spread ruled out this approach. [7] The energy of the electron beam decreases down the length of the undulator due to emission of spontaneous radiation, but we do not expect this to necessitate tapering the undulator when it is accounted for, because it induces an energy loss less than the 0.02% energy spread of the beam. [6]

Superconducting DC and warm pulsed bifilar helical electromagnetic unduators were considered, but were thought to be costly, complicated, and difficult to hold to mechanical tolerances and to provide with steering corrections. Access to the beampipe completely surrounded by undulator would be impaired. Part of this consideration was our earlier belief that we should avoid separations between portions of the undulator; we now consider such separations acceptable, so the bifilar helix may be worth revisiting, because it could be built with about half the saturation length of a plane polarizing device.

Besides the bifilar helix, there are permanent magnet helical undulator concepts, both with permanent magnets only (twistor designs) and with hybrid structures. These do not offer short periods like the electromagnet devices, and are as costly and complex as the plane polarizing devices. A pure permanent magnet plane polarizing undulator would work for the LCLS, but it would have a longer period than a hybrid device, because the field is weaker for a given gap, and errors are harder to control in the pure design. Such a design might have the advantage that steering and focusing fields could be superposed by electromagnets surrounding the undulator, but even this concept is not simple. Instead, we chose a hybrid permanent magnet design.

We expect to be able to mount magnets on the girders to tolerances of a few microns. Each end of the girder will be supported by a pier, and a separation of 23.76 cm will be allowed between undulator segments, to allow the electrons to fall behind the photon beam by one photon wavelength. The separation will be used for diagnostics, and possibly for a steering corrector magnet. The undulator sections can be mounted on high precision mover, if it is necessary to control their positions using feedback.

The separation between two sections of FEL undulator is given by the condition that as the electron and the photon wave travel the distance z, the phase of the wave advances by $\Delta\phi = 2\pi$. A time t is required for the electron to travel the distance z. Then we have generally:

$$\Delta\phi = 2\pi = k\,z - \omega t = k\,z - \omega \int_0^z \frac{dz}{v_z(z)}$$

(1)
In the separation, where v_z is constant:

$$t = z / c\beta_z \approx z\,(1 - 1/2\gamma^2) / c$$
(2)

so that :

$$2\pi \approx k\,z\,(1 - (1 - 1/2\gamma^2)) = \pi\,z / \lambda\gamma^2$$
(3)

and z will be given by:

144

$$z = 2 \gamma^2 \lambda = 2 \gamma^2 \frac{\lambda_u (1 + K^2/2)}{2 \gamma^2} = \lambda_u (1 + K^2/2)$$

(4)

In our case, $z = $ 23.76 cm for $K = 3.72$, and undulator period $\lambda_u = 3$ cm, independent of λ, the radiation wavelength. The electron's velocity v_z is not constant in the undulator or at the end of the undulator, so the exact separation must be calculated numerically. The termination of the magnetic field should be done in manner that causes the least loss of resonant wiggles, probably with a ferromagnetic field clamp.

Simulations of the LCLS show that strong focusing must be added to the undulator lattice in order to maintain a small beam diameter; natural focusing would give a beta function length of more than 50 m; optimal focusing occurs with a beta function of less than 20 m. We intend to operate the LCLS from 5 GeV to 15 GeV. If we focus with variable permanent magnet quadrupoles or electromagnetic quadrupoles, we can obtain optimum focusing at all energies. If we have fixed focusing strength and choose an optimum beta function at 15 GeV, we are not far off optimum at 5 GeV.

In the early stages of the design, we thought that the undulator had to be continuous, with no segmentation. In this case, strong focusing had to be built into the undulator. We would have chosen a combined function lattice, with shaped pole tips to add a 10 T/m field gradient to meter long segments of the undulator. We originally felt that beta modulations $\Delta\beta /{<}\beta{>} = 2(\beta^+ - \beta^-)/(\beta^+ + \beta^-)$ should be kept below 10% for good FEL gain. The strength of beta modulations is dependent on the distance between focusing and defocusing lenses. In a distributed function lattice, this distance could be a meter or less, but such short distances would be impractical in a separated function lattice.

GINGER and FRED-3D simulations are not suited to variations in focusing strength, but a new version of TDA-3D was capable of modeling this effect. The new TDA-3D results, however, showed that much larger beta modulation, even 50%, had very little effect on the gain length. It seems that beta modulation compresses electron density in one dimension when it expands in the other dimension, so that there is little net change of the electron density, and hence little change in Pierce parameter ρ that governs gain length. The lumped focusing strategy substantially relaxed the very demanding tolerances on the undulator required in the distributed focusing design.

Lumped focusing allowed us to adopt a separated function FODO lattice. Since we wish to run at 5 GeV, with a 6 m beta function, it is desirable that the FODO cell be less than 6 m long. Trajectory simulations show that a 4 m spacing between correctors and BPM's should give less than 5 μm RMS/gain length of walkoff. If the spacing were 2 m, we could experience unacceptable trajectory angle kinking. However, for beam based alignment, simulations show that there is a 3/2 power law increase of alignment error as a function of distance between correctors and BPM's; a 2 m spacing would be desirable. [8] Therefore, we nominally adopt a 1.92 m (64 periods) undulator segment length with a BPM and corrector in each separation between segments. For alignment, we will use them all, but for running we will impose 'soft constraints' that will straighten the trajectory optimally but will not attempt to thread the beam through all the BPM's exactly.

145

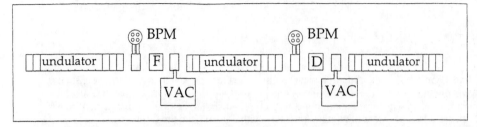

FIGURE 1: A schematic side view of the undulator structure, showing the FODO lattice with separations between 2m undulator sections for diagnostics, focusing correctors, and vacuum ports. The undulator magnets are mounted on aluminum girders whose temperature is stabilized by flowing water regulated to 0.1K

The quadrupole magnets are also serve two other functions. They are used in the initial alignment to establish a straight line trajectory. They will be moved mechanically on vertical and horizontal slides to a point where they cause the least deflection of the electron beam when the energy of the electron beam is varied. At this point, the quadrupole axes will coincide. They are also used as run time correctors; they will be moved transversely to offset small trajectory errors that are detected by the BPM systems.

The quadrupole focusing magnets could use electromagnetic or permanent magnet designs. Because electromagnets dissipate heat, and because there is some danger that a short circuit in a coil could missteer the beam, we prefer permanent magnet quadrupoles. Assuming a field gradient of 50 T/m over a length of 0.1 m, a 25 µm motion would result in a 5 µm deflection at a BPM that is 2 m downstream. In addition to errors caused by the non-zero field integrals of the undulators, the earth's field will need to be corrected for. A 1 Gauss environmental field will cause a 15 GeV beam to deflect by 8 µm every 2 m.

The fields inside a Halbach 'orange slice' pure permanent magnet quadrupole are given to leading order by [9]:

$$B^*(z) = B_x - i B_y = B_0 e^{-i\phi} = \frac{M B_r}{\pi} \frac{z(r_2 - r_1)}{r_1 r_2} \cos^2(\frac{\pi}{M}) \sin(\frac{2\pi}{M})$$

(5)

where $z = x + iy = r e^{i\theta}$, M is the number of 'orange slices', r_1 is the radius of the bore, and r_2 is the outer radius. In the case where M = 4, the we have:

$$B_0 = \frac{2 Br}{\pi} \frac{r(r_2 - r_1)}{r_1 r_2}$$

(6)

The field gradient will therefore be:

$$\frac{dB_0}{dr} = \frac{2 Br}{\pi} \frac{(r_2 - r_1)}{r_1 r_2}$$

(7)

A gradient of 50 T/m is far below the capacity of pure permanent magnet design; we would need only a simple quadrupole design with four trapezoidal pole pieces of NdFeB 2.7 cm outer radius and a 1 cm bore. The LCLS is very insensitive to higher harmonics of the B field. [8] The first higher harmonic is the dodecapole, whose magnitude is about 6% of the quadrupole field; the tolerance is a factor of 380000.

TOLERANCES

For large scales, such as the length of an undulator segment, the principle that governs the tolerances is that the emitted radiation meet the resonance condition, so that it can create more coherent electron bunching. Errors that cause off resonance radiation should be no larger than the resonance width defined by the energy spread of the electron beam, which is 0.02% at 15 GeV, and 0.068% at 5 GeV. For small scales, such as the length of one period of the undulator, simulations show that random magnetic field errors are tolerable up to $\Delta B_{max}/B_{max} < 0.1\%$, after which there is a degradation of the FEL gain.

The resonance condition for the energy of the first harmonic FEL radiation is:

$$E_1 \text{ (eV)} = \frac{950 \ E_e^2 \text{ (GeV)}}{\lambda_u \text{(cm)} \left(1 + K^2/2\right)} \tag{8}$$

where E_e is the energy of the electron beam, and $K = 0.934 \ B_{max} \text{ (T)} \ \lambda_u \text{ (cm)}$. In the case of the LCLS, $K = 3.7$ and $B_{max} = 1.32$ T.

For an NdFeB hybrid, the maximum value of the magnetic field is [9]

$$B(\text{ x }) \text{ (T)} = 3.44 \ e^{(-gap/\lambda_u(5.08 \ -1.54 \ gap/\lambda_u)} \cosh \left(2 \ \pi \ x / \lambda_u \right) \tag{9}$$

The undulator period is λ_u and x is in the direction across the gap. The pre-exponential, 3.44, in the formula for B_{max} is based on NdFeB magnetic materials with $B_r = 1.1$ T. Stronger materials are now available, which would increase the pre-exponential proportionally. [10] However, after high temperature thermal stabilization against radiation induced damage, this value of B_r is more realistic.

For the optimum 15 GeV LCLS values, $\lambda_u = 30$ mm, gap = 6 mm, we can calculate the sensitivities of the parameters nearby:

$$E_1 \sim \frac{E_e^2 \ gap^{1.56}}{\lambda_u^{4.31}} \tag{10}$$

An electron beam energy shift equal to the 15 GeV energy spread of $0.0002 \ E_e$ would cause a resonance shift of $0.0004 \ E_1$. The B field varies as $\cosh(2 \ \pi \ x / \lambda_u)$ acrosss the gap, so the beam could be off axis in this direction by about 100 μm (190 μm) before it was off resonance by more than the energy spread at 15 GeV (5 GeV). It could be off axis by an even larger amount parallel to the gap, because the undulator field can be made 'wide' in this direction.

The energy spread criterion would limit local gap and period errors to about 1 μm which are extremely severe tolerances. However, FRED-3D simulations show that FEL efficiency drops off only if random local $\Delta B_{max}/B_{max}$ is greater than 0.1%. [11] This tolerance is reached when we have a local period error of 35 μm, or a local gap error of 7 μm. These are merely tight tolerances, which we can meet with machining and assembly and magnetic shimming techniques.

It is possible to order magnets with $\Delta B/B_{rms} < 0.5\%$, and an angular error of $0.5°$ from the nominal easy axis, but 1.0%, $1°$ magnets are much less costly. We have found in the case of pure permanent magnet devices that we can use computerized sorting techniques to improve undulator errors by a factor of about 5 over the case of a randomly assembled undulator; we would hope to be able to do as well with a hybrid structure. [12] If we are able to sort to 0.2% magnetic errors, it may be preferable to specify mechanical tolerances only to this level, since we have to resort to magnetic shimming to get to 0.1% $\Delta B/B_{rms}$ anyway. This would double the gap and period errors specified above to 14 and 70 μm, respectively,

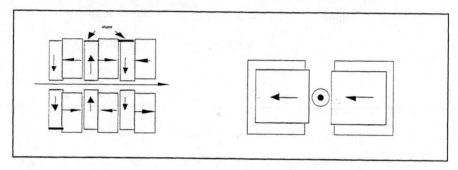

FIGURE 2: Left: Top view of undulator, showing shims on poles and small gaps between pole assemblies. Shims may also be applied to the gapside faces of the NdFeB. Right: End view of undulator, showing NdFeB overlapping vanadium permendur pole pieces. The field of the undulator is horizontal, so that the radiation is vertially polarized.

The undulator pole assembly comprises vanadium permendur poles with an NdFeB magnet closely attached to one face, as shown above. The NdFeB magnet opposite the other face will be separated from the pole by a small gap, in order to prevent mechanical stackup error. This is not ideal magnetically, as it can lead to some direct fields from the NdFeB reaching the gap. But if the separation can be kept to the 50 μm level, these fields should be small.

Even with precision machining, sorting of magnets, tight specifications, and precise assembly, we expect measurable field errors. A standard Hall probe can, in principle, measure fields in the Tesla range with 20 μT (0.2 Gauss) resolution. [13] It is our strategy to shim each pole, to minimize the trajectory and photon resonance errors. We have a coordinate measuring machine with 0.8 μm absolute positioning accuracy over a volume of 1.2 m x 1 m x 0.6 m. [14] We have fitted it with a temperature compensated Hall probe (calibration drift = 80 ppm/K), and we should be able to characterize the magnetic field precisely.

We have modeled the effect of placing small shims on the ends of the pole pieces away from the gap. The effect is to lower the field in the gap by a few Gauss, or a few 10's of Gauss. If we purchase magnets with 1% (125 Gauss) errors, and lower the errors to 0.2% (25 Gauss) with sorting, we are well within the range of this shimming strategy. A desirable feature of this shimming strategy is that the effect is local to one pole, with little fanout. Thin ferromagnetic shims may also be placed on the gapside faces of the NdFeB magnets to control phase errors.

Detailed design of the undulator has been carried out using the AMPERES 3-D magnet modeling code [15]. These simulations show us that the vanadium permendur poles should be 25-30 mm wide (parallel to gap) 5 mm thick, and 30 mm tall (normal to gap). The NdFeB should be 35-40 mm wide, 10 mm thick (longitudinally) and 35 mm tall. It might be possible to reduce these dimensions somewhat, if significant economies could be achieved.

ALIGNMENT

The basic specification for alignment is that the beam trajectory must be straight to within about 5 mm over each field gain length (~ 10 m). High performance laser tracker survey instruments could be used to position absolute beam position monitors (such as carbon wires) to within about 50 μm over this distance, perhaps somewhat less with great care and thermal stability. Straightness interferometry could probably reduce this figure, but 5 μm would be very difficult. It is possible to use a hydrostatic level to establish vertical position to 1 μm [16]. The hydrostatic level may then be used to measure the vertical position of the wires in a few positions, and a mathematical fit may be established for the wire. If the horizontal position of the wires is planar to a small tolerance, the wires then become an absolute reference with a few microns uncertainty.

Beam based alignment is preferred to external mechanical techniques in high energy physics machines, such as the SLC and FFTB projects at SLAC. Our project is different from high energy physics machines, in that we position the beam over a long distance, not just at a point. However, we cannot expect to use mechanical techniques for the higher tolerances at higher beam energies where beam based techniques will be required.

Once the undulator is aligned, we plan a suspended wire and mechanical actuator feedback system to maintain the BPM mechanical alignment at the micron level, with a response time on the order of seconds. Two stretched wires, suspended between monuments at either end of the undulator, act as position references. See figure 4. The wires sag hyperbolically, but are very stable; one end runs over a pulley and has a constant weight attached to it, for constant tension. Optical or inductive pickups attached to the electron beam position monitors detect changes in BPM position with respect to the wires. These changes will be accounted for when the BPM information processing system calculates electron beam position. If the electron beam moves, correctors will be repositioned to null the beam motion as measured by the BPM system downstream.

The initial alignment of the undulator is a major challenge. We have examined four beam based approaches, and are pursuing those described below. We will narrow the number of approaches during the R&D phase of the LCLS project.

Our first approach is simply to lower the alignment tolerances by running the LCLS at the lower limit of energy, 5 GeV. We will have located carbon wire position monitors to within the 50 μm survey tolerance, and we can then steer the electron beam so that it strikes the carbon wires. From our simulations, we expect 15 Å FEL radiation with this trajectory. The saturation length is nominally about 40 m so we should get to saturation at 100 m even under non-optimal conditions. It may be possible to bootstrap the alignment from this condition, by progressively raising the energy. However, with 50 correctors that can be moved in two planes, our configuration space is rich, and our output has few independent parameters.

A second alignment strategy is to use the quadrupole focusing magnets. As the energy of the beam is varied, the quadrupoles are moved to positions where the beam is deflected least; this should be the center of the quadrupole. There can be small dipole errors between each quadrupole, so we correct the position of the magnet or the trajectory of the beam to find a trajectory that has no dispersion with energy. Such a trajectory must be a straight line, at high enough energy. The concept here is that we do not measure straightness directly, which is hard, but subtsitute a measurement of a change in position with energy, which is much easier.

A third technique is to allow the high energy electron beam from the linac to strike a target and produce bremsstrahlung which could be used for alignment. We would place pinholes in line between a small source at the upstream end of the undulator and a small detector at the downstream end. When bremsstrahlung pased through the pinholes, we would then move quadrupoles to precise positions with respect to the pinhole. High energy bremsstrahlung diffracts negligibly, but does scatter from collimators and pinholes, which could compromise precision. It would be necessary to create a small source (by using a wire on axis, for example), and to have a small detector [17]

The fourth strategy for beam based alignment is to use a beam position monitor that is sensitive both to photons and electrons. Ultimately, it is overlap between the photon and electron beams that we want, and this is the most direct technique to ensure this. We propose to place a carbon wire in the beampipe, and find the electron beam from its bremsstrahlung, and the photon beam by diffraction. A carbon wire like this can withstand the electron and x-ray beams with very long lifetime. We measured the powder diffraction pattern from 7 mm carbon wires, and saw a 1.5 Å scattering peak at about 28 degrees. By moving the wire and the electron beam, we can superpose the electron beam onto the 1.5 Å photon beam.

The spontaneous radiation at the 1.5 Å fundamental diverges by $\sigma_{r'} = \sqrt{\lambda_{rad}/L_u} = 1.2$ μrad, or 120 μm at 100 meters. The FEL beam diverges by $\varepsilon / \sigma_r = 0.74$ μradians, or 74 μm at 100 meters. On axis, the FEL power is expected to be greater than the spontaneous power after about 7 gain lengths. The size of the electron beam is only 65 μm diamter RMS everywhere. Therefore, sweeping a 20-30 μm carbon wire (rms effective size = diameter/4) through the beam should give us the ability to find the beam centroid to 5 μm resolution, as required.

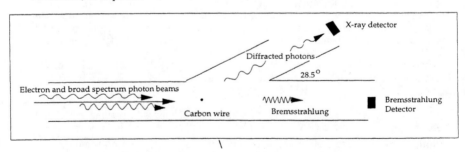

FIGURE 3: The combined electron and photon beam position monitor. Electrons are detected by the bremsstrahlung monitor, and photons are detected by the diffracted x-ray detector.

150

BEAM POSITION MONITORS

Simulation codes similar to those for circular machines were run in order to determine what beam position monitors and correctors would be needed to control the trajectory walkoff to 5 μm RMS per gain length. Beam position monitors have uncertainty as to their position, noise in their signals, and limited resolution. Correctors may also have uncertainty as to position and limited resolution. We require a resolution of 1 μm or better for both BPM's and correctors. As stated previously, we have adopted a 4.32 m FODO cell length, with two 1.92 m undulators and an F and a D quadrupole and corrector magnet in each of the 24 cm separations between undulator segments.

We examined many beam position monitor technolgies, including intercepting carbon wires, Compton scattering from laser 'wires', striplines and wall current monitors, diffraction radiation monitors, RF cavity monitors, fluorescent materials such as YAG crystals, and transition radiation monitors. We required simplicity, stability, low drift, low impedence, high resolution. and fabricability. The candidate technologies we are presently entertaining are carbon wires for initial beam finding, and striplines or RF cavity monitors for feedback control. We have advanced the design of a stripline for this particular application with MAFIA modeling; it would be used if we had no separations and had to fit the BPM into the undulator gap. The CERN (CLIC) and SLAC (FFTB) RF cavity designs are somewhat more costly than stripline BPM's, but are less sensitive to cabling and conncector problems, their electrical center is easier to establish, and their modeling and development is more mature. [18] We have decided to choose the RF cavity as our candidate BPM technology.

Since FEL performance is a sensitive function of beta function, it will be important to have the ability to measure the beam size. The FFTB experiments used carbon wires for very precise beam size measurements, and we could do the same, by detecting bremsstrahlung.

GEOPHYSICAL, THERMAL, AND MECHANICAL DESIGN

The LCLS undulator will be placed in the existing Final Focus Test Beam (FFTB) facility at SLAC. The first third of the tunnel for the unduator is underground; the last two thirds are in an above-ground heavily shielded concrete structure. The substrate throughout is miocene sandstone. We propose to place the undulator on piers made of a proprietary sand and expoxy material [19], which is much more damping than metal piers, and much more stable against water related swelling and shrinking than ordinary concrete. In the range below 100 Hz, there is cultural and ocean generated vibration, but at the top of existing sand/epoxy FFTB piers, the measured amplitude of these vibrations is less than 50 nm.

Diurnal thermal distortions of the tunnel outside the hilliside can be as much as 100 μm /day, but they can be reduced by isolating the tunnel with trenches cut into the substrate alongside it. The slow diffusive ground motion can separate points 100 m apart by about 100 μm /year. [20] These measurements indicate that our feedback system needs to have a dynamic range of some fraction of a millimeter. Presently the tunnel air temperature is stable to about 1K, but we could install a separate isothermal shield for the undulator.

We intend to stabilize the undulator structure with flowing water whose temperature is controlled to 0.1K.

We propose to detect and correct diurnal, thermal, and ground motions with a system that was successfully implemented on the FFTB. [21] Two wires will be suspended inside hollow chambers that run the length of the undulator. The chambers are fixed to the magnets, and one end of the wire runs over a pulley and is tensioned by a fixed hanging weight. The exact shape of the curve described by the wire is not a catenary, but is distorted by imperfections in the wire at the micron level. However, a wire suspended this way is very stable. We will use either inductive position monitors sensitive to pulses of current through the wire, or optical LED and split photodiode position sensors; these systems are capable of resolving changes in the position of the wire with submicron resolution.

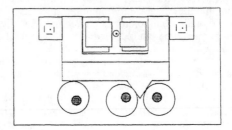

FIGURE 4. End view of the undulator magnet and mover showing three of the cams on one end of the girder (there are two on the other end) and the wire position monitors (upper left and right). The beampipe is 6 mm OD stainless steel tube.

The magnets may be mounted on 5 eccentric cams, which are driven by stepping motors and harmonic drive gear reduction units. [22] The stepping motors might have 200 steps/turn, the harmonic drives might have 100:1 gear ratios, and the cam might have an eccentricity of 1 mm, for a net resolution of 10 steps/μm. In the FFTB, these units have been shown to have submicron resolution and repeatability, which matches that of the wire position monitoring system. Rotation of the 5 cams controls pitch, yaw, roll, x, and y motion; an kinematic algorithm is used to control the motors; all 5 may have to move to correct an error in any of these 5 degrees of freedom. It is our intention to energize the stepping motors only when they move, so as not to create unwanted heat. Each motor dissipates about 10 W in motion. If, on average, one motor is running at a time, only this 10 W has to be removed from the system.

RADIATION AND BEAM EFFECTS, AND WAKEFIELDS

We do not think that radiation below a few MeV contributes significantly to magnet damage, since it is below the threshhold for efficient pair production; tests are under way at the Advanced Photon Source to test this quantitatively. The LCLS can be thought of as a wiggler with a critical energy of about 200 keV, so only bremsstrahlung and direct e-beam should affect the magnets. With suitable collimation and beam missteering protection, we should not have a problem with radiation induced damage. If our FEL were run at 120 Hz, with pulses only 280 fsec long, the continuous equivalent beamtime is only 1 msec/year.

If we have only a 6 mm gap between vanadium permendur pole pieces, and perhaps 7 mm between NdFeB magnets, radiation might pose a threat to the magnetization. At the

ESRF, demagnetization of NdFeB magnets occurred due to beam missteering. However, that material was not thermally stabilized by being baked after magnetization. This procedure allows marginally magnetized domains to demagnetize, so it causes a loss of some magnetization of the material, but this loss is homogenous. Radiation induced demagnetization is modeled like a thermal demagnetization process, and thermal stabilization is apparently a defense against radiation induced demagnetization. [23] High coercivity NdFeB [8] used in the KEK in-vacuum undulator has been exposed to circulating beam for 7 years at a 12 mm gap with no apparent demagnetization. [24] The material was thermally stabilized to 125C, with a loss of about 2% of Br. If NdFeB is not robust enough against radiation damage, we could use Sm_2Co_{17}, which costs much more than NdFeB, has lower Br, but has much greater radiation resistance.

Calculations show that our electron beam will not endanger the beampipe, but a single electron beam strike in the magnet material could fracture it, so we will protect the undulator with collimators, which will also scrape the halo off the beam. We calculate that 2-3 radiation lengths of Ti, followed by 20-30 radiation lengths of Cu is an appropriate collimator, with a possible sacrificial stopper at the entry to the undulator, for safety. [25]

The vacuum in the LCLS beampipe need not be very good; 100 nTorr should be adequate. The main disadvantage of higher pressure is the creation of gas bremsstahlung. However, we did need to examine the 5 mm ID beampipe for the effects of wakefields. For stainless steel, resistive wall losses would generate an energy spread in the electron beam of $\Delta E/E = 0.0016$ and an emittance growth of $\Delta \varepsilon/\varepsilon = 100\%$, which is unacceptable, but it is simple enough to plate the inner wall with a skin depth of copper, which reduces $\Delta E/E$ to 0.0003 and $\Delta\varepsilon/\varepsilon$ to 3%, which are acceptable. [26] The total heat generated by resistive wall losses is on the order of a watt for 120 Hz operation, so heat dissipation will not be a problem. We have also calculated the expected effects of pumping ports, flange joints, etc and find them to be much smaller than the wall effects. The effect of longitudinal wakefields from surface roughness of the beampipe is a cause for concern; a high polish on the inside of the beampipe will be required.

SUMMARY

We have briefly described above some of the problems that we addressed in studying the design of an x-ray FEL at SLAC. Among the problems that any such design must address are: 1) tight mechanical tolerances, 2) geophysical and thermal environmental problems 3) beam position monitoring 4) initial alignment strategy 5) stability of alignmnent 6) radiation dose managment and 7) wakefield effects. Of these, we consider the initial alignment the most challenging, because we have no precedent for it. The other issues have been addressed in existing SLC and FFTB machines at SLAC, and in work at other laboratories.

ACKNOWLEDGEMENTS

The author is pleased to thank the many contributors to the LCLS undulator design study, including John Arthur, Richard Boyce, Max Cornacchia, Jeff Corbett, Robert Hettel, Don Martin, Heinz-Dieter Nuhn, Jim Sebek, Roman Tatchyn, (SSRL), Vinod Bharadwaj, Karl Bane, Gordon Bowden, Paul Emma, Alberto Fasso, Clive Field, Cho Ng, Pantaleo Raimondi, Sayed Rokni, Robert Ruland, John Sheppard, Vaclav Vylet, Dieter Walz, (SLAC), Klaus Halbach, Dave Humphries, Kwang-je Kim, Steve Lidia, Ross Schlueter, Ming Xie (LBNL), Lou Bertolini, Lee Griffith, Marcus Libkind,

(LLNL), Ilan Ben-Zvi (BNL), and Claudio Pellegrini (UCLA). This work was supported by the US Department of Energy, Office of Basic Energy Sciences under contract number DE-AC03-76SF00515.

REFERENCES

1   H. Winick, et. al. Nucl. Inst. & Methods A347 (1994) p. 199
2   M. Cornacchia, SPIE Proceedings # 2988, (1997) p. 2
3   R. Bonifacio, C. Pellegrini, L.M. Narducci, Opt. Commun. 50 (1985)
4   H-D Nuhn, to be published
5   M. Xie, Proceedings of the Particle Accelerator Conference (1995)
6   H-D Nuhn, private communication
7   H-D Nuhn and C. Pellegrini, to be published
8   P. Emma, private communication
9   K. Halbach, Nucl. Inst. & Methods 187 (1981) p. 109
10 Sumitomo Special Metals Co. Neomax 28SH NdFeB material (1993)
11 H-D Nuhn, to be published
12 S. Lidia and R. Carr, Rev. Sci. Inst. 66 (1995) p. 1865
13 Group 3 Technology Ltd. model DTM 130 Teslameter
14 Leitz Enhanced Accuracy Model PMM 12-10-6 Coordinate Measuring Machine
15 Integrated Engineering Software Inc. Winnepeg, Canada
16 Fogale Nanotech 190 Parc Georges Besse, 30000 Nimes, France
17 C. Field, Nucl. Instr.& Methods. A 360 (1995) 467.
18 W. Schnell, J.P.H. Sladen, I Wilson, and W. Wuensch, CLIC note 170, CERN SL/92-33 (1992)
19 Anocast Division of Anorad Corportation, Chagrin Falls, Ohio
20 Zero Order Design Report for the Next Linear Collider, Appendix C. SLAC report 474
21 R. Ruland - private communication
22 G. Bowden, P. Holik, S.R. Wagner, G. Heimlinger, R. Settles Nucl. Instr.& Methods A368 (1996)p. 579
23 O.P. Kahkonen, M. Talvitie, E. Kauutto and M. Mannigen, Phys. Rev. B49 (1994) p. 6052
24 S. Yamamoto, private communication
25 D. Walz, private communication
26 K. Bane, private communication

154

Photocathode Guns for Single Pass X-ray FELs[*]

D. T. Palmer
Stanford Linear Accelerator Center
2575 Sand Hill Road
Stanford, California 94309

September 23, 1997

ABSTRACT

The present state of the art in photoinjector designs will be presented in this review. We will discuss both proposed and operational photoinjectors with operating frequencies from L-band (1.424 GHz) to X-band (11.424 GHz). Also a novel pulsed DC gun will be presented. All the RF photoinjector discussed here use an emittance compensation scheme[12] to align the different slices of the the electron beam to decrease the beams normalized rms emittance.

1 Proposed Photoinjectors

1.1 Pulsed DC Photoinjector

A pulsed DC photoinjector[3] has been proposed with an accelerating gradient of $1 \frac{GV}{m}$. Beam dynamics simulations indicate that this design can achieve substantial lower normalized rms emittance than is achievable with rf photoinjectors.

[*] Work supported by the Department of Energy, contracts DE-AC03-76SF00515, DE-AC02-76CH00016 and DE-FG03-92ER40793

1.2   L-Band Radio Frequency Photoinjectors

A proposed injector for the TESLA Test Facility(TTF) has eliminate the RF power feed constraint by used a door knob coupler.[4] This novel power coupler design therefore allows for the positioning of the emittance compensation magnet in the optimum position for its design. A second photoinjector designed, built and tested for the TTF is the FERMI Lab A0 photoinjector which has worked around the rf power feed constraint problem by using a split emittance compensation magnet design.[5]

1.3   X-Band Radio Frequency Photoinjectors

The push to higher gradients in RF guns has spurred the development of a 11.5 cell integrated photoinjector based on the PWT accelerator design at UCLA.[6] This gun runs at a peak field of $240 \frac{MV}{m}$ and is capable of attaining a normalized rms emittance of less than 1.00π mm mrad for a total bunch charge of 1 nC.

2   Operational Photoinjectors

2.1   S-Band Photoinjectors

The proposed photoinjector[7] for the Linear Coherent Light Source(LCLS)[8] at the Stanford Linear Accelerator Center is the BNL/SLAC/UCLA 1.6 cell Symmetrized S-Band Emittance Compensated RF gun and emittance compensation magnet. This RF gun has been designed to reduce the RF dipole mode contribution in the electron beam normalized rms emittance, $\epsilon_{n,rms}$.[9] The RF power feed into the full cell of this gun constrains the positioning of the emittance compensation magnet. PARMELA[10] simulations indicate that there can be a 15% improvement in injector performance with the elimination of this mechanical constraint.[11]

3   Beam Dynamics Studies

We shall present beam dynamics studies of the BNL/SLAC/UCLA 1.6 cell Symmetrized S-Band Emittance Compensated RF Gun, which is the prototype for the LCLS RF gun. The beam dynamics results presented here were conducted at the Brookhaven National Laboratory Accelerator Test Facility with a final beam energy of 40 MeV.

3.1 Low Charge Emittance Measurements[12]

PARMELA was used to simulate the emittance compensation process and the subsequent acceleration to 40 MeV. A correlation of the minimum spot size with an emittance minimum was noted during these simulations. This was experimentally verified during the commissioning of the 1.6 cell rf gun, using the beam profile monitor located at the output of the second linac section, as can be seen in figure 1. The result in figure 1 is for a total bunch charge of 0.329 ± 0.012 nC, an electron bunch length of $\tau_{95\%} = 10.9$ psec with an $\epsilon_{n,rms} = 1.17 \pm 0.16 \, \pi$ mm mrad.

The dependence of transverse emittance on the bunch charge under two different experimental conditions are presented in figure 2. The linear dependent emittance versus charge was conducted under constant solenoidal magnet field that was optimized for a total charge of 390 pC and only the laser energy on the cathode was varied. For the quadratic dependent emittance versus charge, as the laser energy was varied the beam was optimized using the solenoidal magnet and steering magnets to produce the smallest symmetric beam profile at the first high energy profile screen. Note that even under these diverse experimental condition the measured emittance are consistent with the data in figure 1.

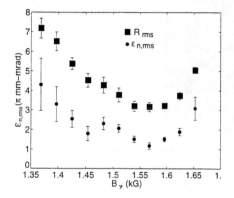

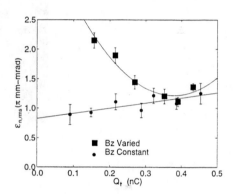

Figure 1: $\epsilon_{n,rms}$ and R_{rms} versus $B_{z,max}$

Figure 2: $\epsilon_{n,rms}$ versus Q_t

For the quadratic dependent emittance versus charge experiment it is clear that the beam line tune is not satisfactory. This experiment points out an defect in beam line tuning, in that tuning for best emittance is not that same as tuning for the best spot size. Advanced diagnostics must tune for emittance and not spot size quality which is a qualitative beam feature.

3.2 Cathode Magnet Field Emittance Contribution[11]

Due to the single emittance compensation solenoidal magnet a small but finite field at the rf gun cathode the electron bunch are produced with a finite angular momentum. We have measured the relative angular rotation due to this finite field. In this experiment a 8-fold symmetric mask[13][14] was inserted into the the laser beam thereby producing the laser profile on the cathode shown in figure 3.

The smallest beamlet located at 45° is used to break the symmetry. In this way we were able to measure the betatron rotation of the beam thru the solenoidal magnet which was found to be approximately 90°. The 8-fold beamlet relative angular rotation was measured as a function of bucking magnet field for point to point imaging of the 8-fold beamlets from the cathode to a beam profile monitor located 66.4 cm from the cathode. Figure 4 represents the relative angular rotation of the 8-fold symmetric beamlets due to the cathode magnetic field and has a linear dependence.

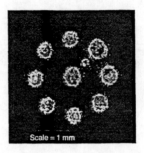

Figure 3: 8 fold symmetric beamlets

Experimental Result	$0.010 \frac{\pi \ mm-mrad}{G}$
PARMELA	$0.006 \frac{\pi \ mm-mrad}{G}$
THEORY	$0.007 \frac{\pi \ mm-mrad}{G}$

Table 1: Comparison of experimental results, simulation and theory for the emittance growth due to the cathode magnetic field

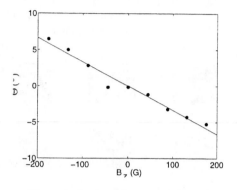

Figure 4: $\Phi_{rel}(z = 0)$ versus $B_z(z = 0)$

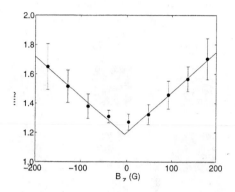

Figure 5: High energy R_{rms} versus $B_z(z = 0)$

The rms spot size as a function of cathode magnet field was measured with a beam energy of 40 MeV. The dependence of the high energy spot size with respect to cathode magnetic field is shown in figure 5. This data indicates that the minimum spot size occurs with a cathode magnetic field of -5 G. The cathode field with the bucking magnet off is +4 G. These results indicate that the cathode field can not be zeroed better than the mechanical construction and alignment of the photoinjector itself.

Due to experimental instabilities it was not possible to measure the emittance growth due to the induced cathode magnetic field. From the spot size variation at high energy due to the induced cathode magnetic field and the correlation between spot size and measured $\epsilon_{n,rms}$,[11] the emittance growth as a function of cathode magnetic field can be estimated. In table 1 the correlated experimental results of $\epsilon_{n,rms}$ growth as a function of induced cathode magnetic field is presented along with PARMELA simulation and theoretical[15] predications.

3.3   RF Field Asymmetry Emittance Contribution[16]

Phase variations in the rf gun, due to rf breakdown in the TEM coaxial conductor, prevented measurement of emittance growth due to gun disymmetrization from being conducted.

In the either mode of operation, the 8-fold beamlets were point to point focused, using the single emittance compensation magnet,[11] to a beam-profile monitor screen located 66.4 cm from the cathode plane. The center beamlet profile in figures 6 and 7 represent the beamlet profiles in the symmetric and asymmetric mode of operation respectively, with the central beamlet on the geometric center of the rf gun. Mechanical alignment and dark current studies indicate that the geometrical and integrated electrical centers of the rf gun are with in $50\mu m$ of each other in the symmetrized mode. Steering the laser beam $\pm$ 1 mm on the x and y axis we are able to probe the electromagnetic field in the 1.6 cell rf gun out to a radius of approximately 2.4 mm and produce the eight additional images in both figures 6 and 7. It should be noted that the betatron rotation of the solenoidal magnet has been removed numerical from these images so that a direct comparison of the individual beamlet position and distortion can be directly compared to the full cells cavity penetrations.

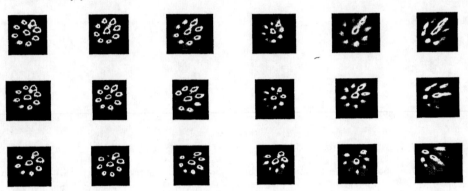

Figure 6: Beamlets profile with rf gun symmetrized

Figure 7: Beamlets profile with the rf gun unsymmetrized

In the analysis of the center 8-fold beamlets in figures 6 and 7 for the symmetrized and unsymmetrized mode of operation we have seem no centroid deflection for the outer ring of beamlets with respect to the on axis beamlet, as is expected since the TM_{110} magnetic field is uniform in the beam region of the cavity. This effect is seem in figure 8. The difference in total deflection between the different modes of operation is due to the peak rf field that the rf gun could hold off was lower in the unsymmetrized cases, since the vacuum/plunger TEM coaxial conductor would breakdown at the higher rf fields used in the symmetrized mode of operation. The normalized deflection in both modes are approximately the same.

The effect of the symmetrized versus unsymmetrized mode of operation can be seem in figure 9. This figure represents the individual beamlets Full Width Tenth Maximum (FWTM) profiles. The data that represents the unsymmetrized mode of operation has a dominate dipole mode contribution as would be expected. In the symmetrized case the dipole contribution is minimized with the dominate mode being the quadrupole contribution, as would be expected from theory. Analyses is ongoing to fully understand this effect.

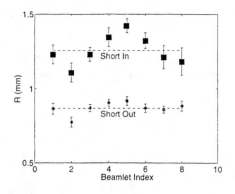
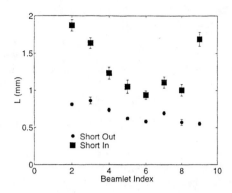

Figure 8: Beamlet centroid position from center beamlet

Figure 9: Individual beamlet radial extent

Multi-pole field effects were also studied by decreasing the laser spot size to 400 μm and setting the laser injection phase to the Schottky peak. This injection phase causes an effective electron bunch-lengthening and a noticeable energy-spread tail was observed. By adjusting the laser spot position we were able to eliminate this energy spread tail. This alignment minimizes the integrated higher-order-mode contribution to the beam distortion. Analysis indicates that the symmetrized BNL/SLAC/UCLA 1.6 cell photocathode rf gun's electrical and geometric center are within 170 μm of each other, which is within the laser spot alignment error of 250 μm. Compared to similar experimental results of the 1.5 cell BNL gun[17] whose electrical and geometric centers differ by 1.0 mm, the 1.6 cell gun has fulfilled the symmetrization criteria.

3.4 Thermal emittance Contribution

We have noted that at Q=0 for the linear dependence emittance versus charge there is a residual emittance term of 0.8 π mm-mrad. This term is due to ϵ_o, ϵ_{rf} and ϵ_{mag}. The magnetic term can be neglected since the initial cathode spot size and the small magnetic field at the cathode will contribute only 0.03 π mm mrad.[11] The rf contribution should be small, since there is no transverse space charge forces to cause beam size expansions in the rf gun. The beam, in this limiting charge case, will only sample a small portions of the rf fields which decrease the rf contribution by σ_x^2.[18] The beams intrinsic thermal emittance including the Schottky correction is given in equation 1. A theoretical estimate of the thermal emittance at 125 $\frac{MV}{m}$ for a copper cathode was found to be 0.44 π mm-mrad.

$$\epsilon_o \leq 0.8\ \pi \text{ mm mrad} \tag{1}$$

The beam thermal emittance is given by equation 2.

161

$$\epsilon_0 = \frac{r}{2}\sqrt{\frac{kT}{mc^2}} \tag{2}$$

Where the term kT is given by the difference between the laser photon energy and the effective work function of the cathode material. The effective work function of the cathode material is given by

$$\phi = \phi_0 - \sqrt{\beta E_0 \sin(\theta)} \tag{3}$$

where ϕ is the work function of the material under high gradient, ϕ_0 is the zero field work function and β is the optically polished photo-emitting areas field enhancement factor. The dependence of the beams thermal emittance is shown in figure 10 as a function of field gradient.

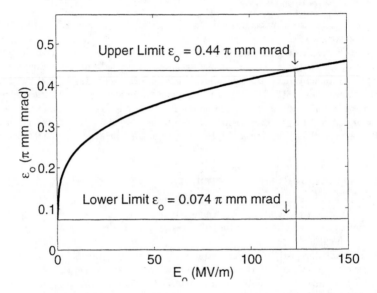

Figure 10: Thermal emittance, ϵ_o, as a function of cathode field

We have estimated the thermal emittance contribution from the Cu cathode to the total normalized rms emittance to be $\epsilon_o \approx 0.45 \pi$ mm mrad for a field gradient at the cathode of 140 $\frac{MV}{m}$.

Where this result is compared to our experimental result in the limit of zero charge which will caused the emittance term due to space charge to tend to zero we are only left with ϵ_0, ϵ_{rf}, ϵ_{mp}

Emittance Terms	Value πmmmrad
ϵ_o	0.44
ϵ_{sc}	0.00
ϵ_{rf}	0.20
ϵ_{mp}	0.15
ϵ_{B_z}	0.04

Table 2: Emittance contributes terms to the total normalized rms emittance $\epsilon_{n,rms}$

and ϵ_{mag}. These terms can be estimated from theory and compared to our experimental results in the limit of zero charge.

Using the calculated emittance values in table 2, we find that the measured emittance in the zero charge limit should be between $0.51 \leq \epsilon_{n,rms} \leq 0.83$ (π mm mrad). The zero charge intercept of figure 2 is $\epsilon_{n,rms} = 0.84$ π mm mrad. The agreement of the experimental data in the zero charge limit with the upper limit of the theoretical calculation lends credence to our Schottky corrected thermal emittance value for the Cu cathode of $\epsilon_o = 0.44$ π mm mrad.

An interesting ideas to eliminate the thermal emittance contribution or at least decrease its effect is to design the photoemission process in such a way to take advantage of the Schottky effect. By this I mean the system should be designed with a cathode material with a higher work function than that of the laser photon energy. With an appropriate cathode field that has been chosen to optimized the emittance compensation process, the effective work function of the material would be less than that of the laser photon energy by some small but finite amount. In this way the excess electron can be tuned by the accelerating field such that this excess energy is zero.

Also the use of a spin polarized rf gun operating in the emittance compensation regime would allow for the production of an emittance compensation electron beam with a zero thermal emittance contribution. Since the production of a spin polarized electron beam necessitates that the photon energy be tune to the band gap energy there should be no excess electron energy that could manifest itself into a thermal emittance term. It should be pointed out that this would not be the case is a negative electron affinity photocathode were used since in these cathode materials the vacuum energy level is less than that of the conduction band minimum.

It should be pointed out that both of these proposed methods make the available laser power an important parameter since we are not depending on the Schottky effect or the NEA to increase the quantum efficiency of the photocathode material. Also of possible importance is the time dependent variation of the band structure of the GaAs due the large applied time depend electromagnetic field.

163

Parameter	LCLS	Experimental Results
Q_T	1 nC	1.020 ± 0.059 nC
$\tau_{95\%}$	10 psec	14.7 psec
$\epsilon_{\text{n,rms}}$	$1.00\ \pi$ mm mrad	$4.74 \pm 0.24\ \pi$ mm mrad

Table 3: 1 nC emittance results

3.5   High Charge Emittance Measurements[12]

For X-ray free electron laser application such as the Linear Coherent Light Source(LCLS) the relevant photoinjector parameters are 1 nC, $\tau_{95\%} = 10$ psec and $1.00\ \pi$ mm mrad normalized rms emittance. The measured and LCLS values are presented in table 3.

Simulation results using PARMELA with a longitudinal flat top electron pulse of $\tau_{95\%} = 10$ psec, full width tenth maximum (FWTM), indicate that a $\epsilon_{\text{n,rms}} \leq 1\ \pi$ mm mrad is attainable.[19]

The results of our experimental studies present two challenges. First is the increased $\tau_{95\%}$ and second is the larger than expected $\epsilon_{\text{n,rms}}$. These are not unrelated since the emittance contribution due to rf scales as σ_z^2. Therefore an important issue is to find the mechanism causing the electron bunch length to increase. Two possible mechanisms were studied; laser pulse bunch lengthening due to laser power saturation in a doubling crystal, or space charge bunch lengthening. Experimental studies of the correlation of the laser power and $\tau_{95\%}$ were undertaken. Figure 11 shows the results of these studies. While keeping the total electron bunch charge constant at 360 pC, the laser intensity in the green($\lambda = 532$ nm) was increased. The results is that $\tau_{95\%} = 10.9$ psec. Keeping the laser intensity in the green constant and varying the uv on the cathode to increase the electron bunch charge from 400 pC to 1nC results in an increase in the $\tau_{95\%}$ from 10.9 psec to 14.7 psec (Figure 11). This clearly shows that the 1 nC electron bunch length is not due to the laser pulse length but due to some beam dynamics issue, almost surely longitudinal space charge forces.

PARMELA simulation do not show the space charge bunch lengthening seen in these experiments and this simulation error is being investigated. A group was formed at PAC97 to cross check the validty of different space charge dominated beam dynamics codes with respect to each other and with the experimental results of the BNL/SLAC/UCLA 1.6 cell S-band emittance compensated RF gun.

This debunching effect due to space charge forces can be corrected by decreasing the laser pulse length and using the debunching effect to lengthen the $\tau_{95\%}$ to the required 10 psec. This technique has the drawback the longitudinal phase mixing of the electron pulse could possible occur thereby degrading the emittance compensation process. A second possibility is to increase the laser spot size. But this would increase the emittance due to the rf emittance contribution, the remnant magnetic field at the cathode and also increase the intrinsic thermal emittance.

164

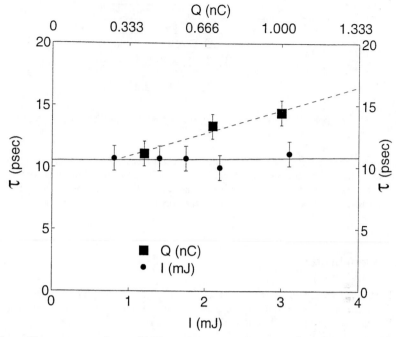

Figure 11: $\tau_{95\%}$ versus I_{green} with Q_t constant. $\tau_{95\%}$ versus Q_t with I_{green} constant.

Using a longitudinally flat top laser pulse would seem to be the best solution since the magnetic and thermal emittance would not increase and the longitudinal space charge forces would be decreased in the central portion of the bunch.

Due to the non-uniformity of the transverse laser pulse and transverse variation of the cathode quantum efficiency, the electron bunch transverse charge density $\rho(x, y)$ is not uniform. Laser assisted explosive electron emission(LAEEE)[20] has been used to smooth out this effect. The measured improvement in $\epsilon_{\text{n,rms}}$ due to LAEEE are presented in table 4. Qualitative data for QE(x, y) of the copper cathode is not available, but for the magnesium cathode LASEEE improved the QE(x, y) by a factor of two, from an order of magnitude variation across the cathode spot to a 50% after LASEEE.

A proposed correction for the cathode QE variation problem has been suggested by the author. By measuring the energy variation of the transverse laser profile with a calibrated CCD and the transverse electron charge distribution with a n-pole strip line, a correction signal can be sent to an array of liquid crystal pixels, that is inserted into the laser beamline, to distort the lasers transverse energy distribution in such a manner that a given electron charge distribution is produced. This non-invasive feedback system is presented in figure 12.

Before Laser Cleaning	
Q_t	918 ± 77.5 pC
$\epsilon_{n,rms}$	5.18 ± 0.25 π mm mrad
After Laser Cleaning	
Q_t	1020 ± 58.7 pC
$\epsilon_{n,rms}$	4.74 ± 0.24 π mm mrad

Table 4: High charge emittance study results before and after LAEEE

It should be noted that by using another CCD to measure the 2D variation of the electron beam as is strikes a beam profile screen, we could eliminate the need for the feedback control system, but this method would be invasive and would require periodic transverse charge distribution to insure that the cathode properties are not varying on the time scale of the FEL run.

3.6 Conclusion

The experimental results from the BNL/SLAC/UCLA 1.6 cell S-band emittance compensated RF gun has pointed out areas for future work. These include the experimental investigation of the thermal emittance contribution to the normalized rms emittance from Cu cathode materials. The photocathode materials group at INFN LASE Milano lead by Paolo Michelato, has agreeed to conduct zero field electron velocity distribution measurements on CU (100). These results will allow for an exact calculation of the thermal emittance contribution in the zero field limit. Two novel techniques for the reduction or elimination of the thermal emittance contribution to

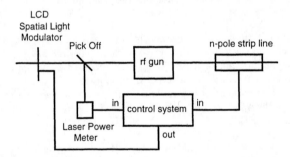

Figure 12: A non-invasive feedback system to correct for the 2D variation of the electron charge distribution.

the normalized rms emittance have been presented. Also comparing different beam dynamics simulation codes for consistence with each other and with the experimental results is of critical importance and is ongoing.

4 Acknowledgment

The work presented here is the the product of years of work and I would like to thank Drs. Ilan Ben-Zvi, Bruce Carlsten, Kwang-Je Kim, Roger Miller, Claudio Pellegrini, James Rosenzweig, Luca Serafini, Richard Sheffield, XiJie Wang and Herman Winick for their intellectual effort in the development of RF Photoinjector technology. The author would like to thank the technical staff at BNL, SLAC and UCLA for all their dedicated work during the design, construction, installation and commissioning of the BNL/SLAC/UCLA 1.6 cell symmetrized S-band emittance compensated RF photoinjector.

5 REFERENCES

[1] B. E. Carlsten, *NIM*, **A285**, 313 (1989)

[2] L. Serafini and J. B. Rosenzweig, Phys. Rev. E, **55**, 7565, (1997)

[3] T. Srinivasan-Rao *et al.*, Proc. 1996 Adv. Accel. Concepts Workshop

[4] K. Flotmann *et al.*, TTF-FEL Design Report

[5] E. Colby, *et al.*, Proc. 1997 Part. Accel. Conf., 4W.20

[6] J.B. Rosenzweig *et al.*, Proc. 1997 Part. Accel. Conf., 9V.33

[7] A. D. Yeremian *et al.*, Proc. 1997 Part. Accel. Conf., 4W.14

[8] H. Winick, *Electron Spectrosc. Relat. Phenom.*, **75**, 1-8 (1995)

[9] D. T. Palmer *et al.*, Proc. 1995 Part. Accel. Conf., SLAC-PUB-6799

[10] L. M. Young, private communications

[11] D. T. Palmer *et al.*, Proc. 1997 Part. Accel. Conf., 4W.10

[12] D. T. Palmer *et al.*, Proc. 1997 Part. Accel. Conf., 2C.10

[13] Z. Li, Ph.D. Thesis

[14] D. W. Feldman *et al.*, IEEE J. Quantum Electronic, **27**, 12, 2636-2643 (1991)

[15] K. Flotmann, private communications

[16] D. T. Palmer *et al.*, Proc. 1997 Part. Accel. Conf., 4W.11

[17] X. J. Wang *et al.*, Proc. 1995 Part. Accel. Conf. (1995) p. 890

[18] K. J. Kim, *NIM*, **A275**, 201 (1989)

[19] D. T. Palmer *et al.*, Proc. 1995 Part. Accel. Conf., SLAC-PUB-6800

[20] X. J. Wang *et al.*, J. Appl. Phys. 72(3), 888-894

Working Groups Summary

Report from the Working Group on FEL Physics and Numerical Modeling[1]

W. M. Fawley* and E. T. Scharlemann[†]

*Lawrence Berkeley National Laboratory
Berkeley, California 94720
†Lawrence Livermore National Laboratory
Livermore, California 94550

Abstract. We summarize the presentations and conclusions of the Working Group on FEL Physics and Numerical Modeling.

INTRODUCTION

Recent interest in SASE-driven x-ray FELs has developed both through an improved understanding of the basic physics and through performance predictions from numerical modeling. Accordingly, the workshop organizers felt it beneficial to have one of the three working groups focus on FEL physics and numerical modeling. Approximately twenty of the workshop participants attended the afternoon meetings of this group, in which the time was about evenly spent between semi-formal presentations and general discussion. In addition, a number of joint sessions were held with the other topical working groups on "Beam Production" and "FEL Components and Diagnostics" to discuss common issues.

To organize this written summary, we have split the material into two broad sections: 1) physical processes that affect the start-up and performance of SASE-driven FELs, and 2) key diagnostic measurements and needed modeling improvements – the former to satisfy theoreticians, the latter to give experimentalists added confidence in the theoretical predictions.

---

[1] This work was supported by the U.S. Department of Energy under Contracts DE–AC03-76SF00098 (LBNL) and W-7405-ENG-48 (LLNL).

STARTUP AND PERFORMANCE OF SASE FELS

Observations of Coherent Spontaneous Emission at Existing FELs

Most, if not all, of the proposed single-pass, short-wavelength FEL facilities rely on an initial optical signal generated by low-level bunching in the electron beam when it enters the wiggler. The bare minimum of such bunching will be due to incoherent "shot-noise" arising from the discrete nature of electrons. In the absence of any additional bunching, the effective input signal power will then scale linearly with the input current. However, in the last five years or so, there have been quite a number of reports from facilities with oscillator FELs (*e.g.*, CLIO, FOM, Stanford) suggesting that the effective initial input signal, attributed to coherent spontaneous emission (CSE), can in practice be many orders of magnitude larger than that expected solely from shot noise. The source of the CSE is generally believed to be the Fourier-component of the longitudinal beam current shape corresponding to the resonant FEL wavelength. Even though the amplitude of this Fourier component may be orders of magnitude less than the dominant, very long wavelength mode, its coherence implies that the corresponding effective input signal scales quadratically with the number of particles per wavelength N_λ (and thus with beam current). Inasmuch as N_λ often exceeds 10^6, coherent bunching at a level as small as 10^{-4} can dominate the incoherent terms.

D. Jarosynski and J. Ortega gave presentations during the morning oral sessions concerning the presence of CSE in the CLIO experiments in France. In our working group, D. Oepts reported on observations of CSE that was both quite strong and highly repeatable in the startup of the FOM IR FEL facility in Holland. Although the operating wavelength of the FOM FEL was at the ~5000th harmonic of the accelerator's RF frequency, the electron beam pulse shape had a significant Fourier component at the lasing wavelength, producing CSE 10^4 times stronger than the expected incoherent spontaneous emission. The optical phase of the CSE was essentially fixed relative to the electron bunch: Oepts presented cross-correlation studies (the FEL has ~40 independent micropulses in the oscillator cavity at any one time) that showed nearly 100% correlation between adjacent micropulses and ~50% correlation for pulses separated by 30 neighbors. From this dependence, he estimated that there is approximately 15 fs of random phase jitter on the individual electron bunches (fuller details are given by Oepts in his written summary elsewhere in these Proceedings).

The key factor here in explaining the strength of the CSE is that the FOM electron micropulses have a nearly *triangular* longitudinal shape due to the particular choice of accelerator tuning algorithm (which, for example, may be attempting to minimize coherent longitudinal energy sweep due to wakefields). The Fourier amplitudes $f(k)$ of a triangle drop off only as k^{-2}, much more slowly than the exponential fall-off for the Gaussian pulse shape often adopted by theorists. In fact, it is probably extremely difficult in practice to obtain a current pulse shape that totally avoids any portions with triangular or parabolic ($f(k) \propto k^{-3}$) shape. However, since the wavelength of an x-ray FEL is of order nanometers while a 100-fs electron beam pulse duration corresponds to 30 microns (30,000 times longer), the question is whether $f(k)$ will fall off by less than a factor of 10^5 (in which case CSE might be important) or 10^{10} (in which case one would expect shot noise to dominate any shape-factor induced CSE).

Following discussions on CSE, the working group asked A. Renieri of ENEA Frascati to discuss whether storage ring measurements of FEL startup at short wavelengths could shed some light (so to speak) on the startup physics of SASE-driven x-ray FELs. He indicated that due to the low gain of most present-day storage ring FELs it is difficult to obtain information that is obviously relevant. However, he pointed out that there is a proposal for a moderate-to-high gain storage ring (FELICITA II at Dortmund) which, if built, would have much larger gain and thus possibly produce useful data about startup. Renieri also mentioned that the Super-ACO ring in some cases operates with a theoretical gain less than the losses and nonetheless successfully starts up in what he termed a "chaotic pulsed mode" state. Explanations at ENEA for this apparent discrepancy include fluctuations in the bunch length that might generate a higher peak current or a microwave instability that might generate a kind of internal turbulence with Fourier components close to the operating FEL wavelength. A separate question arose as to whether quantum effects might be seen in a storage ring FEL but Renieri believed this was unlikely since the necessary critical parameter scales as γ^{-3}.

There was also lively discussion in the working group as to the possible importance of CSE in experimental results from the UCLA 10-μm SASE FEL reported by M. Hogan during a morning oral session. Due to the low overall gain, some felt that CSE (whose power scales as L_w^2 for low currents) might be the dominant source of the output radiation, while the UCLA group felt that they had sufficient data to rule this out. Since the UCLA experiment has recently undergone a Galilean transformation to the Los Alamos FEL lab at which the expected single pass gain is in the 10^5 to 10^6 range, one hopes that true exponential SASE will be obvious rather than (in the words of one astute observer) "Slightly-Amplified Spontaneous Emission".

A Scheme to reduce the output spectral bandpass

Both theory and numerical simulation indicate that an output signal developing solely from incoherent shot noise will have a temporal and frequency structure that is extremely "spiky", with a relatively large bandwidth

173

$(\Delta\omega/\omega_o \approx 10^{-3})$. This may be acceptable for many applications, but others may require a much narrower bandwidth. E. Saldin of Samara, Russia presented a proposal for splitting the TESLA FEL undulator into two sections. The first section would generate "normal" SASE-generated radiation. It would be immediately followed by a monochromator to select out a narrow band of wavelengths that would then be amplified in the second undulator. There would also be a chicane in which the (accelerator-produced) energy spread of the e-beam would wipe out any coherent bunching induced by SASE in the first undulator. In theory, this scheme should work well if the first undulator is sufficiently long to allow the radiation to build up three or four orders of magnitude over the equivalent start-up power in spontaneous emission (which should have the same value in the second undulator as in the first if the coherent bunching is eliminated). We note that if the passband of the monochromator is made sufficiently narrow, the output power of the amplified radiation in the second wiggler should follow the negative-exponential statistics suggested by R. Bonifacio, even though the electron beam micropulse is many coherence lengths long.

Effects of Incoherent Synchrotron Emission

As the operating wavelength of an FEL gets shorter and shorter, the effects of incoherent synchrotron emission become more and more important. These effects include both an overall energy loss (which can be compensated by tapering the wiggler) and an increase in instantaneous energy spread (which cannot be compensated and reduces the overall gain). S. Reiche of DESY reported on his modifications to the TDA3D simulation code made to study the effects of incoherent synchrotron emission, and separately to improve the "quiet start" in the presence of an initial energy spread (a fuller exposition by Reiche and collaborators appears elsewhere in these Proceedings). In order to prevent unphysical bunching downstream due to an initial energy spread, he divides his simulation macroparticles into a set of bands in γ. The macroparticle positions are modified such that the average bunching (and one or more higher moments) are forced to zero in each band. This then prevents the unwanted bunching from developing. His simulation results indicate that the synchrotron emission has a small effect on the overall predicted TESLA FEL performance. During discussion, some pointed out that synchrotron emission might have a different angular dependence than CSE, possibly providing a useful diagnostic tool for separating the two phenomena.

Quantum Effects

In addition to the quantum phenomena that may appear in storage rings (see above) and in synchrotron emission, quantum effects should be very strong in a proposed experiment at Vanderbilt University summarized by C. Brau. This experiment involves a Compton FEL with very low energy electrons, in which recoil effects from the radiation emission should reduce the gain significantly. Brau pointed out that no detailed theory exists for FEL performance in this regime (although he recognized it would have little applicability to an x-ray FEL employing GeV-energy electrons).

New FEL simulation codes

N. Vinokurov (Argonne National Laboratory and Novosibirsk) gave a short summary of his new FEL simulation code which pre-calculates the trajectories of electrons and then uses the retarded fields of one electron acting upon another to calculate the source terms for the FEL response. He has used the code to calculate the effects of putting diagnostic gaps between different undulator sections of the LEUTL project at ANL. His results suggested that for reasonable gaps there are no serious effects (*e.g.*, debunching or diffraction losses) on the overall gain other than a slight increase in the effective gain length. We note that the mechanics of this code are quite different from most other FEL particle simulation codes, thus providing an independent benchmark for FEL performance predictions.

Another new FEL code based on a different approach was described by N. Piovella (Univsity of Milan). The electromagnetic field is followed in Fourier space [*i.e.*, $E(z,\omega)$ rather than $E(z,t)$] and one does not need to adopt the wiggler-averaged source term of the standard Kroll-Morton-Rosenbluth (KMR) formulation. Piovella's method should work extremely well for FELs with relatively broad bandwidths such as those based on short electron micropulses with bunch lengths comparable to the slippage length. In general, x-ray FELs do not fall in this regime but longer wavelength experiments (*e.g.*, the UCLA 10-μm SASE experiment) often do.

JOINT MEETINGS WITH THE OTHER WORKING GROUPS

The joint sessions with the other working groups concentrated on what type of electron beams might exist at the end of the accelerator and what diagnostics are needed to understand and optimize FEL performance. In the former area, the issue of electron beam halos was prominent, with concern for whether it will be necessary to scrape them off before entrance into the wiggler. We note that in both the ELF and Paladin experiments in the 1980's at LLNL it was felt extremely important to eliminate low brightness beam components (often $\geq 75\%$ of the original beam current) in order to maximize FEL gain. It is not yet clear as to how important low brightness halos will be in the various proposed short wavelength experiments (LCLS, TESLA, LEUTL) and – if it is necessary to scrape them – where best to do so. From the group discussion, it became obvious that answers to the scrapeoff question strongly depended upon one's assumption about halo parameters. DESY workers had done simulations in which halos strongly reduced gain and saturated power levels while their LCLS counterparts had numerical results in which halos had the same effect as a slightly reduced current. We note that our experience simulating halos has been that the exact morphology of a halo's 6-D phase space density can play as much of a role as its actual integrated fractional magnitude.

In general, experimentalists from the other working groups desired more flexibility in FEL simulation codes with regard to overall input and output capabilities. S. Milton (ANL) believed it important to understand the sensitivity of x-ray/VUV FEL performance to various parameters such as input electron phase space distributions, wiggler alignment tolerances, gaps in wigglers, and the alternate-gradient focusing flutter and possible mismatch sausage oscillations which occur if the wiggler employs external quadrupoles. In the category of focusing options, a z−dependent β-function capability should also be implemented. He also proposed that the simulation codes extend their output diagnostics to produce beam phase space snapshots at user-selected longitudinal locations.

Theorists indicated that they wanted diagnostic data concerning both electron beam properties (including phase space distribution) at the entrance to the undulator and FEL output quantities such as average power, spectrum (including harmonic emission), and temporal pulse-to-pulse statistical fluctuations vs wiggler length. The dependence of FEL gain on input current, emittance and energy spread is also important. Other information, such as a detailed map of the actual undulator errors, will also be needed to make point-to-point comparisons between simulation predictions and experimental results. M. Cornacchia (SLAC) pointed out that these desired diagnostics could be subdivided into two distinct categories: 1) those needed to make the FEL work, and 2) those needed to confirm details of FEL simulation – with this category (such as beam properties on a sub-femtosecond time scale) far more difficult to obtain. D. Yeremian (SLAC) suggested that the simulation codes can provide indications of trends and sensitivities to various beam and undulator parameters. These trends could and should guide experimentalists as they tune the FELs. Many experimentalists suggested that previous experience with FEL oscillators indicated that in the absence of strong exponential gain, spontaneous emission provided the most important diagnostic for evaluating the effects of "tweaking" available knobs.

CONCLUSIONS

Our overall impression is that the SASE field needs data from a short wave-length experiment whose single pass gain is so strong that any CSE component is demonstrably negligible in comparison. From a gain point of view, it is probably not important if the initial signal is predominantly shot-noise or produced by a bunch shape factor. However, the output photon statistics will be quite different in the two cases, with the latter probably irrelevant to x-ray FEL performance.

It does not appear that in the SASE field there is an overabundance of theorists chasing insignificant problems, but we do hope that the various experimental groups continue their friendly cooperation to ensure that at least one sub-optical SASE experiment is actually built (as opposed to "destructive interference" from five or more proposals preventing any one from getting construction and operating funds).

SUMMARY REPORT OF WORKING GROUP 2 ON "BEAM PRODUCTION"

Luca Serafini[a] and Max Cornacchia[b]

[a]INFN and Universita' di Milano
Via Celoria 16, 20133 Milano, Italy
[b]SSRL - Stanford Linear Accelerator Center
Stanford, CA 94309

Abstract. We summarize here the discussions and the communications presented in the working group on "Beam Production", whose main aim was actually focused on identification of main processes limiting the maximum achievable electron beam brightness at injection into the undulator of a generic linac-based X-Ray single pass FEL. These processes have been divided into three main categories: stochastic effects in the photoelectric emission at the cathode, emittance degradation due to space charge effects in the injector, coherent spontaneous emission in bends (*e.g.* magnetic compressors) and its related impact on the emittance and/or peak current budget.

I. INTRODUCTION

The quest for high brightness electron beams has been traditionally driven in the past by the requirements of short wavelength single pass Free Electron Lasers[1]. The recent development of two major projects (TTF-FEL at DESY[2] and LCLS at SLAC[3]), aiming at producing coherent radiation in the VUV-X ray domain, makes such a quest even more demanding in terms of emittances smaller than 1 mm·mrad with beam peak currents in excess of a few kA.

The activity of the working group was mainly dedicated to a general analysis on whether the present technology for bright electron beam production, essentially based on RF laser-driven photo-injectors coupled to multi-staged magnetic compressors and further linac acceleration, is presently able to meet these highly demanding needs.

For the purpose of a better organization, the discussion was basically split into three main subjects:

i) characterization of the beam emittance at the cathode

ii) beam quality control through the injector, herein defined as the region of beam laminarity

iii) beam dynamics in the linac with distributed magnetic compression, *e.g.* with two or more compressors located at different energies along the linac

CP413, *Towards X-Ray Free Electron Lasers*
edited by R. Bonifacio and W. A. Barletta
© 1997 The American Institute of Physics 1-56396-744-8/97/$10.00

During the working group sessions 13 oral communications were presented, covering theoretical aspects of beam dynamics, status report of various projects, planned and/or experiments in progress on beam diagnostics and characterization, development of new simulation tools.

A joint session with the other working groups was devoted to discuss the relevant but challenging aspects of a possible integrated analysis of a X-ray FEL experiment, *e.g.* with a full beam simulation from the cathode surface up to the FEL radiation detector.

A final general discussion among the working group participants provided a list of relevant issues for future studies and/or R&D activities.

II. PHOTOCATHODES FOR BRIGHT BEAMS

An absolute minimum limit for the rms normalized transverse emittance of an electron beam in a linac is set by the normalized rms emittance at the cathode surface, which is a measure of the rms area spanned by the photoelectron distribution: this quantity scales linearly with the cathode spot size and as the square root of the rms photoelectron energy, with an additional dependence on the characteristics of the photoelectric emission[4]. In fact, it may be reduced, for example, by channelling effects on the photoelectrons drifting through the cathode toward the surface, producing a forward oriented emission instead of an isotropic one.

In absence of damping rings - or any other emittance damping device - as usual in any FEL driver linac, this quantity, usually known as cathode temperature emittance, cannot be further reduced because of its liouvillian invariance: the best one can do is to minimize all the phase space dilution effects taking place along the acceleration through the injector and the linac, so to make the beam emittance at the undulator entrance as close as possible to the temperature emittance.

This clearly points out the relevance of measuring the temperature emittance at the cathode and, at the same time, understanding all possible mechanisms capable to reduce its value beyond the natural scalings, which do imply the use of smaller laser spot sizes at the cathode (favouring higher frequencies) and lower photoelectron energies (favouring semi-conducting cathode materials).

In the contest of this subject, two communications were presented in the working group, by C.A. Brau (Vanderbilt Univ.), who described the experiments in progress at his laboratory on photo field-emission from needle cathodes: these are potentially able to upgrade the brightness by even 5 orders of magnitude with respect to usual photocathodes[5], mainly because of the capability to generate very high current densities at the tip of the needle - in excess of 10^{12} A/cm^2 . The use of ultra small cathode spot sizes allows to achieve very low temperature emittances, down in the range of 10^{-9} m·rad; however, the strong non-linearity of the electric field in the tip region induces a severe phase space area curvature (*i.e.* a spherical aberration) which, unless corrected, blows up the effective rms beam emittance by a huge factor[6]. Investigations are under way to assess the real potentialities of these cathodes.

A second important line of activity, presented in a talk by P. Michelato (INFN-Milan), deals with the experimental characterization of the photo-electron distribution in phase space at the cathode surface: an experiment is being presently set up in Milan based on a TOF (time of flight) technique to collect the low energy (0.1-5 eV) photoelectrons emitted by a short laser pulse (250 fs to 1.5 ps) with no further acceleration, and an energy resolution below 30 meV [5]. A properly controlled drift with selection of the emission angle allows to map the phase space density distribution at the cathode surface, and therefore to evaluate the temperature emittance. Although the first measurements are planned using a CsTe cathode, different kind of cathode materials will be tested, in search of possible channelling effects capable to focus the emission in a forward cone.

To this respect, a discussion in the working group, following a note in the invited talk by D.Palmer (plenary session), pointed out the contribution from Schottky noise effect in metallic photocathodes[7] in enhancing the temperature emittance via an increase of the photoelectron energy. This is thought to be a potentially dangerous effect which should be experimentally better investigated.

III. LAMINAR BEAM DYNAMICS

In order to avoid further emittance increase, above the temperature emittance, due to beam dynamics effects along the injector as well as the linac, one should be able to control properly the beam optics, avoiding phase space dilutions which may be driven by collective and non-linear behaviors of space charge forces, in the quasi-laminar relativistic beam regime, and/or coherent radiative effects associated with spontaneous synchrotron radiation emission, in bends, which in turns produce phase space correlations and, eventually, blow up the projected rms normalized emittance.

Regarding the laminar beam regime in the injector region, in order to meet the needs of a X-Ray FEL, which is pointing toward emittances in the sub mm·mrad range, we need both a complete understanding of the dynamics of cold plasma oscillations, which are the basis of emittance correction, and the development of new challenging beam diagnostic tools able to measure such low emittances.

The basic theoretical understanding and explanation of the emittance correction technique, originally envisaged by Carlsten[8], and now fully integrated into the rms envelope equation formalism formerly developed by Lapostolle and Sacherer[9], has been derived by Serafini and Rosenzweig[10], who found a new equilibrium mode for a quasi-laminar accelerated beam called *invariant envelope* , characterized by vanishing longitudinal correlations in the transverse space charge field, which is the basic condition to achieve emittance correction.

One of the interesting predictions of this analysis is the possibility to perform emittance correction in many steps, connecting each other different kind of beam equilibria, *i.e.* Brillouin flow[11] in drifts and invariant envelopes in accelerating sections: the matching between these equilibria is achieved automatically once the accelerating sections are set at the proper accelerating gradient and the input beam in the first section is properly matched onto the invariant envelope. The process is halted, and the residual emittance is set at the value achieved up to that point, when the dimensionless parameter

$$\rho_{INV} = \frac{\langle g^2 \rangle}{1 + 4\Omega^2} \left[\frac{(I_p/I_0)}{\gamma \varepsilon_{n,th} \gamma'} \right]^2$$, usually very large in the injector region[12], reaches

values close to 1, which represents the threshold between the laminar space charge dominated regime and the emittance thermal flow regime. Here I_p is the beam peak current ($I_0 = 17$ kA), $\varepsilon_{n,th}$ the cathode normalized temperature emittance, γ the beam normalized energy, γ' the accelerating gradient, while Ω^2 is the normalized focusing frequency[13] ($\Omega^2 \cong 1/8$ in standing wave linacs, $\Omega^2 \cong 0$ in travelling wave linacs) and $\langle g^2 \rangle$ is the rms average of the transverse space charge field form factor, which comes out to be 1 for a uniform current distribution in the bunch, $1/\sqrt{3}$ for a gaussian[12,14].

The laminar beam dynamics region is therefore characterized by the regime $\rho_{INV} \gg 1$: whenever a source delivers a beam which is still in this regime, its rms projected normalized transverse emittance is still subject to possible degradations or further corrections, since the single particle motion is not dominated by incoherent betatron motion but by the collective cold plasma envelope oscillations.

This behavior has been confirmed by many simulations and is now under active experimental investigations: during the working group sessions we had a presentation by D. Yeremian (SLAC) describing the preliminary lay-out of the LCLS injector[15] (S-band), which extends from the cathode up to 150 MeV energy point. This is an unusually high energy for the front-end of the injector, but the minimum energy needed to leave the laminar regime, as shown by that analysis. Yeremian and co-workers are also studying the option to adiabatically connect groups of accelerating sections at different accelerating gradient, preserving anyway the laminar flow behavior as necessary to achieve a full emittance correction and control up to the matching into the first compressor.

At DESY (TTF-FEL) they are following a similar approach, although using an L-band gun with special axi-symmetric on-axis coupler[16], injecting into a TESLA SC multi-cell cavity acting as a booster[17]: their emittance curves clearly show the characteristic damping profile of a beam close to invariant envelope in the laminar flow.

For both designs a crucial issue is the time profile of the laser pulse illuminating the cathode: it is well known that a uniform profile gives better performances due to the minimum longitudinal dependence of the geometric factor $g(\zeta)$ on the bunch internal coordinate ζ. Nevertheless the costs associated to such a requirement (sub-ps rise times required to the laser pulse) are quite an issue, hence most of the projects are planning to use multi-gaussian pulses with rise time of about 1 ps, as originally proposed by Coacolo et al. [18]: the preliminary predictions from the simulations give emittance growths around 1 mm·mrad (discarding from 5 to 10 % of the bunch particles in the head and tail, to avoid contamination from the bifurcated tails) versus a record of 0.3 mm·mrad with a uniform bunch[14] (100% particles, including the tails). One should note that the simulations are usually performed without taking into account the cathode temperature emittance, which should be added in quadrature to the emittance growth.

An interesting alternative approach was presented by D. Yeremian (SLAC) who discussed the possible upgrade of the NLCTA injector to allow possibility of an FEL test stand[19]: this would be based on an integrated multi-cell (11+1/2) RF gun structure working on a π-mode standing wave, according to what suggested recently by Rosenzweig *et al.*[20] on the basis of the invariant envelope theory. Preliminary simulations indicate the possibility to produce a 20 MeV beam out of the gun at 1 kA peak current and 1 mm·mrad emittance, a quite higher brightness than that of split injectors (short 1+1/2 cell gun + booster). The clear advantages of this option are originated mainly by the better longitudinal dynamics behavior, because of the continuous acceleration in the gun up to a relevant exit energy, and by a more effective focusing from the ponderomotive RF effect in the gun, which allows a better control on the beam envelope with lower magnetic peak fields required to the solenoid lens.

Several other laboratories are designing and/or upgrading present injectors to be used as FEL drivers: K.Masuda (IAE-Kyoto Univ.) presented the performances of a 4+1/2 cell S-band thermoionic RF gun to be converted into a laser-driven gun, while K.Yanagida (SPring-8) discussed the lay-out and upgrade of the SPring-8 injector and linac system, equipped with magnetic compressors, to produce a beam for the planned single-pass FEL experiments at 20 and 4 nm radiation wavelength.

The design of the APS low energy undulator test linac (LEUTL) was presented by S.Milton (ANL), who gave a general talk in the plenary session, and discussed in more details, in the working group, a new design for a thermoionic gun under study at ANL, based on a split two (and three) cell cavity for the gun with independently phased cells: the anticipated performances are far superior than the usual design of a thermoionic RF gun (π-mode standing wave in the whole multi-cell cavity), mainly because of the better control of the longitudinal dynamics which shows a quite strong bunching effect in the second cell. Peak currents above 200 A at 2-3 mm·mrad rms tranverse normalized emittance are envisaged, comparable to a good laser-driven RF gun design.

The predictions of the beam brigthness achievable with a complex system like the one under discussion (injector-linac-compressors) requires, to be reliable, the capability of numerical simulations to model properly in a 3D environment the relevant beam dynamics processes, which, in this case, are basically space charge and coherent radiation induced emittance blow-up. The community working on photo-injectors was missing for quite a long time the availability of a fully 3D code able to model the beam dynamics from the cathode surface up to the injector exit taking into account a series of possible 3D effects like transverse multipole RF fields, non axi-symmetric RF field components due to structure non-symmetries and/or misalignments, photocathode non-uniformities, flat beams, quadrupole and sextupole fields in the optics, etc.

L.Giannessi (ENEA) presented the preliminary results achieved with the code TREDI as well as the algorithms and the basics of field calculations which are part of the code engine. This is a 3D code based on the calculation of Lienard-Wiechert retarded potential, which should be able to model all the effects listed above plus the coherent synchrotron radiation effects in bends (magnetic compressors). For the moment it has been bench-marked versus ITACA[21] for a standard 2D case (round beam in a BNL-like RF gun[22]), obtaining a quite promising agreement. So far this code seems to be the best candidate for future 3D simulations of RF photo-injectors, although the cpu-time scalings are not particularly favourable, implying the need of parallel computing whenever the modeling of a real 3D beam and structure is requested.

IV. COHERENT RADIATIVE EFFECTS

Another simulation code which is under development, although in a quite advanced stage, is the one in progress at DESY by T.Limberg and co-workers, aimed at the detailed modelling of coherent synchrotron radiation (CSR) effects in bends, mainly focused at the evaluation of beam quality degradation through magnetic compressors, and based on the wake-field of gaussian line charges for any arbitrary orbit and observation points.

This coherent synchrotron radiation effect, first proposed by Carlsten *et al*. [23], is usually thought to be consisting of two contributions, namely that due to non inertial space charge forces (near field effect) and that due to coherent synchrotron radiation forces (far field effect, wake-field like). The first arises by lack of cancellation of the electric and magnetic components in the transverse space charge field of an electron bunch which is bent along a trajectory whose curvature radius is not very much larger than the bunch length. The second is produced by the synchrotron radiation produced by the bunch tail and detected by the bunch head due to the shorter straight path followed by the radiation along the cord of the arc of circle locally defined by the trajectory.

Both these contributions can lead to growth of energy spread and emittance dilution because are clearly longitudinally correlated, *i.e.* they act in different ways on various slices (longitudinal position) of the bunch. Since the typical bright electron beams under discussion are almost laminar in the longitudinal plane, even through magnetic compressors unless they are over-compressed, so that the mixing of neighbour slices is typically weak, such a longitudinal correlation in the transverse forces of various nature leads naturally to a dilution of the transverse phase space because different slices are projected over different angles, causing an increase of the total rms projected normalized emittance. The so called slice emittance, *i.e.* the emittance associated with a single slice, may remain unchanged during the process: actually this turns out to be the same mechanism producing emittance growth in RF photoinjectors[10].

As discussed in a presentation by T.Limberg, the CSR effect can produce also slice emittance growth because of the non-linear radial behavior of the wake-field, which is known to be one of the most serious issue in FEL performance degradation.

E.Schneidmiller (ASC) reported about a detailed 1D analytical treatment of energy spread induced by CSR on electrons travelling on finite arcs of circle, as opposed to previous evaluations based on steady-state regime (particle moving on a circular orbit), which allows the evaluation of CSR induced effects on beams going through finite length magnets, keeping into consideration also the effects along the straight part of the trajectory downstream the bending magnet. While previous treatments were valid for $R/\gamma^3 \ll \ell_{bunch} \ll R\phi_{magnet}^3 /24$, where R is the curvature radius, Schneidmiller's approach is valid in general for any small magnet bending angle ($\phi_{magnet} \ll 1$) and free space environment (no shielding effect from the beam pipe is taken into account). The reported set of formulas for the energy spread are in very good agreement with preliminary test runs of the DESY code, making them a quick tool for upper estimations of the impact of CSR effects.

The preliminary results from simulations of the TTF compressors concerning the emittance growth indicate that the shielding applied by the vacuum chamber in the magnetic compressors is mandatory to avoid an unbearable beam quality degradation: a reduced vertical size for the beam pipe avoids indeed the propagation of most of the CSR spectrum because of the obvious cut-off effect on the wake field. However this is not enough to prevent serious damages to the beam, so that the curvature radii of the compressor magnets must be kept large enough to damp the residual CSR effects (which tend to increase dramatically at small radii). This, in turns, sets up a lower limit in the overall length of the two magnetic compressors, within the linac, which, in case of TESLA-FEL, seem to be in excess of 5 m and 18 m, respectively.

J.C. Sheppard (SLAC) summarized the new design of the SLAC linac necessary to produce the requested beam brightness by the LCLS-FEL experiment at energies ranging between 5 and 15 GeV. Main issues of this design in preserving through the linac the nominal beam quality provided by the injector, *i.e.* 1.5 mm·mrad rms emittance at 1 nC bunch charge, achieving at the same time a final bunch rms length of 20 μm, are the wake-field induced effects coupled with CSR effects in the two magnetic compressors. Applying a careful adjustment of the compressor lay-out Sheppard reported a predicted emittance dilution due to CSR confined within a few percent of the initial value: further emittance bumps are applied to keep under control the emittance dilution through the linac.

Although the predictions based on these simulations are clearly very encouraging, there is some concern on the exact knowledge of the short bunch wake potentials that have been used and their reliability, making this issue one of the topics for future development and concern. Photocathode laser stabilities, both in time and amplitude, as well as variations in cathode performances, are of course other crucial issues.

It is interesting noticing that the final longitudinal distribution of the current in the bunch, as predicted by these simulations, shows sharp peaks at the bunch head and tail, with an almost uniform density for the bunch central part.

Investigations are in progress to understand the dependence of such a current distribution on the initial electron bunch shape in the RF photoinjector, which turns out to be dependent on the laser pulse time profile illuminating the photocathode. One of the problem to be clarified should be how much the final performances of the compressor-linac system are sensitive to the details of the initial electron bunch shape as well as the time jitters of the whole RF photoinjector and linac complex.

As presented by L. Serafini during the working group, there is some degree of freedom to tailor the electron bunch shape at the photoinjector exit: this is based on a strong longitudinal blow-out expansion of the bunch under the effect of the longitudinal space charge field, which, in the domain of ultra-short laser pulses, is linear and gives rise to uniform elliptical charge density distribution in the bunch, a well known ideal distribution assuring linear transverse and longitudinal space charge fields.

V. CONCLUSIONS AND FUTURE PERSPECTIVES

During a joint session with working group 1 (FEL theory and simulations) and working group 3 (FEL diagnostics and applications), we had an extensive discussion concerning the issue of evaluating the dependence of the FEL performances (in terms not only of total radiation power in the micro-pulse but also longitudinal and transverse coherence) on the details of the electron bunch density distribution. This deals not only with possible current modulation along the bunch, but also with halos in transverse phase space mainly caused by bifurcated tails which go through cross-overs during the emittance correction process, and eventually mix with the bunch core due to the rotation applied in the longitudinal phase space by the magnetic compression.

On one hand, the number of FEL radiation wavelengths spanned usually by the electron bunch is so large that even current spikes in the bunch macro-distribution are ineffective on the scale of the FEL wavelength, i.e. the bunching on the FEL scale is negligible - a simulation of such a distribution would need the modelling of a huge number of FEL buckets, making the simulation time unbearable. Nevertheless, we learned during the joint session that even a very weak Fourier-component of the beam current modulation corresponding to the resonant FEL wavelength can increase by many orders of magnitude the effective input signal of a SASE-FEL (due to Coherent Spontaneous Emission), hence this mechanism should be better analyzed form the point of view of the associated beam dynamics.

On the other hand, an FEL simulation starting with a real distribution in phase space as the one produced by the injector-linac-compressor complex, again, would require a modelling of the beam from the photo-cathode surface all the way down to the undulator exit (and possibly the FEL detector) - unfortunately this looks hard because of the different degree of statistics applied for injector-linac simulations (thousands of particle over the whole bunch) compared to that of FEL simulations (thousands of particle over one or a few buckets). Although some test runs have been performed both at SLAC and at DESY to evaluate the impact of halos in the FEL dynamics [24,25], the results seem to be still not conclusive and quite dependent on the type of halo modelling used in the simulation: this issue definitely deserves a much deeper attention and analysis in the future - it is clearly of critical importance to the FEL performances.

The final session of our working group was dedicated to a general discussion about the recommended topics of relevance for future investigations and R&D activity. These came out to be:

i)   Measure of the cathode thermal emittance and
procedures for its minimization

ii)   Preservation of quantum efficiency uniformity on the cathode surface
(reasonable goal: deviations < 10%)

iii)   Derivation of the scaling law of emittance at the injector exit versus
the beam current

iv)   Measure of CSR effects on beam quality in transients

v)   Analysis and Measure of wake-fields generated by ultra-short bunches
(those with rms length < 20 μm)

A special attention should be paid to the problem of longitudinal space charge de-bunching at the cathode as evaluated by present simulation codes, some of them being not able to predict the correct behavior: this issue is of great relevance after the recent measurements with the Next Generation RF gun at BNL[7] (BNL-SLAC-UCLA collaboration), performed at high charge (1 nC), showing an evidence of bunch lengthening not predicted by the simulations. To this purpose, D.T. Palmer set up an international network of RF gun simulation checkers which, in the near future, will address this as well as other problems of relevance in RF gun beam dynamics modelling.

The working group also updated the general table[26] originally set up by R. Sheffield[27] collecting the main characteristics and performances of electron beam sources in operation and/or development around the world.

ACKNOWLEDGEMENTS

We are very grateful to the organizers and the chairman for hosting this stimulating and fruitful workshop, and for setting it in such a beautiful place as Villa Feltrinelli on the Garda Lake shore.

This work was supported in part by U.S. Dept. of Energy grants DE-FG03-93ER40796 and DE-FG03-92ER40693, contract DE-AC03-76SF00515, and Offices of Basic Energy Sciences.

We also thank all our group participants for their contributions to the discussions and for their presentations: C.A. Brau, J. Clendenin, M. Ferrario, K. Flottmann, L. Giannessi, D. Giove, A. Jackson, T. Limberg, K. Masuda, P. Michelato, S. Milton, D. Palmer, F. Parmigiani, E. Schneidmiller, D. Sertore, J. Sheppard, S. Suzuki, T. Tomimasu, K. Yanagida, D. Yeremian, K. Yoshikawa

REFERENCES

1. C. Travier, *Nucl. Instr. Methods A* **304**, 285 (1991)
2. "A VUV Free Electron Laser at the TESLA Test Facility at DESY. Conceptual Design Report", June 1995, DESY Print, TESLA-FEL 95-03
3. M. Cornacchia, "Performance and design concepts of a free electron laser operating in the x-ray region", *Proc. SPIE*, Vol. 2998, p.2-14, Free Electron Laser Challenges, Eds. P.G. O'Shea and H.E. Bennet
4. J.S. Fraser *et al.*, *IEEE Trans. Nucl. Sci.* **32**, 1791 (1985)
5. P. Michelato, "Photocathodes for RF Photoinjectors", *Proc. of FEL'96 Conf.*, August 1996, Rome
6. L. Serafini, *Nucl. Instr. Methods A* **340**, 40 (1994)
7. D.T. Palmer, invited talk on "Photocathode Electron Sources for Single Pass X-Ray FEL", *these proceedings*
8. B.E. Carlsten, *Nucl. Instr. Methods A* **285**, 313 (1989)
9. P. Lapostolle, *Proton Linear Accelerators* (Los Alamos, 1980)
10. L. Serafini, J.B. Rosenzweig, *Phys. Rev. E* **55**, 7565 (1997)
11. M. Reiser, *Theory and Design of Charged Particle Beams*, John Wiley & Sons, New York, 1994
12. L. Serafini, *AIP CP* **395**, 47 (1997)
13. J.B. Rosenzweig, L. Serafini, *Phys. Rev. E* **49**, 1599 (1994)
14. L. Serafini, J.B. Rosenzweig, "Optimum Operation of Split RF Photoinjectors", *Proc. Particle Accelerator Conference 1997*, May 1997, Vancouver, Canada
15. D. Yeremian, this workshop
16. Y. Huang *et al.*, "Simulation study of the RF Gun for the TTF-FEL", Jan. 1997, Desy Print, Tesla rep. TESLA-FEL 96-01
17. J. Sekutowitz *et al.*, "Note on the SC Linear Collider TESLA Cavity Design", June 1997, DESY Print, TESLA rep. 97-06
18. J.L. Coacolo *et al.*, "TTF-FEL Photoinjector Simulation giving a High Quality Beam", *Proc. of FEL'96 Conf.*, August 1996, Rome
19. D. Yeremian, "Results of NLCTA first beam", *Proc. VII Workshop on Advanced Accelerator Concepts*, Oct. 1996, Lake Tahoe, CA
20. J.B. Rosenzweig *et al.*, "Beam Dyn. in an Integrated Plane Wave Trans. Photoinj. at S and X band",*Proc. Part. Acc. Conf. 1997*, Vancouver, Canada
21. L. Serafini *et al.*, *Proc. 1st European Particle Accelerator Conf.*, World Scientific, Singapore, 1988, p.866
22. D.T. Palmer *et al.*, *IEEE Cat. No. 95CH35843*, 1996, p.2432
23. B. Carlsten *et al.*, *Phys. Rev. E* **53**, 124 (1996)
24. B. Faatz, communication during the joint session
25. H.D. Nuhn, communication during the joint sessi
26. A postscript file named SOURCETAB.PS containing the table can be found in the Web page of UCLA-PBPL at address http://pbpl.physics.ucla.edu/
27. R. Sheffield, *AIP CP* **395**, 11 (1997)
    L.Serafini, D.Yeremian, "Particle Beam Sources", *Proc. VII Workshop on Advanced Accelerator Concepts*, Oct. 1996, Lake Tahoe, CA

FEL Components and Diagnostics

Roger Carr

Stanford Synchrotron Radiation Laboratory
Stanford Linear Accelerator Center
Stanford, CA 94309

Abstract: FEL hardware includes undulators, alignment systems, electron beam diagnostics, and mechanical and vacuum systems. While most FEL's employ conventional undulators, there is some interest in novel types, particularly where conventional designs cannot be used, as at very short periods and high fields. For these areas, superconducting technology is indicated. The most serious issue facing long FEL undulators is that of alignment; mechanical techniques may not be accurate enough, and beam-based strategies must be considered. To maintain alignment and control the electron trajectory, beam position monitors with micron precision are required. Beam size monitors are also required to assure control of emittance. The talks given in the working group sessions touch on undulators, alignment, and electron beam diagnostics, and they are summarized here.

INTRODUCTION

An FEL system requires an accelerator to provide an electron beam, matching optics, an undulator, and exit optics for the radiation produced. The undulator system might comprise a periodic magnet array, focusing magnets, beam position monitors, beam size monitors and other diagnostics, a component alignment system, a beam position feedback system for trajectory control, a beampipe and vacuum system, a mechanical support system, other instrumentation and controls, and radiation and beam protection systems. In this workshop, only some of these topics were addressed; the contents of the talks given in the working group are summarized below.

Superconducting Undulator Design

Dr. Kiyoshi Yoshikawa of Kyoto University presented a paper describing computational analyses of magnetic fields in a superconducting staggered pole undulator. In this design, the field of a superconducting solenoid is intercepted by ferromagnetic poles in a staggered array, and forced back and forth across a gap. This design keeps the superconducting magnet to its simplest form, a solenoid, thereby reducing cost and technology development. It offers the potential of higher fields at shorter periods than can be obtained in a permanent magnet device. Superconducting technology would be of interest if it is desired to make an FEL more compact. It also offers easy tunability of the magnetic field strength. The workshop session encouraged further work with superconducting designs because of their desirablity for future FEL's.

Beam Size Monitors

Dr. Stefano Marchesini of the CEA Genoble described work done at the ESRF on beam size monitors. With the dramatic reduction of emittance in third generation Synchrotron Radiation (SR) sources the diffraction limit is not far from being reached in the vertical plane. Approaching the diffraction limit, SR becomes nearly independent of the electron beam parameters, and diagnostics based on SR becomes difficult. The

CP413, *Towards X-Ray Free Electron Lasers*
edited by R. Bonifacio and W. A. Barletta
© 1997 The American Institute of Physics 1-56396-744-8/97/$10.00

measurement of the electron beam size based on SR can be performed either by imaging or by measuring the transverse coherence. When the electron beam becomes smaller than the diffraction limited beam, the image of the source becomes independent of the electron beam size, and when the coherence size becomes larger than the SR beam the coherence properties become independent on the electron beam. These two limits were discussed, and some experiments were reviewed. The first was two mirror interferometry, which is analogous to Young experiment with reflecting optics. The second was holography of a fiber is based on the interference produced by the beam diffracted by the fiber. The more the beam is diffracted, the less it is correlated with the reference beam. The third experiment used lenses for hard X rays that are produced by drilling many holes in a material with small refractive index and small absorption. The variation of refractive index gives the effect of a refractive lens in visible optics, and absorption can be only a few percent in a material like beryllium.

The APS FEL

Dr. Efim Gluskin of the Advanced Light Source at Argonne National Laboratory presented the design of an FEL planned for their 700 MeV injector linac. This project would produce soft x-rays in the range of 1500 - 500 Å, an important intermediate wavelenght between the visible and the x-ray range. A demonstration of SASE in this wavelength range should be free of beam structure effects that confuse interpretation at longer wavelengths. This undulator will rely on natural focusing in the vertical, and quadrupole lenses in separations between segments for horizontal focusing. The quadrupoles defocus in the vertical, so they should have half the total focusing strength of the natural focusing to get equal strength in the horizontal and the vertical. This could also be achieved with shaped-pole sextupole focusing, which has the advantage of causing less longitudinal velocity dispersion. One can lessen this effect by using less quadrupole strength below the equal focusing value.

The APS design incorporates a multiple pole 'comb' pole design, where many pole pieces are machined from a single block of iron, and loaded with permanent magnet material in slots. Mechanical assembly errors will be reduced by accurate machining. Four combs are interlaced to make an undulator. The advantage is that the single block will be at one scalar potential, which should reduce errors, and further tuning is allowed by moveable tuning pins.

The APS design is a precursor to higher energy x-ray FEL's, and a valuable experimental test of SASE. Mechanically and magnetically it is an innovative design, which could prove useful for long FEL undulators.

Beam Based Alignment

Dr. Bart Faatz of DESY presented studies of beam based alignment strategies for the TESLA Test Facility FEL undulator. The gain of any FEL depends both on beam and undulator quality. With these qualities given, one stil has to make sure that the electron beam trajectory through the undulator is as straight as possible in order to ensure a maximum overlap of e-beam and field along the entire undulator. The over lap can be reduced by undulator errors, quadrupole misalignment and misalignmnet of an entire module in the case of a multi section undulator. Beam based alignment is proposed for the TTF-FEL in addition to other methods, such as the one describe in the next section. Usually the beam is aligned by varying the quadrupole and corrector strengh, the TTF-FEL has fixed quadrupole. Therefore, the beam energy is changed in order to get a similar result. For different error distributions and disalignments, the results have been checked, assuming a beam position monitor close to the center of each quadrupole and a

190

corrector as close to each BPM as possible. Simulations show that uncorrected trajectories with errors of 240 μm may be corrected to error levels of 35 μm.

The technique of minimizing energy dispersion by adjusting correctors and quadrupoles overcomes the problem of non-zero dipole fields between quadrupoles that would confuse the technique of adjusting quadrupole strengths. In the FFTB at SLAC, the adjustment of quadupoles to obtain a 30 μm tolerance orbit was done in a line with no dipoles between the quadrupoles.

Optical Beam trajectory monitor

The optical trajectory monitor is being developed at DESY for the TESLA TTF FEL by Dr. Johnny Ng and Dr. Ute Müller. Dr. Ng presented the design of this scheme, in which radiation from the electron beam passes through a pinhole and then falls onto a two dimensional silicon area detector that is oriented in a plane transverse to the beam. The radiation comes from a length of the trajectory, so that it enters the pinhole over a range of angles, and thus a range of energies. However, if the trajectory is curved, it will make a curved path on the detector, and it is this curvature that is the measured quantity. If a detector and a pinhole are used in both the horizontal and vertical, trajectory error can be measured in both directions. Then correctors may be used to straighten the trajectory. One set might be used for each 4.5 m undulator module. The system is still in the design stage, and there are some technical difficulties, such as spectral range, diffraction, and wakefields that must be thought through. The detector is being constructed at the Max Planck Institute in Munich.

Dr. Ute Müller has performed simulations to predict the performance of the optical trajectory monitor system. In the TESLA TTF FEL, the quadrupole magnets are part of a combined function scheme; quadrupole magnets are placed between the poles of the undulator in 50 cm F and D lengths. There is a four parallel wire steering corrector for each of these 50 cm lengths that runs alongside the beam. For trajectories that have uncorrected walkoff error magnitudes of ± 200 μm RMS the system can be used to reduce thse errors to about ± 10 μm RMS. The system relies only on the spontaneous undulator radiation, so it will work independently of FEL action. The steering algorithm is sequential; thus it corrects single errors locally, rather than with a global simultaneous solution.

SUMMARY

These talks touched on only a few of the hardware issues that FEL construction require. The most significant uncertainties to reaching x-ray energies with FEL's have to do with establishing that SASE, cleanup, optical guiding, and photon statistical stability work as calculated by theory. But there are also significant problems with building the devices. Conventional undulators must be constructed at the very edge of the state of the art in error control. Superconducting undulators for long FEL's still require development. Alignment tolerances that can be met with mechanical means for low energy FEL's, like the TESLA TTF device, cannot be achieved mechanically for higher energy FEL's like the LCLS and the 30 GeV machine at DESY. Beam based alignment must be used, and it has not been throughly developed in this regime. Beam position monitors technology probably exists to satisfy x-ray FEL requirements, but innovative concepts like the optical trajectory monitor promise better control over errors.

Contributions

Computer Modelling of Statistical Properties of SASE FEL Radiation

E.L. Saldin[1], E.A. Schneidmiller[1], M.V. Yurkov[2]

[1] Automatic Systems Corporation, 443050 Samara, Russia
[2] Joint Institute for Nuclear Research, Dubna, 141980 Moscow Region, Russia

Abstract.
   The paper describes an approach to computer modelling of statistical properties of the radiation from self amplified spontaneous emission free electron laser (SASE FEL). The present approach allows one to calculate the following statistical properties of the SASE FEL radiation: time and spectral field correlation functions, distribution of the fluctuations of the instantaneous radiation power, distribution of the energy in the electron bunch, distribution of the radiation energy after monochromator installed at the FEL amplifier exit and the radiation spectrum.

   All numerical results presented in the paper have been calculated for the 70 nm SASE FEL at the TESLA Test Facility being under construction at DESY.

INTRODUCTION

   Self amplified spontaneous emission free electron laser (SASE FEL) [1–4] is considered now as perspective source of coherent VUV and X-ray radiation. The projects of X-ray SASE FELs are developed at SLAC and DESY [5,6] and 6 nm SASE FEL is under construction at the TESLA Test Facility (TTF) at DESY [7,8]. The first phase of the TTF FEL project will operate at the radiation wavelength 40 – 100 nm [9].

   The process of amplification in the SASE FEL starts from the shot noise in the electron beam having stochastic nature. It means that the SASE FEL radiation is also stochastic object, that is why there exists definite problem for description of the SASE FEL process requiring development of time-dependent theory of the FEL amplifier. Some averaged output characteristics of SASE FEL have been obtained in refs. [1–3,11–15]. Quantum consideration of photon statistics in SASE FEL has been performed in ref. [10]. An approach for time-dependent numerical simulations of SASE FEL radiation has been developed in ref. [15]. Realization of this approach allowed to obtain some statistical

CP413, *Towards X-Ray Free Electron Lasers*
edited by R. Bonifacio and W. A. Barletta

properties of the radiation from SASE FEL operating in linear and nonlinear regime [15,16].

The most comprehensive analysis of statistical properties of SASE FEL radiation has been performed in paper [17]. It has been shown that description of SASE FEL radiation in terms of statistical optics (time and spectral correlation functions, time of coherence, interval of spectral coherence, probability density functions of the instantaneous radiation power and of the finite-time integrals of the instantaneous power, probability density function of the radiation energy after the monochromator installed at the exit of SASE FEL) is adequate for the problem. In particular, analytical formulae for the first and the second order time and spectral correlation functions have been obtained. It has been shown that the radiation from SASE FEL operating in the linear regime possesses all the features corresponding to completely chaotic polarized radiation: the higher order correlation functions (time and spectral) are expressed via the first order correlation function, the probability density distribution of the instantaneous radiation power follows the negative exponential distribution and the probability density function of the finite-time integrals of the instantaneous power and of the energy after monochromator follows the gamma distribution.

In this paper we describe an approach to computer modelling of statistical properties of SASE FEL radiation with time-dependent simulation codes. To be specific, we calculated all the numerical results for the 70 nm SASE FEL at the TESLA Test Facility being under construction at DESY [9]. Nevertheless, they can be simply scaled for calculation of another SASE FELs by means of application of similarity techniques (see, e.g. ref. [18,19]).

METHOD FOR TIME-DEPENDENT SIMULATION

Time-dependent algorithm for the simulation of the FEL amplifier should take into account the slippage effect which connected with the fact that electromagnetic wave moves with the velocity of light c, while the electron beam moves with the longitudinal velocity v_z. Electron motion in the undulator is a periodic one, so the radiation of each electron $E(z,t)$ is also periodic function:

$$E(z,t) = f(z - ct) = f(z - ct + \lambda) \,,$$

with period

$$\lambda = \lambda_{\rm w} \frac{c - v_z}{v_z} \simeq \frac{\lambda_{\rm w}}{2\gamma_z^2} = \lambda_{\rm w} \frac{1 + K^2/2}{2\gamma^2} \,,$$

where $\lambda_{\rm w}$ is the undulator period, K is undulator parameter, γ is relativistic factor and $\gamma_z^2 = \gamma^2/(1 + K^2/2)$.

It seems to be natural to construct the following algorithm [15]. Suppose, we have electron bunch of length l_b. We divide this length into $N_b = l_b/\lambda$ boxes. FEL equations are used in each box for calculation of the motion of the electrons and evolution of the radiation field within one undulator period. The using of steady-state FEL equations averaged over undulator period is justified by the fact that FEL amplifier is resonance device with a narrow bandwidth. Then we should take into account the slippage effect, i.e. that electromagnetic radiation advances the electron beam by the wavelength λ while electron beam passes one undulator period. It means that the radiation which interacted with the electrons in the j th box slips to the electrons located in the next, $j+1$ th box. Then procedure of integration is repeated, etc.

This algorithm allows one to calculate the values of radiation field for each box as function of longitudinal coordinate z. Time dependence of the radiation field has the form:

$$E(z,t) = \tilde{E}(z,t)e^{-i\omega_0(z/c-t)} + C.C. \tag{1}$$

at any position along the undulator. Here we explicitly segregated slowly varying complex amplitude $\tilde{E}(z,t)$. At any fixed point z along the undulator the time interval between the arrival of the radiation connected with adjacent boxes is equal to $\Delta t = t_{j+1} - t_j = \lambda/c$, so we have discrete representation of $\tilde{E}(z,t_j)$. To calculate spectral characteristics we use Fourier transformation:

$$\bar{E}(z,\omega) = \int\limits_{-\infty}^{\infty} e^{i\omega t}[\tilde{E}(z,t)e^{-i\omega_0 t} + C.C.]dt . \tag{2}$$

The first and the second order time and spectral correlation functions, $g_1(t,t')$, $g_2(t,t')$, $g_1(\omega,\omega')$ and $g_2(\omega,\omega')$, are calculated in accordance with the definitions:

$$g_1(t-t') = \frac{\langle \tilde{E}(t)\tilde{E}^*(t')\rangle}{\left[\langle| \tilde{E}(t) |^2\rangle\langle| \tilde{E}(t') |^2\rangle\right]^{1/2}} ,$$

$$g_2(t-t') = \frac{\langle| \tilde{E}(t) |^2| \tilde{E}(t') |^2\rangle}{\langle| \tilde{E}(t) |^2\rangle\langle| \tilde{E}(t') |^2\rangle} ,$$

$$g_1(\omega,\omega') = \frac{\langle \bar{E}(\omega)\bar{E}^*(\omega')\rangle}{\left[\langle| \bar{E}(\omega) |^2\rangle\langle| \bar{E}(\omega') |^2\rangle\right]^{1/2}} ,$$

$$g_2(\omega,\omega') = \frac{\langle| \bar{E}(\omega) |^2| \bar{E}(\omega') |^2\rangle}{\langle| \bar{E}(\omega) |^2\rangle\langle| \bar{E}(\omega') |^2\rangle} . \tag{3}$$

The coherence time τ_c [20] and interval of the spectral coherence $\Delta\omega_c$ [17] are calculated as follows:

$$\tau_c = \int\limits_{-\infty}^{\infty} \mid g_1(\tau) \mid^2 d\tau \, ,$$

$$\Delta\omega_c = \int\limits_{-\infty}^{\infty} \mid g_1(\omega - \omega') \mid^2 d(\omega - \omega') \, . \tag{4}$$

Normalized envelope of the radiation spectrum is reconstructed from the first order time correlation function as follows

$$G(\Delta\omega) = \frac{1}{2\pi} \int\limits_{-\infty}^{\infty} d\tau g_1(\tau) \exp(-i\Delta\omega\tau) \, . \tag{5}$$

To obtain output characteristics of the radiation from SASE FEL one should perform large number of simulation runs with time-dependent simulation code. The result of each run contains parameters of the output radiation (field and phase) stored in the boxes over the full length of the radiation pulse. At the next stage of numerical experiment the arrays of data should be handled to extract information on statistical properties of the radiation (see eq. (3) − (5)). Probability distribution functions of the instantaneous radiation power, of the finite-time integrals of the instantaneous power and of the radiation energy after the monochromator installed at the exit of SASE FEL are calculated by plotting histograms of large number of statistical data.

RESULTS OF NUMERICAL SIMULATIONS

In this section we illustrate with numerical example the results of computer modelling of statistical properties of the radiation from SASE FEL. Simulations have been performed by means of 1-D time-dependent simulation codes [17]. We performed 2400 statistically independent runs with linear simulation code and 100 runs with nonlinear simulation code. At the next stage of numerical experiment the arrays of data have been handled with postprocessor codes to extract statistical properties of the SASE FEL radiation.

To test linear simulation code we used rigorous results of the SASE FEL theory presented in paper [17]. Nonlinear simulation code has been tested by means of linear simulation code at the linear stage of operation. Such a cross-testing have shown full physical consistency of the results obtained by means of analytical approach and by means of numerical codes.

High gain linear mode of operation

We begin study the properties of SASE FEL radiation with the high gain linear regime. Parameters of the radiation at this point correspond to the exit of the second stage of 70 nm option SASE FEL at DESY.

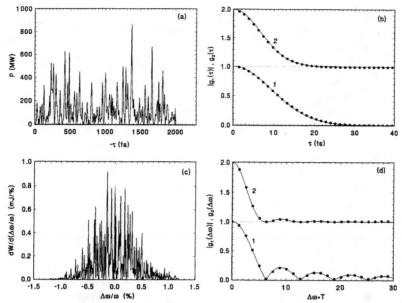

FIGURE 1. Temporal and spectral characteristics of output radiation from SASE FEL operating in high gain linear regime. Here plot (a) – typical temporal structure (one shot), plot (b) – module of the first order (curve 1) and the second order (curve 2) time correlation functions calculated over 5×10^4 independent statistical events, plot (c) – typical spectral structure (one shot) and plot (d) – module of the first order (curve 1) and the second order (curve 2) spectral correlation functions calculated over 5×10^4 independent statistical events. Solid circles in plots (b) and (d) are the results of analytical calculations.

Fig. 1a presents typical time structure of the radiation pulse. Fig. 1c presents the spectrum of the radiation pulse corresponding to this shot. The time and the spectral structure of the radiation pulse change from shot to shot and the information about the properties of the radiation can be obtained only using the statistical analysis of series of shots.

In Fig. 1b we present the results of calculations of the first and the second order time correlation functions. The circles on these plots are calculated with analytical asymptotical formulae [16,17]:

$$g_1(\tau) = \exp\left(-\frac{9\rho^2\omega_0^2\tau^2}{\sqrt{3}\hat{z}}\right) , \qquad g_2(\tau) = 1 + |\, g_1(\tau)\, |^2 , \tag{6}$$

where ω_0 is resonance frequency, ρ is saturation parameter [3] and $\hat{z}$ is the reduced undulator length [3,19].

At large number of statistically independent runs there is the possibility to perform precise calculations of the first and the second order spectral cor-

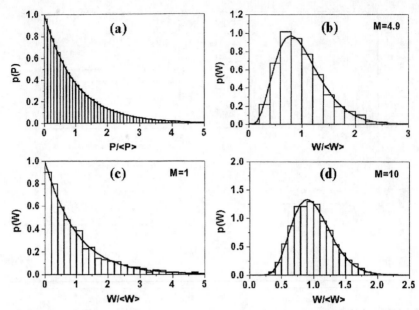

FIGURE 2. Probability density functions of the instantaneous radiation power (plot (a)), of the finite-time integrals of the instantaneous power (plot (b)), of the radiation energy after the monochromator installed at the exit of SASE FEL (plots (c) and (d)). SASE FEL operates in the high gain linear regime. Plot (a) has been calculated over 2×10^5 independent statistical events and plots (b) – (d) have been calculated over 2400 shots. Solid curves are the results of analytical predictions (negative exponential distribution in plots (a) and (c) and gamma distribution in plots (b) and (d)).

relation functions. Fig. 1d presents the results of the calculations of these functions. It is seen that there is excellent agreement between numerical and analytical results. In the modelled situation the bunch profile was a rectangular one of duration T which defined the shape of the spectral correlation functions [17]:

$$g_1(\omega, \omega') = \frac{\sin \frac{(\omega - \omega')T}{2}}{\frac{(\omega - \omega')T}{2}} \ , \qquad g_2(\omega, \omega') = 1 + \frac{\sin^2 \left[\frac{(\omega - \omega')T}{2} \right]}{\frac{(\omega - \omega')^2 T^2}{4}} \ . \tag{7}$$

A brief view on Fig. 1a indicates that instantaneous values of the radiation power are random values and practically important problem is to find the law for the probability density function. It has been shown in paper [17] that in the linear regime of SASE FEL operation instantaneous values of the radiation power P must be distributed in accordance with the negative exponential law:

$$p(P) = \frac{1}{\langle P \rangle} \exp \left(-\frac{P}{\langle P \rangle} \right) \ . \tag{8}$$

Histogram presented in Fig. 2a agrees very well with this prediction.

We performed numerical study of the fluctuations of the energy in the radiation pulse integrated over finite time. In accordance with theoretical predictions [17] the probability density function of the energy, $p(W)$, should be described with the gamma distribution:

$$p(W) = \frac{M^M}{\Gamma(M)} \left(\frac{W}{\langle W \rangle} \right)^{M-1} \frac{1}{\langle W \rangle} \exp \left(-M \frac{W}{\langle W \rangle} \right) , \qquad (9)$$

where $\Gamma(M)$ is gamma function of argument M and

$$M = 1/\sigma_W^2 , \qquad \sigma_W^2 = \langle (W - \langle W \rangle)^2 \rangle / \langle W \rangle^2 . \qquad (10)$$

In Fig. 2b we present a histogram calculated with 2400 independent shots at the value of parameter $M = 4.9$. It is seen that within statistical accuracy the gamma distribution describes rather well the probability distribution of the energy in the radiation pulse.

For a number of planned experiments a monochromator should be installed at the exit of SASE FEL and the problem arises about the shot-to-shot distribution of the radiation energy after the monochromator. Theoretical analysis predicts that energy at the exit of the monochromator will be approximately distributed in accordance with the gamma distribution (9) [17]. Histograms presented in Figs. 2c and 2d show that gamma distribution fits rather well the numerical data. It is important to notice that the gamma distribution describes the fluctuations of the energy after monochromator with any arbitrary line shape.

Nonlinear mode of operation

Using statistical technique described above, we study statistical properties of the radiation from SASE FEL operating in nonlinear mode. In Fig. 3a we present the dependence of the average radiation power on the reduced undulator length (this plot is in full agreement with previous calculations presented in ref [15]). It is seen that in the case of SASE FEL the process of the field growth does not stop at saturation point. Coherence time, τ_c, decreases drastically in the nonlinear regime (see Fig. 3b).

In Fig. 3c we present the results of calculation of the normalized rms deviation of the instantaneous fluctuations of the radiation power, $\sigma_p = \langle (P - \langle P \rangle)^2 \rangle^{1/2} / \langle P \rangle$, as function of the undulator length. One can see from this plot that at the linear stage of the SASE FEL operation the value of the deviation is equal to unity. It has been shown above that in this case the probability density distribution of the radiation power is described by the negative exponential law (8). In the nonlinear mode of operation the deviation of the power fluctuations differs significantly from unity. It indicates that

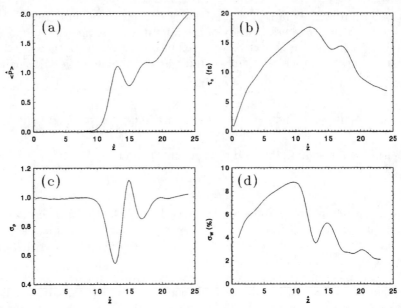

FIGURE 3. Averaged characteristics of SASE FEL as function of undulator length: averaged output power (plot (a)), coherence time (plot (b)), normalized rms deviation of the instantaneous fluctuations of the radiation power (plot (c)) and normalized rms deviation of the energy fluctuations in the radiation pulse (plot (d)).

the probability density function should differ from the negative exponential law. Fig. 4a presents the distributions of the instantaneous radiation power at saturation point. It is seen that this distribution differs cardinally from a negative exponential one. Nevertheless, in the deep nonlinear regime the power distribution tends again to the negative exponential one and the value of the rms deviation also tends to unity (see Fig. 3c) [17].

We performed numerical study of the fluctuations of the energy in the radiation pulse integrated over finite time. Fig. 3c presents the normalized rms deviation of the energy fluctuations in the radiation pulse as function of the undulator length. It is seen that the fluctuations achieve their maximum in the end of linear regime. The first local minimum corresponds to the saturation point. This is in a good agreement with Fig. 3c showing that relative fluctuations of the instantaneous radiation power achieve their minimum at the saturation point.

From practical point of view it is important to know characteristics of the output radiation from SASE FEL operating at saturation. Fig. 4a presents distribution of the instantaneous power at saturation point. In Fig. 4b we present the results of calculations of the first and the second order time cor-

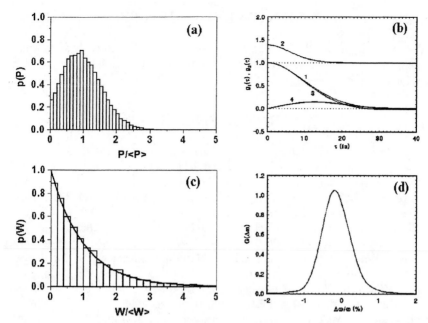

FIGURE 4. Characteristics of the output radiation from SASE FEL operating at saturation: probability density function of the instantaneous radiation power (plot (a)), the first
and the second order time correlation functions, $|g_1(\tau)|$ (curve 1), $g_2(\tau)$ (curve 2), $\mathrm{Re}(g_1(\tau))$
(curve 3) and $\mathrm{Im}(g_1(\tau))$ (curve 4) (plot (b)), a histogram of the probability density distribution, $p(W)$, of the radiation energy after narrow bandwidth monochromator (plot (c)) and
normalized spectrum of the radiation (plot (d)). Spectrum has been reconstructed from the
first order time correlation function.

relation functions. It is seen that there is significant difference with respect
to the linear mode of operation (see Fig. 1b). In particular, we obtain that
relation $g_2(\tau) = 1 + |g_1(\tau)|^2$ does not takes place in the nonlinear regime.

In Fig. 4c we present the histogram of probability density distribution of the
radiation energy after the monochromator. The width of the monochromator
is less than characteristic width of the spikes in the spectrum. It is seen from
this plot that at a small window of the monochromator the fluctuations of
the radiation energy follow the negative exponential law. We also performed
calculations of the first and the second order spectral correlation functions for
the radiation of SASE FEL operating at saturation. With the accuracy of
performed calculation we can state that spectral correlation functions are the
same as those for the linear mode of the SASE FEL operation (see Fig. 1d).

In Fig. 4d we present the envelope of the spectrum when SASE FEL operates
at saturation. Spectrums have been reconstructed using calculated values of
the first order time correlation function (see Fig. 4b).

ACKNOWLEDGMENTS

We wish to thank R. Bonifacio, J. Feldhaus, P. Gürtler, J. Krzywinski, B. Lengeler, G. Materlik, J. Pflüger, J. Roßbach, J.R. Schneider and B. Sonntag for useful discussions.

REFERENCES

1. Kondratenko A.M., and Saldin E.L., *Part. Acc.* **10**, 207 (1980).
2. Derbenev Ya.S., Kondratenko A.M., and Saldin E.L., *Nucl. Instrum. and Methods* **193**, 415 (1982).
3. Bonifacio R., Pellegrini C., and Narducci L., *Opt. Commun.* **50**, 373 (1984).
4. Murphy J.B., and Pellegrini C., *Nucl. Instrum. and Methods* **A237**, 159 (1985).
5. Tatchyn R. et al., *Nucl. Instrum. and Methods* **A375**, 274 (1996).
6. *Conceptual Design of a 500 GeV e+e− Linear Collider with Integrated X-ray Laser Facility* (Editors Brinkmann R., Materlik G., Rossbach J., Wagner A.), **DESY 97-048**, Hamburg, 1997.
7. *"A VUV Free Electron Laser at the TESLA Test Facility: Conceptual Design Report"*, **DESY TESLA-FEL 95-03**, Hamburg, 1995.
8. Rossbach J., *Nucl. Instrum. and Methods* **A375**, 269 (1996).
9. Brefeld W. et al., *"Parameter Study on Phase I of the VUV FEL at the TESLA Test Facility"*, **DESY TESLA-FEL 96-13**; *Nucl. Instrum. and Methods* (Proc. of 19th FEL Conf.), in press.
10. Bonifacio R., and Casagrande F., *Nucl. Instrum. and Methods* **A237**, 168 (1985).
11. Kim k.J., *Nucl. Instrum. and Methods* **A250**, 396 (1986).
12. Kim K.J., *Phys. Rev. Lett.* **57**, 1871 (1986).
13. Wang J.M., and Yu L.H., *Nucl. Instrum. and Methods* **A250**, 484 (1986).
14. Krinsky S., and Yu L.H., *Phys. Rev.* **A35**, 3406 (1987).
15. Bonifacio R. et al., *Phys. Rev. Lett.* **73**, 70 (1994).
16. Pierini P., and Fawley W., *Nucl. Instrum. and Methods* **A375**, 332 (1996).
17. Saldin E.L., Schneidmiller E.A., and Yurkov M.V., *"Statistical properties of radiation from VUV and X-ray free electron laser"*, **DESY TESLA-FEL 97-02**, Hamburg, 1997; *Optics Communications*, in press.
18. Bonifacio R. et al., *Riv. Nuovo Cimento* **13**, No.9 (1990).
19. Saldin E.L., Schneidmiller E.A., and Yurkov M.V., *Phys. Rep.* **260**, 187 (1995).
20. Mandel L., *Proc. Phys. Soc.* **74**, 223 (1959).

A Multi-frequency Approach to Free Electron Lasers Driven by Short Electron Bunches

Nicola Piovella

Dipartimento di Fisica dell'Università di Milano[1]
via Celoria 16, 20133 Milano, Italy

Abstract. A multi-frequency model for free electron lasers (FELs), based on the Fourier decomposition of the radiation field coupled with the beam electrons, is discussed. We show that the multi-frequency approach allows for an accurate description of the evolution of the radiation spectrum, also when the FEL is driven by short electron bunches, of arbitrary longitudinal profile. We derive from the multi-frequency model, by averaging over one radiation period, the usual FEL equations modelling the slippage between radiation and particles and describing the super-radiant regime in high-gain FELs. As an example of application of the multi-frequency model, we discuss the coherent spontaneous emission (CSE) from short electron bunches.

INTRODUCTION

Previous works studying Self-Amplified Spontaneous Emission (SASE) in Free Electron Lasers (FEL) starting from noise [1,2] are based on an approximated model of partial-differential equations [3,4] whose validity is suspected when the electron bunch is only few radiation wavelengths long. In fact, when the electron beam length is comparable to the radiation wavelength, the phase of the electrons entering the undulator is correlated with that of the optical pulse, and it can not be assumed initially distributed uniformly between 0 and 2π. The electrons distributed over a length comparable with the radiation wavelength cause a coherent enhancement of the spontaneous emission radiation, which can be several orders of magnitude larger than the shot-noise contribution, the only source of noise usually taken into account. The coherent enhancement of emission can not be described by the previous models, i.e. with the particles uniformly (or randomly) distributed over a resonant radiation wavelength at the undulator entrance. In this paper we discuss

---

[1] Work supported by the Istituto Nazionale di Fisica Nucleare

CP413, *Towards X-Ray Free Electron Lasers*
edited by R. Bonifacio and W. A. Barletta
© 1997 The American Institute of Physics 1-56396-744-8/97/$10.00

a different approach for the description of the coherent emission from short electron beams, i.e. from beams that are not very long with respect to the radiation wavelength. This study is motivated by the recent experiments [5–7] of FELs operating in the SASE configuration, i.e. single-pass, high-gain regime without input signal. These FELs operate in the IR or FIR spectrum and are driven by low-emittance, high-charge and short-duration electron bunches from RF linear accelerators. Moreover, this study can be important also for the more challenging realisation of X-ray FEL sources in the SASE regime starting from noise [8].

The approach consists in describing the interaction between the electrons and the Fourier components of the radiation field. We transform the Maxwell's equations by Fourier transform in time, reducing the space and time partial differential equations problem to a set of differential equations with respect to the spatial coordinate along the undulator. Then, we neglect the second-order derivative in space (generalised SVEA limit). This last approximation consists in neglecting the field emitted by the electrons in the backward direction with respect to the beam propagation. The particle dynamics is calculated from the Newton equation with the force due to the undulator magnetic and the self-consistent electromagnetic fields. No transverse effects are taken into account in this approach, i.e. a plane-wave form is assumed for the e.m. field. Then, the self-consistent evolution of all the Fourier components of the optical field and the N electrons in the beam is calculated along the undulator axis. At a chosen undulator length, the temporal profile of the optical field is obtained by the inverse Fourier transform of the spectral field, solution of the multi-frequency equations.

THE MULTIFREQUENCY MODEL

The field equation

We start from the Maxwell equation for a plane wave propagating along the undulator axis z:

$$\left(\frac{\partial^2}{\partial t^2} - c^2 \frac{\partial^2}{\partial z^2}\right) \vec{E}_L(z,t) = -4\pi \frac{\partial \vec{J}(z,t)}{\partial t}. \tag{1}$$

We assume a helical undulator and an electric field circularly polarised,

$$\vec{E}_L(z,t) = \frac{1}{\sqrt{2}} \left[\hat{e} E(z,t) + \text{c.c.}\right], \tag{2}$$

with $\hat{e} = (\hat{x} + i\hat{y})/\sqrt{2}$, and a transverse beam current density

$$\vec{J}(z,t) = -\frac{ecn_e(z,t)}{S_e} \vec{\beta}_\perp(z) \tag{3}$$

where $n_e(z,t)$ is the electron number per unit of length, S_e is the uniform cross section, $-e$ is the electric charge and $c\vec{\beta}_\perp$ is the transverse electron velocity. Neglecting the effect of the e.m. field on the electron trajectory,

$$\vec{\beta}_\perp \approx \frac{a_w}{\sqrt{2}\gamma} \left[\hat{e}\, e^{-ik_w z} + \text{c.c.} \right], \qquad (4)$$

where $a_w = eB_w/mc^2 k_w$ is the undulator parameter, B_w is the rms magnetic field in the undulator, $k_w = 2\pi/\lambda_w$ is the undulator wave-number and γ is the electron energy in rest mass units. Using eqs. (2), (3) and (4) and projecting over the versor $\hat{e}^*$, eq.(1) yields

$$\left(\frac{\partial^2}{\partial t^2} - c^2 \frac{\partial^2}{\partial z^2} \right) E(z,t) = \frac{4\pi e c a_w}{\gamma S_e} \frac{\partial n_e(z,t)}{\partial t} e^{-ik_w z}. \qquad (5)$$

We introduce the Fourier transform of the electric field,

$$\tilde{E}(z,k) = c \int_{-\infty}^{+\infty} dt\, E(z,t) e^{-ik(z-ct)}, \qquad (6)$$

and the inverse Fourier transform,

$$E(z,t) = \frac{1}{2\pi} \int_{-\infty}^{+\infty} dk\, \tilde{E}(z,k) e^{ik(z-ct)}. \qquad (7)$$

Using eq.(7) in eq.(5) one obtains:

$$\frac{d\tilde{E}(z,k)}{dz} + \frac{1}{2ik} \frac{d^2 \tilde{E}(z,k)}{dz^2} = \frac{2\pi e c a_w}{\gamma S_e} \int_{-\infty}^{+\infty} dt\, n_e(z,t) e^{-i[(k_w+k)z-ckt]}. \qquad (8)$$

We note that the source term in the field equation is proportional to the Fourier transform of the longitudinal electron density $n_e(z,t)$. Modelling the electron beam as a set of N particles with positions $z_j(t)$,

$$n_e(z,t) = \sum_{j=1}^{N} \delta(z - z_j(t)) = \sum_{j=1}^{N} |c\beta_j(t)|^{-1} \delta(t - t_j(z)) \qquad (9)$$

where $c\beta_j(t)$ is the longitudinal velocity, with $\beta_j = 1 + O(1/\gamma_j^2)$. Inserting eq.(9) in eq.(8), the equation for the k-th Fourier component of the electric field is:

$$\frac{d\tilde{E}(z,k)}{dz} + \frac{1}{2ik} \frac{d^2 \tilde{E}(z,k)}{dz^2} = 2\pi e a_w n_o L_e N^{-1} \sum_{j=1}^{N} \gamma_j(z)^{-1} e^{-i\theta_j^{(k)}(z)} \qquad (10)$$

where we have introduced the electron phase relative to the k-th mode, $\theta_j^{(k)}(z) = (k_w + k)z - ckt_j(z)$ and the constant electron density $n_o = N/L_e S_e$ in the volume of length L_e and area S_e, where L_e is the electron beam length.

The particle equations

The evolution of the particle variables $t_j(z)$ and $\gamma_j(z)$ are determined by the Newton equation, with the force due to the undulator magnetic field and the radiation field:

$$c\frac{dt_j}{dz} = \beta_j^{-1} \simeq 1 + (1 + a_w^2)/2\gamma_j^2 \tag{11}$$

$$mc^2\beta_j\frac{d\gamma_j}{dz} = -e\vec{E}_L \cdot \vec{\beta}_\perp$$

$$= -(ea_w/4\pi\gamma_j)\left[\int_{-\infty}^{+\infty} dk\tilde{E}(z,k)e^{i\theta_j^{(k)}(z)} + \text{c.c.}\right], \tag{12}$$

It is easy to verify that eqs.(10) and (12) conserve the total energy $\mathcal{E}_b + \mathcal{E}_s$, where $\mathcal{E}_b = mc^2\sum_j \gamma_j$ is the beam energy and

$$\mathcal{E}_s = \frac{S_e}{16\pi^2}\int dk(\tilde{E}\tilde{B}^* + \text{c.c.}) = \int dt\vec{S}(z,t)\cdot\hat{z},$$

where $\tilde{B} = \tilde{E} + (1/ik)(d\tilde{E}/dz)$ is the Fourier transform of the magnetic field and $\vec{S}$ is the Poynting vector,

$$\vec{S}(z,t) = \frac{c}{4\pi}[\vec{E}\times\vec{B}] = \frac{c}{8\pi}[EB^* + \text{c.c.}]\hat{z}.$$

In the usual SVEA approximation $\tilde{B} \simeq \tilde{E}$ is assumed, and

$$\mathcal{E}_s = \frac{S_e}{8\pi^2}\int dk|\tilde{E}|^2 = \frac{cS_e}{4\pi}\int_{-\infty}^{+\infty} dt|E(z,t)|^2$$

As the emission is peaked around a main frequency $ck_L = c\beta_R k_w/(1-\beta_R)$, where $c\beta_R$ is the resonant longitudinal velocity associated to the frequency ck_L, we introduce $\bar{k} = (k - k_L)/k_L$ and the electron phase relative to k_L,

$$\theta_j(z) = (k_w + k_L)z - ck_Lt_j(z). \tag{13}$$

Then, $\theta_j^{(k)} = (1 + \bar{k})\theta_j - \Delta z$, where $\Delta = k_w\bar{k} = k_w(k - k_L)/k_L$ is the detuning from k_L. Introducing

$$E_k = \tilde{E}(z,k)e^{-i\Delta z}, \tag{14}$$

in eqs.(10) and (12) and using (13), we obtain the new set of equations:

$$\frac{d\theta_j}{dz} = k_w\left(1 - \frac{\gamma_R^2}{\gamma_j^2}\right) \tag{15}$$

$$\frac{d\gamma_j}{dz} = -\frac{ea_wk_L}{4\pi mc^2\gamma_j}\left[\int_{-\infty}^{+\infty} d\bar{k}\,E_ke^{i(1+\bar{k})\theta_j(z)} + \text{c.c.}\right] \tag{16}$$

$$\frac{dE_k}{dz} + \frac{1}{2ik}\left(\frac{d}{dz} + i\Delta\right)^2 E_k = \frac{2\pi ea_wn_oL_e}{N}\sum_{j=1}^{N}\frac{e^{-i(1+\bar{k})\theta_j(z)}}{\gamma_j(z)} - i\Delta E_k, \tag{17}$$

where $\gamma_R = (1 - \beta_R^2)^{-1/2} = \sqrt{(1 + a_w^2)k_L/2k_w}$ is the resonant electron energy in rest mass units. The second term in the l.h.s of eq.(17) is usually neglected in the Slowly Varying Envelope Approximation (SVEA). An analysis of the spontaneous emission, i.e. assuming constant resonant phases $\theta_j = \theta_{j0}$, shows that eq.(17) admits the existence of a second resonant frequency $ck_b = ck_w/(2 + k_w/k_L) = c\beta_R k_w/(1 + \beta_R)$, corresponding to a radiation field propagating in the backward direction with respect to the electron beam. In the following, we will drop the non-SVEA term, whose effect is negligible in most of the cases of interest [9], considering henceforth only forward propagating radiation.

Universal scaling and link to the previous models

Assuming the Compton limit, $|\gamma_j - \gamma_R| \ll \gamma_R$, and introducing the variables [3] $\bar{z} = 2k_w\rho z$, $p_j = (\gamma_j - \gamma_R)/\rho\gamma_R$ and $A_k = E_k/\sqrt{4\pi mc^2\gamma_R n_o\rho}$, where $\rho = \gamma_R^{-1}(a_w\omega_p/4ck_w)^{2/3}$ and $\omega_p = (4\pi e^2 n_o/m)^{1/2}$, eqs. (15), (16) and (17) become

$$\frac{d\theta_j}{d\bar{z}} = p_j \tag{18}$$

$$\frac{dp_j}{d\bar{z}} = -\left[\int_{-\infty}^{+\infty} d\bar{k}\, A_k e^{i(1+\bar{k})\theta_j} + \text{c.c.}\right] \tag{19}$$

$$\frac{dA_k}{d\bar{z}} = (\mathcal{N}_e/N)\sum_{j=1}^{N} e^{-i(1+\bar{k})\theta_j} - i(\bar{k}/2\rho)A_k. \tag{20}$$

where the N particles are distributed over $\mathcal{N}_e = L_e/\lambda_L$ wavelengths. We note that the Fourier component A_k is driven by the electron bunching $(\mathcal{N}_e/N)\sum_{j=1}^{N} \exp[-i(1+\bar{k})\theta_j]$ and that all the modes A_k contribute to the particle dynamics. In order to recover the equations used previously for FEL propagation studies [3,4], we introduce the variable $\theta = (k_w + k_L)z - ck_L t = k_L(z - c\beta_R t)/\beta_R$, where $\beta_R = k_L/(k_w + k_L)$, and the field amplitude:

$$a(\bar{z}, \theta) = \int_{-\infty}^{+\infty} d\bar{k}\, A_k e^{i\bar{k}\theta}. \tag{21}$$

Then, eqs. (18), (19) and (20) can be written as:

$$\frac{d\theta_j}{d\bar{z}} = p_j \tag{22}$$

$$\frac{dp_j}{d\bar{z}} = -\left[a(\bar{z}, \theta_j)e^{i\theta_j} + \text{c.c.}\right] \tag{23}$$

$$\frac{\partial a}{\partial\bar{z}} + \frac{1}{2\rho}\frac{\partial a}{\partial\theta} = \frac{2\pi\mathcal{N}_e}{N}\sum_{j=1}^{N} e^{-i\theta_j(z)}\delta(\theta - \theta_j), \tag{24}$$

209

where $a(\bar{z}, \theta_j)$ is the field (21) evaluated at $\theta = \theta_j(\bar{z})$. Finally, a local average on θ over the interval 2π around the average coordinate θ_m, yields:

$$\frac{\partial \tilde{\theta}_j(\bar{z}, z_1)}{\partial \bar{z}} = \tilde{p}_j(\bar{z}, z_1) \tag{25}$$

$$\frac{\partial \tilde{p}_j(\bar{z}, z_1)}{\partial \bar{z}} = -\left[a(\bar{z}, z_1) e^{i\tilde{\theta}_j(\bar{z}, z_1)} + \text{c.c.} \right] \tag{26}$$

$$\frac{\partial a(\bar{z}, z_1)}{\partial \bar{z}} + \frac{\partial a(\bar{z}, z_1)}{\partial z_1} = N_\lambda^{-1} \sum_{j=1}^{N_\lambda} e^{-i\tilde{\theta}_j(\bar{z}, z_1)}, \tag{27}$$

where $z_1 = 2\rho\theta_m$. These equations coincide with these of Ref. [3,4], where $\tilde{\theta}_j(\bar{z}, z_1)$ are the phases of $N_\lambda = N/N_e$ particles in a wavelength at the coordinates $\bar{z}$ and z_1.

LINEARISED MODEL

Assuming a small deviation θ_{1j} from the initial phase θ_{0j}, $\theta_j = \theta_{0j} + \theta_{1j}$, a small deviation for the momentum, $p_j = p_{1j}$, and a small field A_k, we introduce the macroscopic quantities for the bunching and the momentum bunching:

$$B_k = -iN^{-1} \sum_{j=1}^{N} \theta_{1j} e^{-i(1+\bar{k})\theta_{0j}} \tag{28}$$

$$P_k = N^{-1} \sum_{j=1}^{N} p_{1j} e^{-i(1+\bar{k})\theta_{0j}}. \tag{29}$$

Then, at the first order in θ_{1j}, p_{1j} and A_k, eqs. (18), (19) and (20) give [10]

$$\frac{dB_k}{d\bar{z}} = -iP_k \tag{30}$$

$$\frac{dP_k}{d\bar{z}} = -\int_{-\infty}^{+\infty} dk' \left[A_{k'} f_{k-k'} + A_{k'}^* f_{k+k'} \right] \tag{31}$$

$$\frac{dA_k}{d\bar{z}} = N_b \left[f_k + (1+\bar{k})B_k \right] - i(\bar{k}/2\rho)A_k, \tag{32}$$

where

$$f_k = N^{-1} \sum_{j=1}^{N} e^{-i(1+\bar{k})\theta_{0j}} \tag{33}$$

$$f_{k-k'} = N^{-1} \sum_{j=1}^{N} e^{-i(\bar{k}-\bar{k}')\theta_{0j}} \tag{34}$$

$$f_{k+k'} = N^{-1} \sum_{j=1}^{N} e^{-i(2+\bar{k}+\bar{k}')\theta_{0j}}; \tag{35}$$

f_k, as defined in eq.(33), is the Fourier transform of the initial electron distribution. As already noted first by Krinsky [10], the electron beam distribution enters into the linearised equations in two ways. The term f_k acts as a driving term of the laser field A_k. For very short radiation wavelengths the Fourier transform of the bunch density is dominated by the high frequency shot noise due to the discrete nature of the individual electrons. However, the term $f_{k-k'}$ and $f_{k+k'}$ in eq.(32) introduce a coupling between different components of the laser due to the finite bunch length of the electron beam. The magnitude of f_k is largest for small $\overline{k}$, less than the inverse bunch length, so coupling between A_k and $A_{k'}$ will be negligible if $|\overline{k} - \overline{k}'| \gg 1/\mathcal{N}_b$.

COHERENT SPONTANEOUS EMISSION

At the zero-order of approximation we neglect the radiation reaction on the particle, described by the driving term B_k in eq.(32):

$$\frac{dA_k}{d\overline{z}} = \mathcal{N}_b f_k - i(\overline{k}/2\rho)A_k, \tag{36}$$

The driving source for a pre-bunched beam is the form factor f_k that, ignoring the discrete nature of the particles, is

$$f_k = \langle e^{-i(1+\overline{k})\theta_0} \rangle = \int_{-\infty}^{\infty} d\theta_0 P(\theta_0) e^{-i(1+\overline{k})\theta_0}, \tag{37}$$

where $P(\theta_0)$ is the initial longitudinal beam distribution. Eq.(36), integrated with zero input field, $A_k(0) = 0$, yields

$$A_k(\overline{z}) = (2\rho/i\overline{k})\mathcal{N}_b f_k \left[1 - e^{-i\overline{k}\overline{z}/2\rho} \right], \tag{38}$$

and, with $\overline{z} = 4\pi\rho N_w$, $a(\overline{z}, \theta) = a_1(\theta) - a_1(\theta - 2\pi N_w)$, where

$$a_1(\theta) = 2\rho\mathcal{N}_b \int_{-\infty}^{\infty} \frac{d\overline{k}}{i\overline{k}} f_k e^{i\overline{k}\theta}. \tag{39}$$

In the following, we calculate the solution for different beam profiles.

Flat-top profile

For a rectangular profile of length $L_b = \mathcal{N}_b \lambda_L$, $P(\theta_0) = [H(\theta_0) - H(\theta_0 - 2\pi\mathcal{N}_b)]/2\pi\mathcal{N}_b$, where $H(x)$ is the Heaviside function, $H(x) = 1$ for $x > 0$ and $H(x) = 0$ for $x < 0$. The form factor is:

$$f_k = e^{-i(1+\overline{k})\pi N_b}\text{sinc}[(1 + \overline{k})\pi\mathcal{N}_b] \tag{40}$$

where $\text{sinc}(x) = \sin x/x$, and

$$a_1(\theta) = 2i\rho \left\{ (e^{-i\theta} - 1)H(\theta) - (e^{-i\theta} - e^{-2i\pi\mathcal{N}_b})H(\theta - 2\pi\mathcal{N}_b) \right\}.$$

We observe that a train of sinusoidal waves is emitted at the beam edges $\theta = 0$ and $\theta = 2\pi\mathcal{N}_b$. When the slippage $\lambda_L N_w$ is smaller than the beam length $L_b = \lambda_L \mathcal{N}_b$ (i.e. when $N_w < \mathcal{N}_b$), then

$$|a(\theta)|^2 = (4\rho)^2 \sin^2(\theta/2) \tag{41}$$

for $0 < \theta < 2\pi N_w$ and $|a(\theta)|^2 = (4\rho)^2 \sin^2(\pi N_w)$ for $2\pi N_w < \theta < 2\pi\mathcal{N}_b$. In the region in front of the beam $(2\pi\mathcal{N}_b < \theta < 2\pi(\mathcal{N}_b + N_w))$ the intensity is $|a(\theta)|^2 = (4\rho)^2 \sin^2[\theta/2 - \pi(\mathcal{N}_b + N_w)]$. When the slippage length is larger than the beam length, $N_w > \mathcal{N}_b$, the intensity inside the beam $(0 < \theta < 2\pi\mathcal{N}_b)$ is given by eq.(41); in the region in front of the beam $(2\pi\mathcal{N}_b < \theta < 2\pi N_w)$ $|a(\theta)|^2 = (4\rho)^2 \sin^2(\pi\mathcal{N}_b)$ and in the region $2\pi N_w < \theta < 2\pi(N_w + \mathcal{N}_b)$, $|a(\theta)|^2 = (4\rho)^2 \sin^2[\theta/2 - \pi(\mathcal{N}_b + N_w)]$. We observe that the emission is coherent, $|a|^2_{peak} \propto \rho^2$, i.e. the unscaled intensity is proportional to n_0^2, and it is not amplified, as the peak value is independent on N_w. This coherence effect is due to the interference between the different slices of width equal to the radiation wavelength. As $|a|^2_{peak} = 16\rho^2$, the radiation emitted by Coherent Spontaneous Emission (CSE) is very small compared to the FEL radiation and is normally neglected in the high-gain FEL theory. However, in the absence of the input field, it is the only noise source, together with the shot-noise, responsible for the start-up of a FEL operating in the SASE regime. If the beam is much longer than the slippage $N_w\lambda_L$, the coherent emission contribution occurs only on a small fraction of the beam, near the edges. By integrating over the radiation pulse, the total efficiency as $\eta = 16\rho^3[N_w/\mathcal{N}_b + (1 - N_w/\mathcal{N}_b)\sin^2 \pi N_w]$ when the slippage is smaller than the beam length, $N_w < \mathcal{N}_b$, and $\eta = 16\rho^3[1 + (N_w/\mathcal{N}_b - 1)\sin^2 \pi\mathcal{N}_b]$ when the slippage is longer than the beam length, $N_w > \mathcal{N}_b$.

Parabolic profile

If the beam has a parabolic shape of length $L_b = \mathcal{N}_b\lambda_L$, $P(\theta_0) = 6\theta_0(2\pi\mathcal{N}_b - \theta_0)[H(\theta_0) - H(\theta_0 - 2\pi\mathcal{N}_b)]/(2\pi\mathcal{N}_b)^3$. The form factor is:

$$f_k = 3\frac{\sin[(1 + \overline{k})\pi\mathcal{N}_b] - (1 + \overline{k})\pi\mathcal{N}_b \cos[(1 + \overline{k})\pi\mathcal{N}_b]}{[(1 + \overline{k})\pi\mathcal{N}_b]^3} e^{-i(1+\overline{k})\pi N_b}, \tag{42}$$

and

$$a_1(\theta) = \frac{3i\rho}{(\pi\mathcal{N}_b)^2} \left\{ g_+(\theta)H(\theta) - g_-(\theta - 2\pi\mathcal{N}_b)H(\theta - 2\pi\mathcal{N}_b) \right\},$$

where

$$g_\pm(\theta) = (1 \mp i\pi\mathcal{N}_b)(1 + i\theta - e^{i\theta}) - \theta^2/2.$$

For long beam, $|f_k|$ decreases as $1/[(1+\overline{k})\pi\mathcal{N}_b]^2$, i.e. faster than the flat-top case. The solution presents the same features as in the flat-top case, with two pulses propagating from the edges of the electron beam, oscillating at the resonant wavelength.

Gaussian profile

If the longitudinal profile of the beam is Gaussian, centred on $\theta_0 = 0$ and of width σ_θ, $P(\theta_0) = (1/\sqrt{2\pi}\sigma_\theta)\exp(-\theta_0^2/2\sigma_\theta^2)$, the form factor is

$$f_k = e^{-(1+\overline{k})^2\sigma_\theta^2/2}. \tag{43}$$

and

$$a_1(\theta) = \rho\sigma_\theta e^{-\sigma_\theta^2/2}\mathrm{erf}\left[\frac{\theta/\sigma_\theta + i\sigma_\theta}{\sqrt{2}}\right].$$

For the Gaussian profile the form factor is many orders of magnitude smaller than for the flat-top or parabolic profiles, also when the beam is only a radiation wavelength long.

THE DISCRETE MODEL

The multi-frequency FEL model of eqs.(18), (19) and (20) and of eqs.(30), (31) and (32) in their linearised version (in the small-signal regime), has been previously used to study the linear evolution of the radiation intensity in high-gain, single-pass FEL amplifiers [10] or the sideband instability in low-gain, saturated FEL oscillators [11], in the case of a uniform coasting electron beam. In this limit the problem is time-invariant. More precisely, if we consider the linear eqs.(30),(31) and (32), in the continuous beam case, $f_k = f_{k+k'} = 0$ and $f_{k-k'} = \delta(k - k')$ (neglecting the shot-noise contribution). Then, the linear gain on each frequency is independent on the intensity on the other frequencies. The coupling between different frequencies occurs only in the non-linear regime, as for instance at saturation, where a broadening of the spectrum is usually observed.

In order to solve numerically the multi-frequency equations the relative frequencies $\overline{k} = (k - k_L)/k_L$ must be discrete, with a convenient frequency spacing and full spectral range. Following Ref. [11], we simulate $\mathcal{N}$ frequencies around the central frequency, setting $\overline{k} = n/M$ and $(\mathcal{N} - 1)/2 \leq n \leq (\mathcal{N} + 1)/2$. The radiation field is modelled as a signal with a central frequency ck_L, contained in a temporal window of periodicity $M\lambda_L$. The full relative width of the spectrum is $\Delta\overline{k} = \pm\mathcal{N}/M$, whereas the spacing between adjacent relative frequencies is $\delta\overline{k} = 1/M$. Hence, the temporal window has a width $(\Delta z_1)_{max} = 4\pi\rho/\delta\overline{k} = 4\pi\rho M$ and the temporal resolution is $\delta z_1 = 4\pi\rho/\Delta\overline{k} = 4\pi\rho M/\mathcal{N}$, in z_1 units.

213

Introducing the discrete Fourier component relative to the n-th mode, $A_n = A_k/M$, eqs.(18), (19) and (20) become:

$$\frac{d\theta_j}{d\bar{z}} = p_j \tag{44}$$

$$\frac{dp_j}{d\bar{z}} = -\sum_n \left[A_n e^{i(1+n/M)\theta_j(z)} + \text{c.c.} \right] \tag{45}$$

$$\frac{dA_n}{d\bar{z}} = (\mathcal{N}_e/M)N^{-1} \sum_{j=1}^N e^{-i(1+n/M)\theta_j(z)} - i\frac{n}{2\rho M} A_n. \tag{46}$$

The frequency spacing $\delta\bar{k}$ should be chosen such that the temporal window contains the entire radiation pulse, whose length is equal, in a single pass through the undulator, to the sum between the beam length and the slippage length, $4\pi\rho M = \bar{L}_b + \bar{z} = 4\pi\rho(\mathcal{N}_b + \mathcal{N}_w)$, in z_1 units. Chosen M in order to have a minimum window necessary to contain the entire radiation pulse, the number of modes $\mathcal{N}$ is determined by the temporal resolution required. In particular, if a resolution less than a resonant wavelength is asked (i.e. $\delta z_1 < 4\pi\rho$), then $\mathcal{N} > M$, i.e. the relative frequency range $\Delta\bar{k}$ must be larger than one. In the numerical code, we load N electrons equally spaced along $\mathcal{N}_e$ wavelengths λ_L at $\bar{z} = 0$. In the continuous, uniform beam case, $\mathcal{N}_e = M$ and the field is periodic in the window $\Delta z_1 = 4\pi\rho M$. If the electron current is not uniform, at the j-th electron can be assigned a weight determined by the beam profile, s_j, so that we generalise eq.(46) as

$$\frac{dA_n}{d\bar{z}} = (\mathcal{N}_e/M) \sum_{j=1}^N s_j e^{-i(1+n/M)\theta_j(z)} - i\frac{n}{2\rho M} A_n. \tag{47}$$

where $\sum_{j=1}^N s_j = 1$. It reduces to a flat-top beam for $s_j = 1/N$.

As an example, Fig.1 shows the radiation intensity $|a|^2$ vs. z_1 at $\bar{z} = 10$, emitted by a flat-top electron beam of length $4\pi\rho\mathcal{N}_b = 1$, containing $\mathcal{N}_b = 5$ radiation wavelengths and without any input signal. In the simulation, $N = 500$, $M = 100$ and $\mathcal{N} = 501$. The initial CSE radiation, showed in the small window enlarging the leading edge of the optical pulse, has been coherently amplified up to the saturation. The Fig. 2 shows the final distribution of the electrons in the phase space (z_1, p) at $\bar{z} = 10$; the initial electrons where uniformly distributed between $z_1 = 0$ and $z_1 = 1$ and at $p_j = 0$. We observe five buckets in the phase space, with the electrons in the first bucket, close to the trailing edge, slightly shifted to the left with respect their initial positions.

The Fig.3 shows the radiation intensity $|a|^2$ vs. z_1 at $\bar{z} = 10$, emitted by a flat-top electron beam of length $4\pi\rho\mathcal{N}_b = 50$, containing $\mathcal{N}_b = 500$ radiation wavelengths. In the simulation, $N = 1000$, $M = 650$ and $\mathcal{N} = 3251$. The Fig. 4 shows the final distribution of the electrons in the phase space

(z_1, p) at $\bar{z} = 10$; the initial electrons where uniformly distributed between $z_1 = 2.5$ and $z_1 = 52.5$ and at $p_j = 0$. Finally, the Fig. 5 shows the scaled energy $E_s = (1/4\pi\rho\mathcal{N}_b) \int z_1|a|^2$ vs. $\bar{z}$. We observe a large spike growing from the trailing edge of the electron beam, with a peak power three orders of magnitude above the CSE contribution amd a scaled energy still far from the saturation, $E_s \sim 1.4$.

CONCLUSION

A multi-frequency model for the free electron laser has been discussed and compared with the usual propagation model used in the SASE studies [1]. This model has the advantage of a more precise description of the electron-field dynamics, as each Fourier mode of the field is driven by the electron bunching at the proper frequency. In particular, it describes the CSE start-up driven by the form factor, determined by the longitudinal beam profile. A further detailed study of the SASE process, including the statistical description of the initial random distribution of the electrons in the beam is under progress and it will appear in a future publication.

REFERENCES

1. R. Bonifacio, L. De Salvo Souza, P. Pierini, N. Piovella and C. Pellegrini, *Phys. Rev. Lett.* **73**, 70 (1994).
2. K.J. Kim and S.J. Hahn, , *Nucl. Instrum. and Meth.* **A 358**, 93 (1995).
3. R. Bonifacio, C. Pellegrini and L. Narducci, *Optics Commun.* **50**, 373 (1984).
4. R. Bonifacio, B.W.J. McNeil and P. Pierini, *Phys. Rev. A* **40**, 4467 (1989).
5. D. Oepts, D.A. Jaroszynski, H.H. Weits and P.W. van Amersfoort, *Nucl. Instrum. and Meth. A* **358**, 72 (1995).
6. D. Bocek, P. Kung, H.C. Lihn, C. Settarkon and H. Wiedemann, *Nucl. Instrum. and Meth. A* **375**, 13 (1996).
7. R. Prazeres, J.M. Ortega, E. Glotin, D.A. Jaroszynski and O. Marcouillé, *Phys. Rev. Lett.* **78**, 2124 (1997).
8. R. Tatchyn et al., *Nucl. Instrum. and Meth. A* **375**, 274 (1996).
9. R. Bonifacio, R.M. Caloi and C. Maroli, *Opt. Comm.* **101**, 185 (1993).
10. S. Krinsky, *AIP Conf. Proc.* **153**, 1015 (1985).
11. D. Iracane, P. Chaix and J.L. Ferrer, *Phys. Rev. E* **49**, 800 (1994).

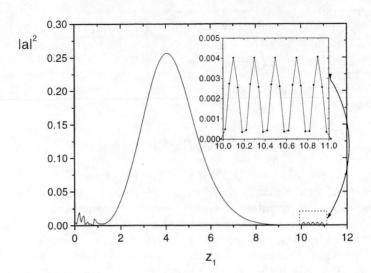

FIGURE 1. Intensity profile emitted by a short electron beam.

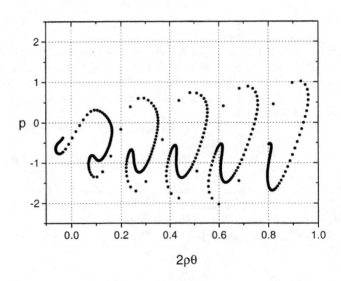

FIGURE 2. Phase-space electron distribution for the case of Fig.1.

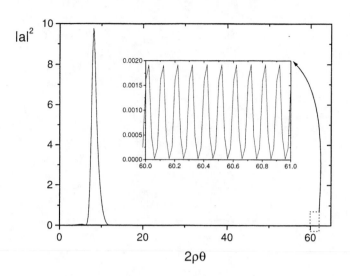

FIGURE 3. Intensity profile emitted by a long electron beam.

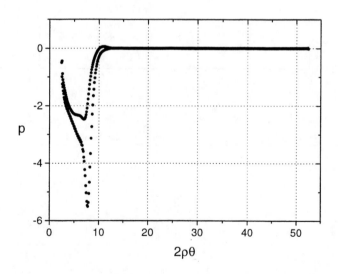

FIGURE 4. Phase-space electron distribution for the case of Fig.3.

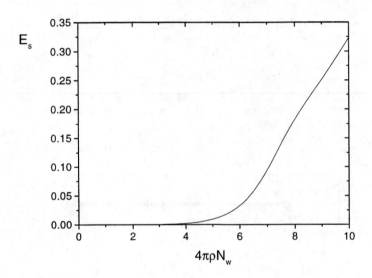

FIGURE 5. Scaled radiated energy for the case of Fig.3.

Method for reducing the radiation bandwidth of an X-ray FEL

J. Feldhaus[1], E.L. Saldin[2], J.R. Schneider[1],
E.A. Schneidmiller[2], M.V. Yurkov[3]

[1] HASYLAB at Deutsches Elektronen-Synchrotron (DESY), D-22607 Hamburg, Germany
[2] Automatic Systems Corporation, 443050 Samara, Russia
[3] Joint Institute for Nuclear Research, Dubna, 141980 Moscow Region, Russia

Abstract.
A novel scheme of a two-stage single-pass SASE FEL has been proposed in papers [1,2]. The scheme consists of two undulators and an X-ray monochromator located between them. The process of amplification in a two stage SASE FEL starts from noise as in the case of conventional SASE FEL, but characteristics of the output radiation differ significantly from those of conventional SASE FEL. It is shown in this paper that the output radiation from a two-stage SASE FEL possesses all the features which usually refer to laser radiation: full transverse and longitudinal coherence of the radiation within the radiation pulse and stability of the output power.

PRINCIPLE OF OPERATION OF A TWO-STAGE SASE FEL WITH MONOCHROMATOR

The FEL scheme consists of two undulators and an X-ray monochromator located between them (see Fig.1). The first undulator operates in the linear regime of amplification starting from noise and the output radiation has the usual SASE properties. After the exit of the first undulator the electron is guided through a bypass and the X-ray beam enters the monochromator which selects a narrow band of radiation. At the entrance of the second undulator the monochromatic X-ray beam is combined with the electron beam and is amplified up to the saturation level.

Necessary and sufficient conditions for the effective operation of a two-stage SASE FEL are as follows:

CP413, *Towards X-Ray Free Electron Lasers*
edited by R. Bonifacio and W. A. Barletta
© 1997 The American Institute of Physics 1-56396-744-8/97/$10.00

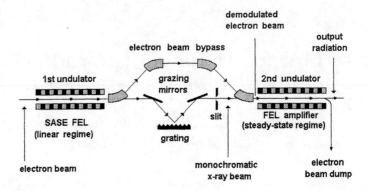

FIGURE 1. The principal scheme of a single-pass two-stage SASE X-ray FEL with monochromator.

$$G^{(1)} \ll G_{\text{sat}}(\text{SASE}) \, , \qquad (1a)$$

$$P_{\text{in}}^{(2)}/P_{\text{shot}} = G^{(1)} R_{\text{m}} (\Delta\lambda/\lambda)_{\text{m}}/(\Delta\lambda/\lambda)_{\text{SASE}} > 10^2 \, , \qquad (1b)$$

$$\lambda/\pi\sigma_{\text{z}} < (\Delta\lambda/\lambda)_{\text{m}} \ll (\Delta\lambda/\lambda)_{\text{SASE}} \, . \qquad (1c)$$

In addition, the modulation of the electron beam induced in the first undulator should be suppressed prior arrival of the electron bunch to the entrance of the second undulator. This is possible because of the finite value of the natural energy spread in the beam and special design of the electron bypass. At the entrance to the second undulator the radiation power from the monochromator dominates significantly over the shot noise and the residual electron bunching. As a result, the second stage of the FEL amplifier operates in the steady-state regime.

Relation (1a) means that the first stage should operate in the high gain linear regime (here $G^{(1)}$ is the power gain in the first stage and $G_{\text{sat}}(\text{SASE})$ is the power gain at saturation).

Relation (1b) requires the radiation power at the entrance to the second undulator, $P_{\text{in}}^{(2)}$, to be much larger than the effective power of shot noise, P_{shot} (here R_{m} is integral reflection coefficient of the mirrors and the dispersive element, $(\Delta\lambda/\lambda)_{\text{m}}$ is the resolution of the monochromator and $(\Delta\lambda/\lambda)_{\text{SASE}}$ is the radiation bandwidth of the SASE FEL at the exit of the first undulator).

Fulfillment of relation (1c) is necessary for obtaining full longitudinal coherence of the radiation over pulse length, σ_{z}.

Operation of a two-stage SASE FEL is illustrated for the example of the 6 nm option SASE FEL at the TESLA Test Facility under construction at DESY [3,4].

TABLE 1. Parameters of the two-stage SASE FEL / TTF-FEL (M) /

Electron beam
Energy, $\mathcal{E}_0$	1000 MeV
Peak current, I_0	2500 A
rms bunch length, σ_z	5×10^{-3} cm
Normalized rms emittance, ϵ_n	2π mm mrad
rms energy spread	0.1 %
External β-function,	300 cm
rms transverse beam size	57 μm
Number of bunches per train	7200
Repetition rate	10 Hz

First stage
Mode of operation	SASE, linear regime
Undulator length, L_w	12 m
Undulator period, λ_w	2.73 cm
Radiation wavelength, λ	6.4 nm
Effective power of shot noise, P_{shot}	100 W
Power gain, G	10^5
Power averaged over pulse,	10 MW
Radiation bandwidth , $\Delta\lambda/\lambda$	0.5 %
Autocorrelation time, $\tau_{1/2}$	2 fs
rms radiation spot size at the undulator exit	40 μm
rms radiation angular divergence	18 μrad
Radiation flash energy	3 μJ
Average power of radiation	0.2 W
Degeneracy parameter	10^9

Monochromator
Integral reflection coefficient, R_m	0.1
Resolution of the monochromator, $(\Delta\lambda/\lambda)_m$	5×10^{-5}

Second stage
Mode of operation	Steady-state, saturation
Input power, $P_{in}^{(2)}$	10 kW
Undulator length, L_w	12 m
Power gain, G	10^6
Radiation bandwidth, $\Delta\lambda/\lambda$	5×10^{-5}
rms radiation angular divergence	15 μrad
rms radiation spot size at the undulator exit	90 μm
Autocorrelation time, $\tau_{1/2}$	200 fs
Peak radiation power	5.3 GW
Radiation flash energy	1.5 mJ
Average radiation power	100 W
Peak spectral brilliance*	3×10^{32}
Average spectral brilliance*	7×10^{24}
Degeneracy parameter	6×10^{13}

*In units of Phot./(sec$\times$ mrad$^2\times$mm$^2\times$ 0.1 % bandw.)

OPERATION OF THE FIRST STAGE

Parameters of the first stage of the SASE FEL are presented in Table 1. It operates in a linear regime with a power gain $G^{(1)} = 10^5$. This value is by 10^3 times less than the power gain at saturation, $G_{\text{sat}}(\text{SASE}) \simeq 10^8$. Output radiation from the first stage operating in the linear regime possesses all the properties of completely chaotic polarized light [5]. In particular, the probability for a certain power $P(t)$ at the time t at the output of the first undulator is given by the negative exponential probability density function (see Fig.2):

$$w(P)dP = \exp(-P/\langle P \rangle)dP/\langle P \rangle . \tag{2}$$

The monochromator does not change this distribution since it is merely a linear filter. However, it changes the characteristic time scale to $(\lambda/c)(\Delta\lambda/\lambda)_{\text{m}}^{-1}$ because its bandwidth $(\Delta\lambda/\lambda)_{\text{m}}$ is considerably smaller than that the FEL amplifier. Fulfillment of condition (1b) ensures that the second stage of the FEL amplifier operates in the steady state regime with the probability close to unity.

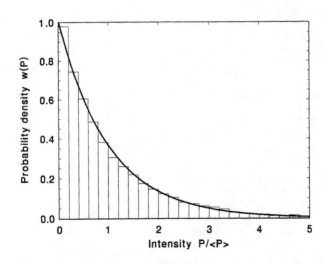

FIGURE 2. A histogram of 600 power measurements. $< P >$ denotes the shot-to-shot averaging. The solid curve represents the negative exponential probability density function $w(P) = \exp(-P/\langle P \rangle)$.

OPERATION OF THE SECOND STAGE OF THE FEL AMPLIFIER

The chosen parameters for the first stage of the SASE FEL and the monochromator satisfy the conditions (1). Therefore, the second stage of the two-stage FEL amplifier operates in the conventional steady-state regime. The parameters of the second stage of the SASE FEL are presented in Table 1. The average value of the input radiation power is 10 kW, which results in a saturation length of 16 m.

The characteristics of the two-stage SASE FEL operating at saturation are presented in Figs.3–10. The FEL process in the second stage has been calculated using the 2-D steady-state code FS2R [6]. Application of the steady-state approach is justified by the fact that the bandwidth of the radiation at the entrance of the second stage is much less than the bandwidth of the FEL amplifier. The only peculiarity which should be taken into account is that the probability density function of the input radiation follows the negative exponential distribution (2).

The dependence of the average (shot-to-shot) output radiation power on the undulator length of the second stage is presented in Fig.3. The input power fluctuates in accordance with the negative exponential distribution (2) with the average value of 10 kW. It is seen from Fig.4 that the fluctuations of the output power reduce significantly when the second stage of the FEL amplifier operates in a nonlinear mode. This feature can be simply understood when one plots the dependence of the output power on the input power for the FEL amplifier operating in the steady-state regime (see Fig.5). It is seen that a higher stability of the output radiation power can be achieved by increasing the length of the undulator. In Fig.6 we present the probability density function for the output radiation power[1]. It is seen that in the nonlinear mode of operation the distribution shrinks. For instance, at a length of the second undulator of 20 m, the mean-squared fluctuations of the output power are below 10 %. In the case under study the length of the longitudinal coherence is about the same as the length of the radiation pulse, so shot-to-shot fluctuations of the energy of the radiation pulse show the same behaviour as the fluctuations of the radiation power.

---

[1] In the case under study the probability density function of the input radiation power is given by the negative exponential distribution, so the probability density function of the output radiation power is given by the expression:

$$w(P_{\text{out}}) = \frac{1}{< P_{\text{in}} >} \sum_{n} \exp\left[-P_{\text{in}}^{(n)}(P_{\text{out}})/ < P_{\text{in}} >\right] |\frac{dP_{\text{in}}^{(n)}}{dP_{\text{out}}}| \, ,$$

where the sum over n must be taken over all branches of the function $P_{\text{in}}(P_{\text{out}})$ (see Fig.5). The singularities of the probability density functions plotted in Fig.6 correspond to the saturation point in Fig.5 where $|dP_{\text{out}}/dP_{\text{in}}| = 0$.

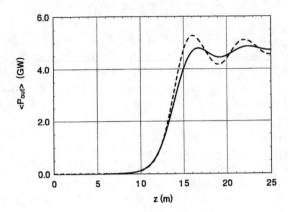

FIGURE 3. Average output power at the exit of the second stage of the two-stage FEL amplifier as a function of undulator length (solid curve). The input power fluctuates in accordance with the negative exponential distribution with the average value of 10 kW. The dashed curve represents the output power of the FEL amplifier operating in the steady-state at a fixed input power $P_{in} = 10$ kW.

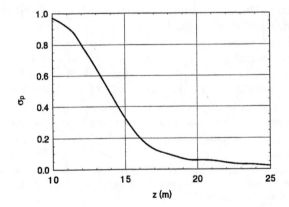

FIGURE 4. The dependence of the standard deviation of the fluctuations of the output power as a function of the undulator length ($\sigma_P^2 = \langle P^2 - \langle P \rangle^2 \rangle / \langle P \rangle^2$).

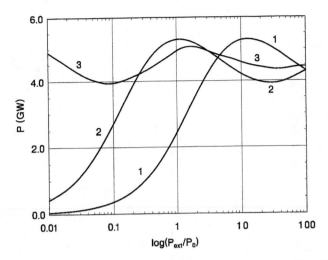

FIGURE 5. Dependence of the output power on the input power for the FEL amplifier operating in the steady-state regime, (1): For an undulator length $L_w = 14$ m, (2): $L_w = 16$ m, (3): $L_w = 20$ m. Nominal input power $P_0 = 10$ kW.

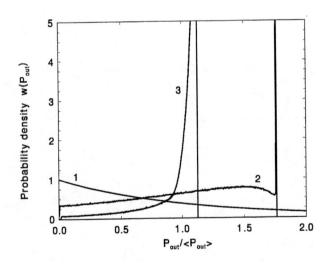

FIGURE 6. The probability density function of the output radiation of the two-stage SASE FEL for different lengths of the second undulator. Curve (1) is the probability density function of the input radiation (negative exponential distribution), and curves (2) and (3) correspond to a length of the second undulator of $L = 14$ m and $L = 16$ m, respectively.

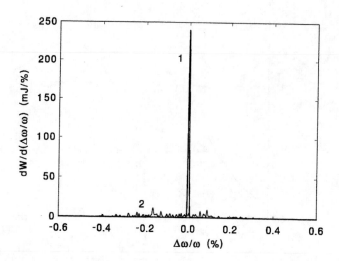

FIGURE 7. Spectral distribution (1) of the energy in one radiation pulse of the FEL amplifier operating in the steady-state regime for an input power $P_{in} = 10$ kW and an undulator length $L_w = 16$ m (saturation point). Curve (2) presents the typical spectrum of a conventional SASE FEL operating at saturation.

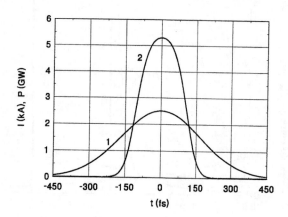

FIGURE 8. Temporal structure of the radiation pulse (2) and the electron beam current (1) at the saturation point of the FEL amplifier operating in the steady-state regime.

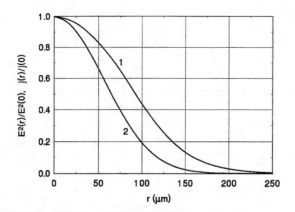

FIGURE 9. Radial distribution of the radiation power density (1) and the electron beam current density (2) at the saturation point of the FEL amplifier operating in the steady-state regime.

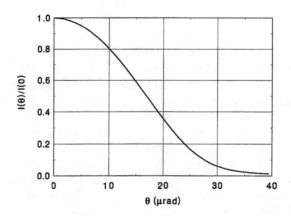

FIGURE 10. Angular distribution of the radiation power in the Fraunhofer diffraction zone at the saturation point of the FEL amplifier operating in the steady-state regime.

Fig.7 presents the spectral distribution of the energy in one radiation pulse of the FEL amplifier operating in the steady-state regime for an input power $P_{in} = 10$ kW and an undulator length $L_w = 16$ m (saturation point). Curve 2 in this plot represents the typical spectrum of a conventional SASE FEL operating at saturation. The output radiation power of the two-stage SASE FEL is close to that of the conventional SASE FEL while the spectral bandwidth is by two orders of magnitude narrower. Thus, the brilliance of the output radiation exceeds the corresponding value of a conventional SASE FEL by two orders of magnitude. In addition, longitudinal coherence of the output radiation over the full radiation pulse is obtained which leads to an autocorrelation time $\tau_{1/2} \simeq 200$ fs (see Fig.8). The spectral bandwidth of $\Delta\lambda/\lambda \simeq 5 \times 10^{-5}$ is close to the limit given by the finite duration of the radiation pulse.

The transverse coherence properties are important for practical applications. As mentioned above, it is expected that the output radiation of the 6 nm SASE FEL at DESY will be fully transverse coherent, although the FEL process starts from noise. The characteristics of the radiation mode at the FEL amplifier exit are intrinsic properties of the FEL amplifier and do not depend on the nature of the input signal [7]. In Fig.9 the distribution of the radiation power density at the FEL amplifier exit (at the saturation point) is plotted as a function of radius. Fig.10 shows the angular distribution of the radiation power in the Fraunhofer diffraction zone. These two figures demonstrate the excellent spatial properties of the output radiation.

CONCLUSION

Free-electron laser techniques provide the possibility to extend the energy range of lasers into the X-ray regime using a single-pass FEL amplifier scheme starting from noise. One particular feature of the FEL amplifier is its rather large amplification bandwidth. This can be considered an advantage when the FEL amplifier amplifies the narrow bandwidth radiation of a master laser, but in the case when the process of amplification starts from noise, it produces relatively wide band output radiation. The bandwidth of the FEL amplifier is an intrinsic feature of the device and is inversely proportional to the power gain length. The latter parameter should be minimized in order to reduce the scale (and the cost) of the device and to decrease the sensitivity of the FEL operation to different imperfections such as undulator errors, etc. As a result, such an optimization leads to an increase of the bandwidth of the SASE FEL radiation. For instance, the bandwidth of the 6 nm conventional single-pass SASE FEL at DESY would be about 0.5 %. The shape of the spectrum is not smooth but spiked. To perform experiments which require a narrow bandwidth of the output radiation, a monochromator has to be installed at the FEL amplifier exit. The shot-to-shot fluctuations of the radiation power after this monochromator will increase with increasing energy resolution and

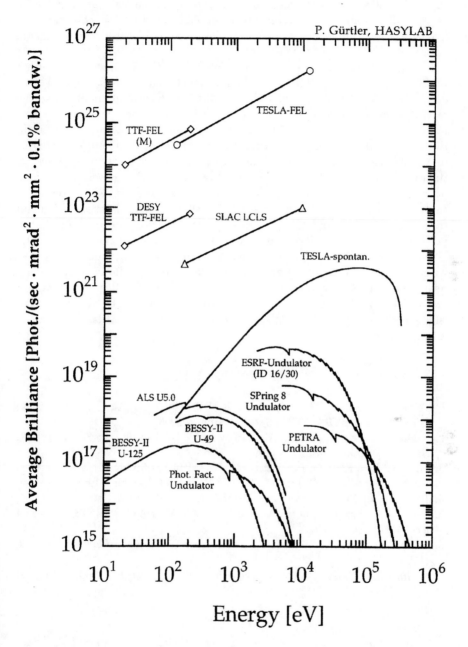

FIGURE 11. Average brilliance for different radiation sources. The brilliance for the FEL sources has been calculated according to $B = 4\dot{N}_{\text{ph}}/\lambda^2/(\Delta\omega/\omega)$.

will be on the order of unity for $\Delta\lambda/\lambda \simeq 5 \times 10^{-5}$. Moreover, conventional X-ray optical elements will suffer from heat load due to the high output radiation power and probably filters have to be installed before the monochromator. As a result, the brilliance of the FEL radiation available at the experimental station might be reduced significantly.

In the considered scheme of a two-stage SASE FEL the monochromator is placed between the undulators. This allows one to reduce the bandwidth of the output radiation (and to increase the brilliance) by two orders of magnitude with respect to a single-pass scheme, while the peak and the average output power are the same (see Fig.11). Shot-to-shot fluctuations of the output radiation power of a two-stage SASE FEL can be reduced to below 10 %. Moreover, in a two-stage scheme the heat load on the monochromator is 10^3 times less than that on a monochromator installed at the exit of a conventional single-pass SASE FEL. The output radiation of a two-stage SASE FEL proposed in this paper possesses all the features which usually refer to laser radiation: full transverse and longitudinal coherence of the radiation within the radiation pulse and stability of the output power. The realization of this scheme at the TESLA Test Facility at DESY would allow to construct a tunable X-ray laser with a minimum wavelength around 6 nm, a micropulse duration of 200 fs, a peak power of 5 GW and an average power of 100 W. The degeneracy parameter of the output radiation of such an X-ray laser would be about 10^{14} and thus have the same order of magnitude as that of quantum lasers operating in the visible.

REFERENCES

1. Feldhaus J., Saldin E.L., Schneider J.R., Schneidmiller E.A., and Yurkov M.V., *DESY Print* **TESLA-FEL 96-10**, Hamburg, 1996.
2. Saldin E.L., Schneidmiller E.A., Yurkov M.V., Feldhaus J., and Schneider J.R., *SPIE Proceedings* **2988**, 145 (1997).
3. "A VUV Free Electron Laser at the TESLA Test Facility at DESY. Conceptual Design Report", *DESY print* **TESLA-FEL 95-03**, Hamburg 1995.
4. Rossbach J., *Nucl. Instrum. and Methods* **A375**, 269 (1996).
5. Saldin E.L., Schneidmiller E.A., and Yurkov M.V., *DESY Print* **TESLA-FEL 97-02**, Hamburg, 1997.
6. Saldin E.L., Schneidmiller E.A., and Yurkov M.V., *Opt. Commun.* **95**, 141 (1993).
7. Saldin E.L., Schneidmiller E.A., and Yurkov M.V., *Phys. Rep.* **260**, 187 (1995).

Prebunching and electron pulse phase stability in FELIX

D. Oepts and H.H. Weits

FOM-Instituut voor Plasmafysica "Rijnhuizen",
P.O.Box 1207, 3430 BE Nieuwegein, The Netherlands
e-mail: oepts@rijnh.nl, weits@rijnh.nl

Abstract.
   The importance of coherent spontaneous emission in a free-electron laser operating with short electron pulses is discussed. A coherent enhancement by many orders of magnitude has been observed in the far infrared with the FELIX device. The associated coherence between independent optical micropulses is used to determine the phase stability of the electron pulses. It is found that the pulse-to-pulse jitter is not more than some tens of femtoseconds on a timescale of nanoseconds.

Introduction

   The Free Electron Laser for Infrared eXperiments, FELIX, operates as a user facility in the mid and far infrared wavelength range from about 5 to 100 μm [1]. It is a multipass oscillator device with relatively low gain (i.e. gain length > undulator length). It is, therefore, only remotely related to single-pass, high-gain, short wavelength devices. Yet, its study can be of interest for projects in the latter field in two respects: (i) as a self-starting oscillator it starts up from spontaneous emission like a SASE device, and (ii) in the highly saturated regime it shows the same superradiant behaviour as can occur in the high gain case [2,3].

   In this contribution we discuss the effects of short electron pulses on the start-up, and the relation with interpulse coherence and electron bunch phase stability.

CP413, *Towards X-Ray Free Electron Lasers*
edited by R. Bonifacio and W. A. Barletta
© 1997 The American Institute of Physics 1-56396-744-8/97/$10.00

231

Coherent Spontaneous Emission

The initial optical field from which a free-electron laser starts up is formed by the spontaneous undulator radiation emitted by the undisturbed input electron beam. For a continuous electron beam this emission results only from the statistical fluctuations in the longitudinal electron density distribution, i.e. shot noise. With a modulated or pulsed beam there is, in addition, a coherent contribution caused by the macroscopic density modulation at the scale of the resonance wavelength. The coherent emission will be strong at the pulse repetition frequency and also at its harmonics, depending on the pulse structure [4].

In FELIX, the standard pulse repetition frequency is 1 GHz, and even a long wavelength of, say, 60 μm corresponds to the 5000th harmonic. It may seem unlikely that emission of such high harmonics is of any importance. However, the radio-frequency linear accelerator system that is used for the electron beam produces pulses with a duration in the order of 5 picoseconds. Therefore, harmonics up to at least the 200th are strongly present. Although the electron pulses are still long compared to the optical wavelengths of interest, the coherent emission due to this coarse prebunching plays an important role in FELIX [5,6]. The spontaneous power emitted by a single pulse of N electrons can be expressed as:

$$P(k) = NP_1(k) + N(N-1)P_1(k)f(k), \tag{1}$$

where $P_1(k)$ is the power emitted by a single electron at the wavenumber $k = \frac{2\pi}{\lambda}$ and $f(k)$ is a form factor determined by the macroscopic longitudinal density distribution, $S(z)$, of the electron pulse,

$$f(k) = | \int e^{ikz} S(z) dz |^2. \tag{2}$$

Here, $S(z)$ is normalized so that $\int S(z)dz = 1$. The first term in Eq. (1) is the incoherent, shot noise, contribution and is proportional to N, while the second, coherent, term is proportional to N^2 because here the fields add in phase. Because N can be a very large number, e.g. of order 10^9 in FELIX, the second term can dominate the first one even when the form factor f is very small, as is usually the case when the pulse length is much larger than the wavelength.

The value of f for wavenumbers much larger than the inverse of the pulse length depends strongly on the precise shape of $S(z)$. The shape factor falls off most rapidly for a Gaussian profile, while a sharp edge greatly increases the importance of the coherent term. The asymptotic behaviour at high wavenumbers is determined by singularities in the derivatives of S [8]. Indeed, it was found in FELIX that the spontaneous emission power was enhanced by a factor of order 10^4 at a wavelength of 80 μm, and still by about 100 at 12 μm

[5,6]. A strong peak was usually seen in the initial phase of the macropulses, where the accelerator has not yet reached steady-state conditions and the pulse shape can differ from the final one. By varying the parameters of the accelerating system, in particular the relative phases of the RF fields in the injector and accelerator sections, the spontaneous emission power could be varied by about two orders of magnitude [7]. The amount of the power enhancement and its dependence on the wavelength could be explained by assuming a pulse shape that seemed not unreasonable for the accelerator system involved [5,9]. Coherent enhancement of the spontaneous emission has also been observed at other infrared FELs [10,11].

Interpulse Coherence

An interesting consequence of the coherent spontaneous emission is that the phase of the radiation is not random, in contrast to the incoherent shot noise case. In the coherent case, the optical phase is determined by the shape of the electron bunch and by its arrival time in the undulator. Thus, if successive electron pulses have identical shape and arrive at fixed intervals of time, then the emitted spontaneous radiation will also show coherence between pulses. Variations in the phase of the electron bunches lead to corresponding variations in the optical micropulse phase. The optical phase, or at least a phase difference, can be determined with sub-wavelength accuracy by interferometric means, and the electron phase stability can therefore be measured on a micrometer scale.

To determine the long range phase coherence of the optical radiation we use a Michelson interferometer. The optical beam is split, and recombined with a path length difference between the two beams. For path length differences equal to a multiple of the interpulse distance, we overlap at the detector optical pulses generated by different electron pulses. When these pulses are coherent, one observes interference fringes when the path difference is scanned over a distance shorter than the pulse length.

Results of such a measurement are shown in Fig. 1. The separate frames correspond to path differences around $0, 1, ...7$ pulse distances (left to right, top to bottom). At each of these settings, the path difference was fine-scanned over a range of about 300 μm in steps of 0.7 μm, and at each position the output power of the interferometer was measured and recorded as a dot in the figure. Because the scan cannot be made in a single macropulse, which has a duration of about 6 μs, each point was measured in a different macropulse, at a fixed time within the macropulse. The time constant of the detection system was such that each point corresponds to an average over about 250 micropulses. The measurements shown were performed at a wavelength of 69 μm.

The data shown in Fig. 1 were taken in the saturated part of the laser

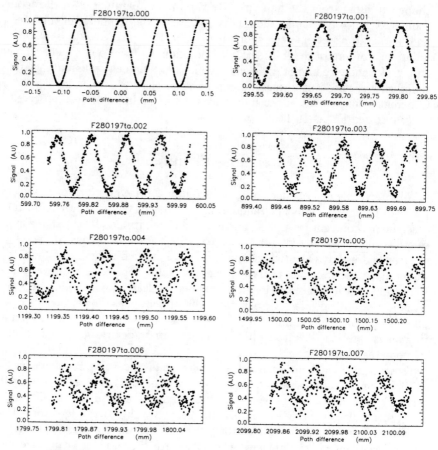

FIGURE 1. Interferometric measurement of interpulse coherence.

macropulses, which means that they do not actually show the spontaneous emission results. These have been measured as well, but show a much lower signal-to-noise ratio. Note that the laser cavity has a roundtrip length of 12 m, while the micropulse separation is 30 cm. This means that 40 separate independent pulses circulate in the cavity. The laser process amplifies the fields coherently, i.e. the phase is preserved, or evolves in the same way for different pulses, so that the relative phases at saturation still represent the phases at start-up. At shorter wavelengths it was possible to reduce the coherent spontaneous emission considerably by manipulating the accelerator settings; in that case the coherence between micropulses was also absent. This shows that the coherence at saturation is indeed determined by the spontaneous emission and does not arise in the laser process itself. Coherence between pulses separated by the 12 m cavity roundtrip length will of course always be present, but this is difficult to observe.

Some observations can be made immediately by inspecting Fig. 1. The presence of strong interference between different micropulses shows that the optical pulses are indeed coherent. This means that the above-mentioned increased spontaneous emission is the result of a reproducible fine structure in the electron pulses and is not caused by excessive electron density noise or randomly fluctuating pulse shapes.

From the almost 100% fringe modulation in the second frame it follows that adjacent micropulses are almost completely coherent. With increasing pulse separation the coherence gradually decreases and the noise in the signals increases. This noise is due to macropulse-to-macropulse variations in the interference, and not to fluctuations of the laser power. The latter are quite small as is seen in the first frame.

The decreasing coherence can be caused by jitter in the arrival time of the electron pulses, or by differences in the pulse shape that increase with their separation. An uncorrelated variation of the pulse arrival times around a fixed interval would reduce the interference in the same way for different interpulse distances. If, on the other hand, the phase varied slowly compared to the pulse repetition rate, then mainly the fringe position would change from micropulse to micropulse, not the amplitude.

No slow drift in the fringe positions has been observed over periods of tens of minutes, which indicates that the stability of the electron repetition frequency on this timescale is certainly better than $1 : 10^5$.

Phase Evolution Modelling

To obtain more quantitative information about the effects of electron pulse jitter on the measured interferogram data, we used an extension of a simple model for the phase evolution of the optical pulses, which was developed previously to simulate induced coherence between pulses [12].

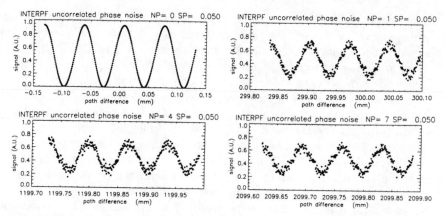

FIGURE 2. Simulation of interferogram data assuming uncorrelated phase jitter.

In this model it is assumed that each optical micropulse has the same shape and can be characterized by a single complex number representing the amplitude and the phase of its carrier wave at a fixed reference position in the pulse. For the spontaneous radiation, we take a constant amplitude, while different models are tried to determine the random part of the phase of each pulse. On successive round trips through the cavity, the amplitude of each pulse is multiplied by a (real) gain factor, and a new spontaneous pulse is added. Saturation is modelled by letting the incremental power gain depend on the pulse power according to the simple $(1 + \frac{I}{I_{sat}})^{-1}$ dependence [13]. In this way, the development of the complex amplitudes for the 40 pulses in the cavity is followed to saturation, and the interferogram signal for a given path difference is calculated from the results. A large number of such runs is performed, corresponding to different macropulses and with different interferometer path differences, to obtain a result that can be compared with the measurements shown in Fig. 1.

Some examples are given in Figs 2-4. Fig. 2 shows the results for 0, 1, 4, and 7 interpulse distances when the electron pulse phases are taken to have a normally distributed (Gaussian) phase variation around a fixed pulse interval. The (rms) magnitude of the phase deviations is taken equal to $0.05 \times 2\pi$ radians of the optical carrier wave. In this case, as was already remarked above, there is essentially no dependence on the pulse separation.

For Fig. 3 it is assumed that there is no short-time jitter in the phases of the electron pulses, but a Gaussian frequency variation on the macropulse time scale.is assumed. The rms variation is taken to be 2.3 kHz around the nominal 1 GHz, which corresponds to $0.01 \times 2\pi$ optical phase shift (rms) per micropulse. In this case, the amplitude of the fringes is basically constant, but the positions vary from macropulse to macropulse by an amount that increases with the separation between the interfering pulses.

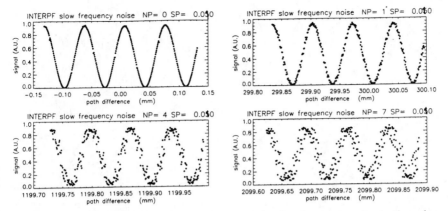

FIGURE 3. Simulation of interferogram data assuming low frequency noise in the pulse repetition rate.

Finally, in Fig. 4 we show the case that best reproduces the measurements. Here, the intervals between successive electron bunches are given a Gaussian deviation from the nominal value, so that the phases of the successive pulses make a random walk around their reference positions. Such a phase diffusion–or white frequency noise–model leads to an exponentially decreasing correlation function, which fits well with the observations such as illustrated in Fig. 1. For the case of Fig. 4 we have taken the rms value of the phase diffusion steps to be $0.05 \times 2\pi$, which corresponds to 11 fs in time, or 3.5 μm in distance. The fact that stable interference fringes have still been observed at a wavelength of 12 μm [6] also agrees with this value.

The decaying coherence and increasing noise seen in the successive frames of Fig. 1 could be caused by other fluctuations in the micropulses than the phase jitter that we considered. In that case, the phase stability would have to be even better to explain the measurements.

The same model has also been used for the case of induced pulse coherence, and reproduces the observations quite faithfully [14].

Conclusion

Coherent spontaneous emission plays an important part in the start-up phase of (far-)infrared free electron lasers like FELIX. The coherent spontaneous emission leads to increased initial power and also to coherence between otherwise independent micropulses. The resulting degree of coherence can be used to characterize the phase stability of the electron pulses. It is found that the micropulse-to-micropulse time jitter can be described as a phase diffusion process in which the rms phase-jumps between adjacent micropulses amount to not more than about 10 fs.

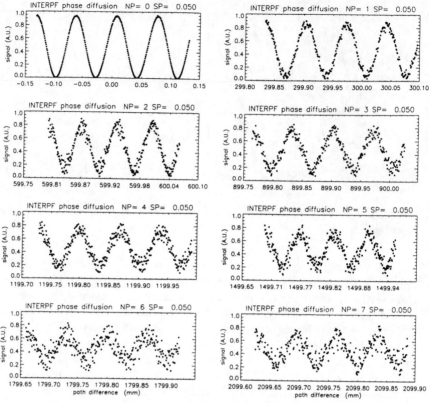

FIGURE 4. Simulation of interferogram data using a phase diffusion model.

Acknowledgment

This work was performed as part of the research programme of the 'Stichting voor Fundamenteel Onderzoek der Materie' (FOM) with financial support from the 'Nederlandse Organisatie voor Wetenschappelijk Onderzoek" (NWO).

REFERENCES

1. Oepts D., van der Meer A.F.G., van Amersfoort P.W., *Infrared Phys. Technol.* **36**, 297 (1995).
2. Jaroszynski D.A., Chaix P., Piovella N., Oepts D., Knippels G.M.H., van der Meer A.F.G., Weits H.H., *Phys. Rev. Lett.* **78**, 1699 (1997).
3. Bonifacio R., McNeil B.W.J., Pierini P.,*Phys. Rev. A* **40**, 4467 (1989).
4. Doria A., Bartolini R., Feinstein J., Gallerano G.P., Pantell R.H., *IEEE J. Quantum Electron.* **29**, 1428 (1993).

5. Jaroszynski D.A., Bakker R.J., van der Meer A.F.G., Oepts D., van Amersfoort P.W., *Phys. Rev. Lett.* **71**, 3798 (1993).

6. Oepts D., Jaroszynski D.A., Weits H.H., van Amersfoort P.W., *Nucl. Instrum. and Meth. A* **358**, 72 (1995).

7. Weits H.H., van der Meer A.F.G., Oepts D., Ding M., *in Proceedings 18th Int. FEL Conf. Rome 1996*, to be published.

8. Best R.W., *Opt. Commun.* **123**, 563 (1996).

9. Weits H.H., Oepts D., to be published.

10. Prazeres R., Ortega J-M., Glotin F., Jaroszynski D.A., Marcouillé O., *Phys. Rev. Lett.* **78**, 2124 (1997) .

11. Berryman K.W., Crosson E.R., Ricci K.N., Smith T.I., *Nucl.Instrum. and Meth. A* **375** 526 (1996)

12. Oepts D., Colson W.B., *IEEE J. Quantum Electron.* **26** , 723 (1990).

13. Dattoli G., Giannessi L., Cabrini S., *IEEE J. Quantum Electron.* **28**, 770 (1992).

14. Oepts D., Weits H.H., *in Proceedings 19th Int. FEL Conf. Beijing 1997*, to be published.

Three-Dimensional Analyses of Magnetic Fields in a Staggered-Array Undulator

Kiyoshi Yoshikawa, Shigeki Shimada, Kouji Okada, Kai Masuda,
Masaaki Sobajima, Jiro Kitagaki, Masami Ohnishi,
Yasushi Yamamoto and Hisayuki Toku

Institute of Advanced Energy, Kyoto University, Gokasho, Uji, Kyoto 611, Japan

Abstract. A staggered-array undulator set inside the superconducting solenoid coils is shown to provide high undulator fields together with original longitudinal magnetic fields, a small undulator period of a few centimeters, easy tunability through the solenoid coil current, as well as compact, less expensive and easy fabrication. These performance characteristics seems very favorable for the single path X-ray FEL generation, since a great number of the undulator period is essential, in general. The previous two-dimensional analyses of electron beam trajectories have shown the necessity of the transverse stabilizing effect, which led to the three-dimensional analyses with successful stabilizing effects for the staggered-array undulator of about one meter for IR FEL. For X-ray FEL generation, electron trajectories with least deviation from the z-axis is rather mandatory for the light to grow over the undulator. Three-dimensional analyses are made to show the effect of parameters on the electron trajectories.

INTRODUCTION

A staggered-array undulator is quite unique, and compact with (a) easy tunability of the wave length by controlling superconducting solenoid coil current, (b) relatively high undulator fields with short undulator periods, (c) low cost and easy fabrication even the superconducting coils be taken into account, and (d) good electron confinement by the longitudinal magnetic field inherent in the solenoid coils, as was first proposed by Stanford group(1,2). It is expected, in particular, to be able to provide a substantial number of undulator periods with a relatively short undulator length, because of the two raws of simple stacks made of iron and aluminum disks, which will be very favorable for the single path X-ray FEL generation, since a great number of the undulator period is essential, in general(3,4).

In the previous study(5,6), spontaneous emission spectra including harmonics from the staggered-array undulator were calculated based on the 3-dimensional electron

CP413, *Towards X-Ray Free Electron Lasers*
edited by R. Bonifacio and W. A. Barletta
© 1997 The American Institute of Physics 1-56396-744-8/97/$10.00

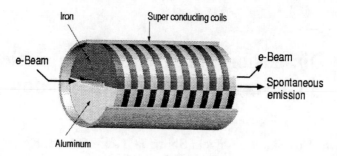

FIGURE 1. Schematic of a staggered-array undulator in the superconducting solenoid coils.

trajectories under two-dimensional magnetic fields to find the optimal f, the ratio of the aluminum disk width to the undulator period λ_u, tunability dependence on the ratio f, as well as the solenoid field B_c, and dependence of the line width on the number of the undulator periods N_w, by solving numerically the two-dimensional magnetic field distributions inside the solenoid for various fractions of iron/ aluminum disk widths. It is found that the transverse stabilizing effects is required to suppress the elctron deviation from the z-axis within an acceptable extent, in particular, for a long staggered-array undulator for X-ray FEL. A three-dimensional simulation code was, consequently, developed to analyze the 3-D effects by the asymmetrical desk structures, which would provide transverse stabilizing effect to the electrons.

UNDULATOR GEOMETRY AND EQUATIONS OF MAGNETIC FIELDS

A cutaway sketch of a staggered-array undulator in the cylindrical super-conducting solenoid coils is shown in Fig.1, where the longitudinal magnetic field B_c without a staggered array is provided by the superconducting solenoid coils of the

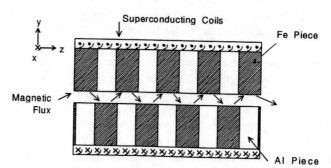

FIGURE 2. Cross sectional view of the staggered-array undulator.

242

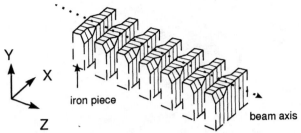

FIGURE 3. A three-dimensional staggered-array undulator with transverse stabilizing effects.

inner radius of a.

With the insertion of the upper and lower disk stacks made of a number of iron/aluminum half disks staggered to each other, the magnetic field components perpendicular to the axis, i.e. undulator fields, are then induced due to the large iron permeability as is shown in Fig.2.

A three-dimensional staggered-array undulator with transverse stabilizing effects is shown in Fig. 3, where the solenoid coils are assumed to have a flat-top axial magnetic field B_c in the central region with sinusoidal field variation from the entrance and to the exit.

MAGNETIC FIELD DISTRIBUTIONS

Table 1 shows reference parameters of the undulator, and electron beams for numerical calculations for SASE X-ray FEL of 4 nm(4). In Fig.4, typical magnetic fields are shown in the vicinity of the beam axis(z axis). It is seen clearly that the undulator field (i.e., $B_y=B_u$) is induced mainly in the region where the upper and lower iron disks face to each other.

TABLE 1. Reference parameters.

Undulator length	$L_u = 700$ cm
Undulator period	$\lambda_u = 1$ cm
Gap length	$g_u = 2$ mm
Ratio of spacer length to λ_u	$f = 0.45$
Number of undulator period	$N_W = 700$
Coil radius	$a = 1.6$ cm
Solenoid field	$B_c = 0.72$ T
Permeability of iron pieces	$\mu = 3000$
Undulator parameter	$K = 0.6$
Beam energy	$E_r = 620$ MeV ($\gamma = 1214$)
Wavelength	$\lambda_L = 4$ nm
Beam radius	$\sigma_r = 0.1$ mm

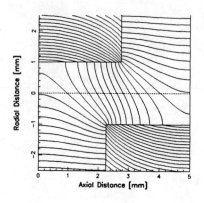

FIGURE 4. Magnetic field lines in the vicinity of the beam axis.

It is found that the solenoid field B_c of 0.7 T can provide the peak undulator field B_y as high as about 0.6 T, and still the longitudinal (axial) field B_z of about 0.3 T remains. The magnetic fields with(3-D), and without(2-D) transverse stabilizing effects are calculated as a function of the ratio of the spacer (aluminum) to the undulator period f, and shown in Fig.5. With increase of f, the undulator field first increases, and after having the maximum field at $f = 0.45$, it starts decreasing. Since the undulator fields strongly depend on the undulator period λ_u, and the gap width g_u, as well, their dependence are calculated and shown in Fig.6. It is seen that the undulator fields can be increased rapidly with increase of the period, and with decrease of the gap. It is to be noted that B_y can be made even larger in excess of B_c.

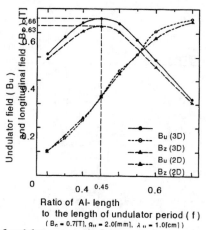

FIGURE 5. Dependence of undulator magnetic fields at the beam axis on the ratio of the spacer.

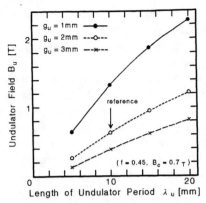

FIGURE 6. Dependence of undulator magnetic field at the beam axis on undulator period and the gap width.

ELECTRON TRAJECTORIES AND SPONTANEOUS EMISSION SPECTRA

Similar to the calculations for 2-D magnetic fields, electron trajectories are calculated under the above-mentioned 3-D magnetic fields by the relativistic equation of motion in the undulator, which results in spontaneous emission spectra through the well known Lienard-Wiehert equation. The Runge-Kutta-Gill method is applied to solve the equations.

Trajectory of the electrons of 0.2mm diameter incident on the z axis ($x_0 = y_0 = 0$ mm), for $K=0.6$ and velocity components are shown, respectively, in Fig.7. It is seen that a number of wiggle motions as many as $N_W = 400$ can be made in the x-y plane,

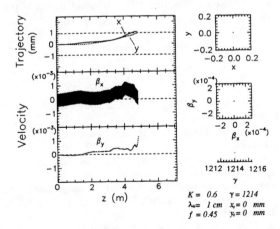

FIGURE 7. Trajectory and velocity components of one electron incident on the z axis($x_0 = y_0 = 0$ mm) for $K=0.6$.

245

until the deviation in the y direction restricts the further wiggle motion in the staggered-array undulator of a gap g_u of 2 mm.

The deviation over the long distance of the undulator is, at present, not fully understood, and being examined.

Although, it is clear that the staggered-array undulator with a very small undulator period seems very promising to SASE X-ray FEL in reducing construction cost and in providing with a number of undulator periods for a short undulator length, as well as excellent tunability, suppression of the deviation of a high-brightness electron beam in the duct due to misalignment of the undulator disks, beam-induced instabilities, or space charge will be the crucial issue in the future, in order the SASE to interact, and grow effectively inside the electron beam over many periods. Careful adjustment of such as the undulator pole structure, and the local background fields to compensate for deviation will be, thus, the next issues to be investigated.

CONCLUSIONS

A staggered-array undulator set inside the superconducting solenoid coils is found to provide high undulator fields together with the original longitudinal magnetic fields, a small undulator period of a few centimeters, easy tunability through the solenoid coil current control, as well as compact, less expensive and easy fabrication. These performance characteristics are very favorable for the single path X-ray FEL generation, since a great number of the undulator period can be made with least construction cost.

Three-dimensional analyses of magnetic fields with transverse stability have predicted favorable features in which the electrons can wiggle over 400 periods with less than 2mm deviation from the beam axis, leading to the possible straight trajectories provided local , and small modification of the magnetic fields be made.

In summary, the staggered-array undulator is found to show potentially excellent capability of generating X-ray FEL with superior compactness.

REFERENCES

1. Huang, Y. C., Wang, H. C., Pantell, R. H., Feinstein, J., and Lewellen, J. W., *Nucl. Instr. Methods in Phys. Res.* **A318**, 765-771 (1992).
2. Huang, Y. C., Wang, H. C., Pantell, R. H., Feinstein, J., and Lewellen, J. W., *IEEE J. of Quant. Elect.* **30**, 1289-1294 (1994).
3. Pellegrini, C., *Nucl. Instr. & Methods in Phys. Res.* **A222**, 364-367 (1988).
4. Kim, K.-J., and Xie, M., *ibid.* **A331**, 359-364 (1993).
5. Okada, K., "Performance characteristics of spontaneous emission spectra from a staggered-array undulator in the superconducting solenoid coil", BS thesis, Dept. of Electrical Engineering, Kyoto University (1996), in Japanese.
6. Shimada, S., Okada, K., Masuda, K., Sobajima, K., Yoshikawa, K., Ohnishi, M., Yamamoto, Y., and Toku, H., *Proc. of Third Asian Symp. on Free Electron Lasers and Fifth Symp. on FEL Applications*, 239-243 (1997).

Studies for a Beam Trajectory Monitor for TTF-FEL at DESY

Ute Carina Müller

DESY, Notkestr. 85, 22607 Hamburg, Germany

Abstract. A beam trajectory monitor for the FEL at the TESLA test facility at DESY has been proposed for the reconstruction of the electron beam trajectory by observing the spontaneous undulator radiation along the beam using the pinhole camera principle. Simulations for this concept have been performed and results are presented here.

INTRODUCTION

For the successful running of a SASE (self amplified spontaneous emission) FEL one of the keypoints is to guarantee a good overlap of the electron and photon beam along the undulator. Photons usually travel on a straight line but the electrons don't because of imperfections of the magnetic components – undulator dipoles as well as focusing quadrupoles. Therefore the requirement for good overlap can only be fulfilled if the electron beam trajectory is corrected to bring it close to the straight line defined by the photon beam. Simulations have shown that in the case of the TTF-FEL at DESY [1] a gain of approximately 100 % can be achieved if the deviation (rms) of the electron beam from a straight trajectory is below $10\,\mu$m [2]. To achieve this tight constraint the current approach for the TTF-FEL is to equip the undulator modules with beam position monitors and correctors [1,3]. As the absolute position of these components is not known beam based alignment correction algorithms have been proposed [4]. They use only the relative position information within each monitor. The idea behind is that a dispersion free trajectory is in first order a straight line. But the electron trajectory itself is not measured directly with this method. Therefore a new type of monitor has been proposed, a beam trajectory monitor (BTM), that can be used to reconstruct the beam trajectory [5,6]. With this information the necessary corrections to bring the electron beam on a straight trajectory can be defined.

The basic idea of this BTM concept is to observe the spontaneous undulator radiation from the electron beam with a pinhole-type camera along the

CP413, *Towards X-Ray Free Electron Lasers*
edited by R. Bonifacio and W. A. Barletta
© 1997 The American Institute of Physics 1-56396-744-8/97/$10.00

undulator. That means only a single detector setup is needed for each undulator module. As sensitive detector behind the pinhole a silicon pixel detector with two columns of 12 pixels each is foreseen. The boundary between the two columns serves as an absolute reference for the straight electron beam trajectory. By combining one camera in the horizontal and one in the vertical direction the position of the electron beam can be reconstructed (for details of the concept see [5,6]). Simulations for certain keypoints of the concept as the radiation signal or systematic effects on position reconstruction have been performed. Results are presented here.

SIGNAL: PHOTONS FROM THE SPONTANEOUS UNDULATOR RADIATION

As a first step the signal, that means in this case the spontaneous undulator radiation, has to be estimated according to the active area and the energy sensitivity of the BTM. The angular range that is covered by a detector of a single camera lies between 0.33 mrad and 1.36 mrad in the direction off axis (that gives information about the longitudinal position) and ± 0.5 mrad on axis in the direction that measures the deviation from the nominal beam axis (see fig. 1 in [6]). To keep diffraction effects as small as possible an energy filter is used. It is transparent for energies above 150 eV whereas the first harmonic of the undulator radiation is at about 20 eV varying slightly for different angles. Although the spectrum of the undulator radiation is well known in the range of the first harmonic, that means the low energy part of the spectrum emitted close to the undulator axis, detailed knowledge of the high energy spectrum far from the axis – that is exactly what is necessary here – is not available. Therefore the amount of signal has to be estimated from simulations. Two approaches have been used:

1. The photon flux has been derived with the help of the widely used Bessel function approximation. It is based on the solution of the general radiation equation in the far field approximation. To derive the flux the contributions from the harmonics has to be added up:

$$\frac{d^3 I}{d\Omega \left(\frac{d\omega}{\omega}\right) dI_B} = 4.55 \cdot 10^{16} \gamma^2 N_U^2 \sum_{n=1}^{\infty} F_n\left(\theta, \phi\right) \left(\frac{\sin N_U \pi \frac{\omega - \omega_n}{\omega}}{N_U \pi \frac{\omega - \omega_n}{\omega}}\right)^2 \qquad (1)$$

with I: photon flux, Ω: unit solid angle, ω: frequency of the undulator radiation, I_B: the electron beam current, $\gamma = E/m$ where E and m are the energy and mass of the electron, N_U: the number of undulator periods, F_n: compositions of the Bessel functions [5,7].

For the numerical calculations, the package URGENT [7] has been used.

248

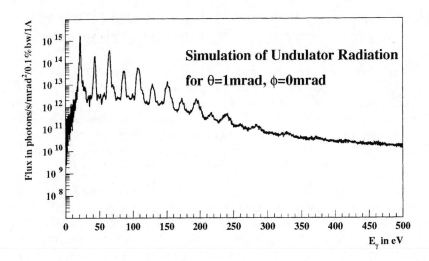

FIGURE 1. Spectrum of the undulator radiation at the angle $\theta = 1$ mrad and $\Phi = 0$ mrad using the simulation program UR [8]. The angle corresponds to an intermediate longitudinal position where the electron beam is on axis. TTF-FEL phase I parameters have been used.

2. The other possibility is to calculate the trajectory for each electron and derive the electromagnetic field from the interaction of the electrons and the magnetic field in the undulator at discrete points along the trajectory. The photon spectrum is the result of a discrete Fourier transformation of this electromagnetic field and a summation for all electrons. The simulation is based on the program UR [8]. One example spectrum at a specific angular position is given in fig. 1.

The spectra of the two approaches have been compared. The peaks of the harmonics agree well but in-between the harmonics the Bessel function approximation is set to zero whereas for the discrete Fourier transformation approach there is also intensity between the harmonics. Also in the high energy region (for energies larger than 300 eV) the level for the second approach is higher than for the first one. Taking these effects into account the discrepancy of the integrated flux that is relevant for the measurement for the two approaches lies in the range of only 10 %. For estimates of the total signal for the different detector pixels see [5]. The amount of signal seems to be high enough for the proposed measurement.

SYSTEMATIC EFFECTS ON THE POSITION AND TRAJECTORY RECONSTRUCTION

Going the next step from the signal to the position measurement, the center of gravity of the undulator radiation emitted from all electrons in a certain longitudinal range can be detected in one pixel row on the detector. No information on the shape of the radiation distribution is measured. Therefore it is very important that the center of gravity of the undulator radiation can be directly related to the center of gravity of the electron beam. Two major effects that might destroy this direct relation have been studied:

1. As the electron beam has a finite width ($\sigma_x \approx 50\,\mu$m [1]) one has to make sure that the angular distribution of the undulator radiation is flat over the full width of the beam and the corresponding angular range. From the simulation of the undulator radiation it has been derived that the distribution is sufficiently flat because of the very small pinhole acceptance, so that the offset between the center of gravity of the undulator radiation and the center of gravity of the electron beam is negligible.

   The variation of the angle with the longitudinal range and the associated variation in the spectral density of the undulator radiation is not critical for the measurement of the center of gravity. It only influences the weighting of different longitudinal areas for one active pixel row.

2. As the undulator radiation for a single electron has to be calculated with respect to the electron direction, divergence or convergence effects of the beam have to be taken into account. For extreme conditions just at the focusing and defocusing elements the shift between the reconstruction of the center of gravity of the undulator radiation compared to the one for the electrons has been estimated to be less than $\pm 1\,\mu$m. As in reality the measurement averages over a certain longitudinal range the effect will be even smaller.

By combining the measured points of all pixel rows the electron beam trajectory can be reconstructed. A simulated trajectory before and after correction is shown in fig. 2. Realistic errors for the magnetic field along the undulator have been assumed [4]. The positions determined with four BTM setups along the whole undulator are overlayed for the trajectory before correction.

CORRECTION PROCEDURE

Having measured the electron beam trajectory without any corrections (see fig. 2) one can start now to apply a correction algorithm to bring the electron beam as close as possible to a straight line. A simple algorithm has been used as a first step. The parameters for quadrupoles and correctors have

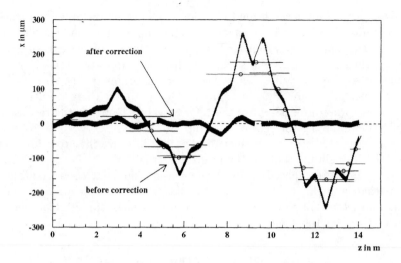

FIGURE 2. Simulation of an electron beam trajectory together with the measured points before correction and the trajectory after the correction algorithm. The rms values of the trajectory are $115\,\mu$m before and $9\,\mu$m after the correction procedure.

been taken from [3,4]. One corrector after the other is adjusted by looking at the measured point that follows the corrector. The best corrector setting is found when the measured point is closest to the nominal beam axis. It is an iterative procedure but usually reaches the requirement of the rms smaller

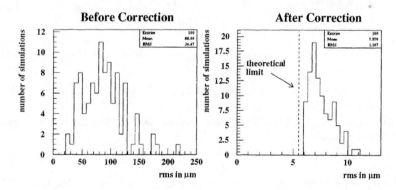

FIGURE 3. Rms values for the electron beam trajectory in the x-plane before and after correction for 100 different error configurations.

251

than $10\,\mu$m already after the second iteration. In fig. 2 an example for a track before and after correction is given. The x coordinate where one can see the wiggling in the undulator is shown. The theoretical limit of the rms because of this wiggling is $5.5\,\mu$m. The simulation has been repeated for 100 error configurations. The rms values before and after correction are shown in fig. 3. The rms mean values are $88.0\,\mu$m before and $7.6\,\mu$m after three iterations of the correction procedure. 98 out of the 100 rms values lie below the requested limit of $10\,\mu$m after the correction procedure.

More sophisticated correction algorithms should improve the procedure but already with this simple concept the result is satisfactory.

A big advantage compared to the beam based alignment correction algorithms is that with this BTM the trajectory of the electron beam is actually seen and that the straight line that one wants to have the electrons on is defined by the detector itself.

SUMMARY

Simulations for a new BTM concept proposed for the TTF-FEL at DESY have been presented concentrating on the signal, systematic effects on the position reconstruction and a first simple correction algorithm. The results look promising. A prototype of this BTM is under construction and is supposed to be tested in TTF-FEL phase I.

ACKNOWLEDGMENT

I would like to thank the organizers for this interesting workshop in a very nice place. I want to thank R. J. Dejus, B. Faatz, and J. S. T. Ng for providing me with the necessary information and useful discussions.

REFERENCES

1. TTF-FEL Conceptual Design Report, TESLA-FEL 95-03, DESY, June 1995; J. Roßbach et al., Nucl. Instrum. and Methods **A375** (1996) 269.
2. B. Faatz et al., TESLA-FEL 96-13, presented at the FEL'96 conference August 26-13,1996 in Rome, Italy.
3. B. Faatz, The SASE-FEL at the TESLA Test Facility at DESY, these proceedings.
4. B. Faatz, Beam based alignment for the TTF-FEL, these proceedings.
5. J. S. T. Ng, TESLA-FEL 96-16, DESY (1996).
6. J. S. T. Ng, these proceedings.
7. R. P. Walker and B. Diviacco, Rev. Sci. Instrum. **63** **(1)** (1992) 392.
8. R. J. Dejus, Nucl. Instrum. and Methods **A347** (1994) 56.

A Beam Trajectory Monitor for the TTF-FEL[*]

Deutsches Elektronen-Synchrotron DESY, 22603 Hamburg, Germany

Abstract. A method to determine the electron beam trajectory inside a long undulator module is described. Three-dimensional information is obtained by imaging the spontaneous radiation off-axis using pinholes and high resolution position sensors. The proposal for such a monitor for the SASE-FEL at the TESLA Test Facility is discussed.

INTRODUCTION

A bright light source with coherent radiation reaching the X-ray range can be created by sending a bright electron beam through a long undulator. This single-pass free electron laser relies on the Self-Amplified Spontaneous Emission (SASE) mechanism which results from the electron bunch interacting with its own radiation from the undulating motion. In order for SASE to take place, the electron and the photon beams trajectories must overlap. Because of the imperfections in the undulator magnets and mis-alignments in the focusing quadrupoles, the average electron beam trajectory tends to deviate from the straight path followed by the photon beam. This deviation must be detected and then corrected using steering coils inside the undulator.

A SASE-FEL operating in the range of 6 nm - 70 nm is under construction at the TESLA Test Facility (TTF) at DESY [1,2]. In the initial stage (TTF-FEL phase-1), the electron beam energy will be in the range of 300 MeV to 500 MeV. Each undulator module is 4.5 m long with a magnetic gap of 12 mm. The beam position inside the undulator will be measured by capacitive pickup- and/or waveguide-type monitors (BPMs), and a beam-based alignment procedure will be used to find the dispersion free trajectory [2]. This approach has the advantage of relying on proven technology, but also the disadvantage of requiring a large number of high precision BPMs (one per 0.5 m) which inevitably translates into cost.

---

[*] Presented at the Workshop on X-Ray Free Electron Lasers, Garda Lake, Italy, June 1997.

CP413, *Towards X-Ray Free Electron Lasers*
edited by R. Bonifacio and W. A. Barletta
© 1997 The American Institute of Physics 1-56396-744-8/97/$10.00

In this paper, we describe a new method to determine the beam position inside long undulator modules [3]. The off-axis spontaneous radiation is detected through a set of pinholes by high resolution silicon pixel detectors. The imaged space points are used to reconstruct the beam trajectory. The monitor is placed near the end of the undulator module, and only one monitor is needed to detect the beam position inside each module. Because the pixel response can be precisely calibrated, this monitor provides a well-defined optical axis. The ability to determine the beam trajectory complements the beam-based alignment approach which relies on the beam dispersion properties but does not measure the actual beam trajectory. In the following, the beam trajectory monitor (BTM) method is described, followed by discussions on a conceptual design. The status and plans of a prototype being developed for the TTF-FEL are then presented.

METHOD

An overview of the beam trajectory monitor is shown in Figure 1. Detailed discussion can be found in [3]. The placement of the position sensors, as well as the sensitive range are shown in Figure 1a. One pair of sensors at orthogonal azimuthal positions is sufficient for beam trajectory reconstruction although two pairs are shown. The two pinhole positions and the pair of measured image points define two lines which are constrained to the same source point (see Figure 1b): $\vec{p} = \vec{v} + a \cdot (\vec{u} - \vec{v})$, where $\vec{u}$ and $\vec{v}$ are the image and pinhole locations, respectively, and a is the slope parameter. Because the two pinholes are at the same distance from the imaging screen, the two slope parameters are identical, and two determinations are available:

$$a_1 = \frac{(h_x - v_x)}{(u_x - w_x) + (h_x - v_x)}, \quad a_2 = \frac{(h_y - v_y)}{(u_y - w_y) + (h_y - v_y)} \tag{1}$$

where $\vec{v}, \vec{u}$ refer to the vertical pinhole and its image, similarly $\vec{h}, \vec{w}$ for the horizontal pinhole. The average can be used. The rest of the equations over-constrain $\vec{p}$. Thus, only one well-measured coordinate in the transverse direction from each pinhole image is required. The x-position of the vertical pinhole image and the y-position of the horizontal pinhole image determine the corresponding components of $\vec{p}$.

The imaging pixels are arranged in two columns with row-wise segmentation, as shown in Figure 1b for the sensor behind the vertical pinhole. In each row, the relative signal in the two pixels is a function of the transverse beam displacement and is measured with high precision. The imaging position accuracy using the charge sharing between the two pixels is expected to be 1 μm, which translates into a 10 μm beam position measurement accuracy at the TTF-FEL for the current design parameters.

a)

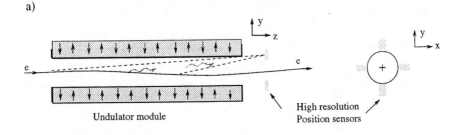

y
z

y
x

↓↑↓↑↓↑↓↑↓↑↓↑↓↑↓↑↓↑↓↑

e   e

+

Undulator module

High resolution
Position sensors

b)

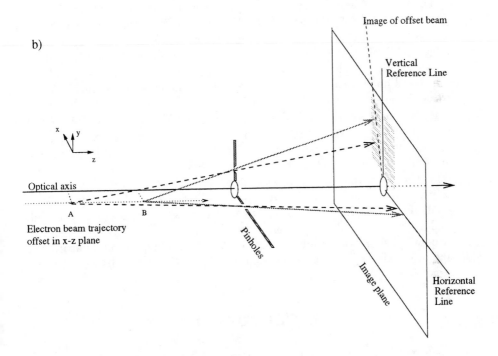

Image of offset beam

Vertical
Reference Line

x   y

z

Optical axis

A   B

Electron beam trajectory
offset in x-z plane

Pinholes

Image plane

Horizontal
Reference
Line

FIGURE 1. An overview of the beam trajectory monitor method. a) The location of the position sensors and the imaging range are shown. b) The method to uniquely determine the source point along the beam trajectory using two pinhole images is illustrated for the simple case of a beam trajectory with an offset. The pixel segmentation and the imaged trajectory for the sensor behind one of the pinholes are also shown.

The row-wise segmentation provides longitudinal resolution along the beam trajectory when it is projected through the pinhole. The longitudinal resolution is determined by the pinhole/pixel angular acceptance, as well as image broadening due to diffraction at the pinhole. For a wavelength of 44 nm, pinhole height of 50 μm, and a distance of 0.5 m between the pinhole and the imaging sensor, the image broadening due to diffraction would be 440 μm in which case the BTM would not be able to resolve the beam trajectory along the beam axis. In order to avoid this effect, a high-pass filter made of a thin layer (0.25 μm) of silver is used. Based on measured optical constants [4], the calculated reduction in the enormous photon flux in the lower harmonics of the spontaneous radiation at TTF-FEL below 150 eV (8 nm) is approximately 10 orders of magnitude.

The results of a trajectory reconstruction simulation are shown in Figure 2. The beam is tracked through the TTF-FEL FODO lattice. The quadrupoles are assumed to have random alignment errors uniformly distributed with a range of $\pm$ 50 μm. (The undulator is not included in this study.) It can be seen that the electron beam deviates from a straight trajectory due to misplaced quadrupole kicks, and the deviation increases as the beam travels towards the end of the undulator. The reconstructed trajectory is also shown. The longitudinal resolution varies from 0.5 m to 2.5 m. A total of four sensors is used to provide two complementing trajectory determinations. For those points with good longitudinal resolution, the transverse beam displacements are well reproduced. When the beam trajectory is averaged over a longer distance, however, the transverse beam position is poorly reconstructed. Despite this apparent shortcoming, an iterative steering correction procedure can be used. For example, the steering dipole strengths can be set one at a time, such that the reconstructed beam displacement near the steerer is minimized. A detailed simulation study using this algorithm, including undulator errors and other effects, shows that the required 10 μm RMS beam displacement can be achieved [5].

CONCEPTUAL DESIGN

The calculation of the undulator photon flux properties and the discussion on the detector requirements are given in [3]. In this section we briefly describe a conceptual design for the BTM. A schematic layout of the BTM is shown in Figure 3.

Silicon pixel detectors are chosen because of the limited available space and the required position accuracy. With a fully-depleted pnCCD with a very thin back-entrance window, a quantum efficiency of more than 50% has been measured for photons in the energy range of 100 eV to 10 keV [6]. The fractional error in the photon signal measurement in each pixel, including electronics noise and fluctuations in the electron-hole pair creation and charge

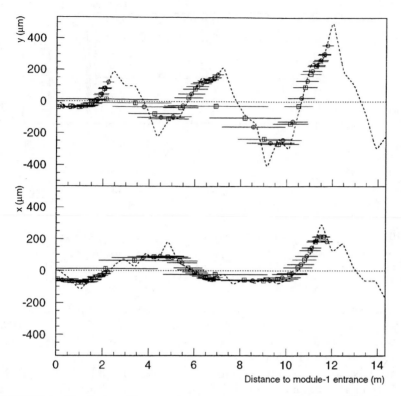

FIGURE 2. Simulation of beam trajectory reconstruction. The TTF-FEL FODO lattice is used, assuming ± 50 μm random quadrupole alignment errors. A typical trajectory is shown as dashed lines in the y-z and x-z planes. The reconstructed trajectory is shown as points (open squares and circles for two pairs of sensors used). The horizontal bars indicate the longitudinal resolution - the range over which the trajectory has been averaged.

collection processes as well as fluctuation in the photon flux, is

$$\left(\frac{\delta S}{S}\right)^2 = \left(\frac{ENC}{S}\right)^2 + \frac{1}{N_\gamma}\left[1 + \frac{1}{\alpha}\left(F + \frac{1}{\epsilon}\right)\right], \tag{2}$$

where ENC is the noise charge at the preamplifier input, N_γ the number of photons in each pixel, F the Fano factor, ϵ the quantum efficiency, and $\alpha = E_\gamma/w$ where E_γ is the photon energy, w the electron-hole pair creation energy. The position error using the charge sharing method is

$$\delta x = \frac{\sqrt{\pi}}{2}\sigma\left(\frac{\delta S}{S}\right) \tag{3}$$

assuming a Gaussian distributed photon flux with RMS width σ centered between the two pixels. For the TTF-FEL phase-1 prototype BTM design,

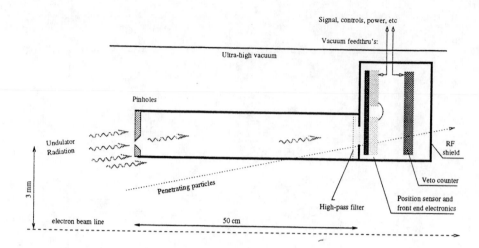

FIGURE 3. A schematic layout of the beam trajectory monitor system showing the detector assembly of pixel detectors, veto counters and front-end electronics, the RF shield, and the pinholes.

$N_\gamma \sim 1000$ and $ENC \sim 300\ e$ would give $\delta x \sim 1\ \mu$m. This condition can be satisfied. In addition, careful attention to the uniformity of the entrance window as well as calibration of the pixel response is required. A pixel detector with an active area of 0.5 mm $\times$ 1.0 mm and a total of 24 channels has been designed with the thin entrance-window technology [7]. Each pixel is directly connected to an on-chip JFET for low noise readout. The pixels are 250 μm long, with widths varying from 25 μm (nearest to the beam) to 100 μm to partially account for the rapid change in the undulator flux angular (θ) distribution and to give an angular acceptance favorable for good longitudinal resolution.

In order to suppress background due to penetrating particles (stray electrons or bremstrahlung X-rays), a veto counter is placed behind the 300 μm thick pixel detector which absorbs completely the VUV signal flux. The veto counter could be either a PIN diode, or another pixel detector.

Each readout electronics channel consists of a charge-sensitive amplifier followed by filter and sample/hold circuits. A remotely controlled switch at the amplifier input is used to vary the signal accumulation time, and the amplifier can be reset by a switch across the feedback capacitor. The photon flux expected in the pixels has a large range. For θ in the range of 0.5 mrad to 2 mrad, the number of expected photons varies by a factor of 4 at $\phi = 0°$ and by a factor of 1000 at $\phi = 90°$. In order to accumulate the required photon signal the integration time is expected to vary from ~ 300 ns to $\sim 200\ \mu$s depending on the radial position of the pixel. To avoid noise pickups, the front-end electronics is placed near the detector in vacuum. Vacuum feed-throughs are needed for connection to the ADC and the control system outside. A buffer/multiplexer is used for serial readout. Precision charge injections into the readout electronics and separately into the detector pixels provide on-line calibration. With this readout system, several measurements of the beam trajectory within one TTF macro-pulse (800 μs long) can be made.

The detectors and the readout electronics are placed inside an RF shield for protection against electromagnetic fields propagating through the beam pipe. For a 5 mm radius beam pipe at the TTF-FEL, the fields above cut-off can be safely shielded with copper approximately 0.5 mm thick. The frequency spectrum of fields excited by a 250 μm long bunch extends up to 190 GHz and then falls off rapidly. Any residual fields leaking through the pinholes (80 μm $\times 100$ μm) are shielded by the high pass filter (0.25 μm thick silver).

The pixel response is precisely calibrated to provide a well-defined optical axis. The pinhole is aligned with respect to the pixel detector using laser light. The relative alignment of two nearby BTMs placed along the beam line is obtained in-situ because there is an over-lap in the beam trajectory measured by the two stations. Absolute alignment of each BTM can be made by referencing to a nearby wire-scanner station which is aligned with respect to the undulator magnetic center.

The variation in the illumination pattern, due to structure in the undulator radiation angular distribution in (θ_x, θ_y), could lead to apparent beam motion. However, because of the very small pinhole width the accepted flux angular variation (in θ_x for the vertical pinhole, for example) is very small, after taking into account the E_γ dependences in filter transmission and quantum efficiency, and the resulting position error is negligible [5].

Further simulation studies are underway to investigate other possible systematic errors. The issue of wakefields due to the BTM as well as the mechanical design of the vacuum chamber need to be studied.

STATUS AND PLANS

The conceptual design phase of a BTM for the TTF-FEL is nearly complete. A prototype is being built and tests are planned. The silicon pixel detector has been designed and the production will start soon. Initial tests in the laboratory will include measurements of noise and position accuracy. Measurements at an undulator beam line will check our calculations of the radiation properties. For the TTF-FEL phase-1, it is proposed to place one BTM at the end of the undulator beam line as a first test of the BTM concept. The initial operation will be parasitic but in a realistic environment to study the effects on wakefield and the background condition. Comparison with the BPMs will provide a cross-check of the optical-axis definition. Steering corrections will demonstrate the effect on the SASE-FEL performance.

SUMMARY

A method to measure the electron beam trajectory inside long undulator modules is described. The conceptual design of a beam trajectory monitor for the TTF-FEL is given, and the required performance is shown to be feasible. A prototype is being built, and a staged test program is starting. With some minor variation, the same concept perhaps could be applied also at other planned SASE-FELs using long undulator modules. From the discussions at this Workshop, it can be concluded that a large number of experimental parameters, such as the electron beam trajectory, needs to be precisely measured in order to understand SASE-FEL in the short wavelength regime. The BTM is not only a useful diagnostic device but will also provide information for the understanding of SASE-FEL physics.

ACKNOWLEDGEMENT

I would like to thank the workshop organizers for their hospitality. I would also like to thank R. Carr, B. Faatz, W.M. Fawley, E. Gluskin, J. Rossbach, and U. Müller for fruitful discussions.

REFERENCES

1. TTF-FEL Conceptual Design Report, TESLA-FEL 95-03, DESY, June 1995; J. Rossbach *et al.*, Nucl. Instrum. and Methods **A375** (1996) 269.
2. B. Faatz, these proceedings.
3. J.S.T. Ng, TESLA-FEL 96-16, DESY, 1996. This report is also available through WWW at http://www.desy.de/~ng/TTF/btm_report.html.
4. B.L. Henke *et al.*, Atomic Data and Nucl. Data Tables, **54(2)** (1993) 181.
5. U. Müller, these proceedings.
6. H. Soltau *et al.*, Nucl. Instrum. and Methods **A377** (1996) 340.
7. MPI Halbleitorlabor, Munich, and KETEK GmbH, Oberschleißheim.

Electron Beam Size Diagnostic by Coherent Measurement and Imaging of Synchrotron Radiation

S. Marchesini

*Departement pour la Recherche Fondamental sur la Matiere Condense SP2M-IRS,
CEA-Grenoble, France E-mail marchesini@cea.fr*

Abstract. A comparison of electron beam size diagnostic in a storage ring by coherence measurement and imaging of Synchrotron radiation is made. Several experiments are reviewed: imaging by hard X-ray refractive lenses, interferometry by fresnel mirrors, Gabor holography of a fiber.

INTRODUCTION

With the dramatic reduction of emittance in third generation Synchrotron Radiation (SR) sources the diffraction limit is not far from being reached in the vertical plane [1]. Approaching the diffraction limit (DL), SR becomes nearly independent from the electron beam parameters, and diagnostic based on the SR becomes difficult. The measurement of the electron beam size based on SR can be performed either by imaging or by measuring the transverse coherence. When the electron beam becomes smaller than the diffraction limited beam the image of the source becomes independent on the electron beam size, and when the coherence size becomes larger than the SR beam the coherence properties become independent on the electron beam. These two limits are analysed. Some experiments are reviewed: Two mirrors interferometry is similar to Young experiment with reflecting optics, Holography of a fiber is based on the interference produced by the beam diffracted by the fiber, the more the beam is diffracted the less is correlated with the reference beam; refractive lenses for hard X rays are produced by drilling many holes in a materialy with the small refractive index and small absorption.

CP413, *Towards X-Ray Free Electron Lasers*
edited by R. Bonifacio and W. A. Barletta
© 1997 The American Institute of Physics 1-56396-744-8/97/$10.00

THEORY

The phase space distribution of the source (size σ_I, angular divergence [1] $k\sigma_J$, $k = 2\pi/\lambda$) is [kim] the convolution of the electron beam (σ_{IE}, $k\sigma_{JE}$) and the Synchrotron radiation from a single electron (σ_{IR}, $k\sigma_J$ if approximated to a gaussian beam):

$$\sigma_I = \sqrt{\sigma_{IR}^2 + \sigma_{IE}^2} \tag{1}$$

$$\sigma_J = \sqrt{\sigma_{JR}^2 + \sigma_{JE}^2} \tag{2}$$

The indices I and J stand for intensity in the near and far field. $\sigma_{IR}\sigma_{JR} = 1/2$ is the Diffraction Limit (DL). σ_{IE} can be measured by imaging if $\sigma_{IE} \gtrsim \sigma_{IR}$. The coherence near the Diffraction Limit needs some more words. The correlation function called in optics *Mutual Intensity* is the Fourier Transform of the phase space distribution, the *Wigner Function* (WF):

$$MI(x,u) = FT[WF](x, k \to u)$$
$$= I_0 \exp\{-x/2\sigma_I)^2\} \exp\{-(u/2\sigma_m)^2\}, \tag{3}$$

where the correlation distance is $\sigma_m = 1/\sigma_J$. Near the DL it is necessary to take into account the fact that the beam size is of the same order of the coherence size using the coherence factor. The coherence factor is the normalised version of the correlation function [$I(x) = MI(x,0)$ is the intensity]:

$$\mu(u) = MI(x,u)/[I(x + u/2)I(x - u/2)]^{1/2}]$$

The coherence distance σ_c is then related to the correlation distance by [R. Coïsson et al.]

$$\frac{1}{\sigma_c^2} = \frac{1}{\sigma_m^2} - \frac{1}{4\sigma_I^2} \tag{4}$$

In terms of electron and diffraction beam properties, the coherence distance in the far field ($\tilde{\sigma}_c$ is [R. Coïsson])

$$\tilde{\sigma}_c = \frac{1}{\sqrt{\sigma_{IE}^2 + \sigma_{IR}^2 \frac{\sigma_{JE}^2}{\sigma_{JR}^2 + \sigma_{JE}^2}}} \tag{5}$$

far from DL the Van Cittert-Zernike theorem holds and $\tilde{\sigma}_c = \frac{1}{\sigma_{IE}}$, at low beta values (electron beam more collimated and divergent than the DL beam) $\tilde{\sigma}_c = \frac{1}{\sigma_{IR}}$.

---

[1] the phase space in these units is typical of cristallography, notice that the phase space volume is dimensionless

The coherence distance can be well determined when it is smaller than the beam itself $\tilde{\sigma}_c \lesssim \sigma_J$.

The two limits are then

$$1 - \frac{\sigma_{IE}}{\sigma_I} = 1 - \frac{\sigma_{IE}^2}{\sqrt{\sigma_{IR}^2 + \sigma_{IE}^2}} \ll 1$$

$$1 - \frac{\sigma_c}{\sigma_J} = 1 - \frac{1}{\sqrt{\sigma_{IE}^2(\sigma_{JR}^2 + \sigma_{JE}^2) + \sigma_{IR}^2 \sigma_{JE}^2}} \ll 1$$

In practice, as long as the beam is far from the DL, it easier to study small beams with coherence measurement since the coherence size is $\propto \frac{1}{\sigma_{IE}}$.

EXPERIMENTAL

Two Mirrors interferometry [K. Fezzaa, et al.]

The spatial coherence of an optical wavefront, defined quantitatively by the "mutual intensity function" (MI), can be experimentally determined (and is in fact operationally defined) by Young's experiment with a variable slit separation. For an X-ray beam it might be more convenient to use reflecting optics, because a thick double slit with a short and variable distance between them is not easy to realize and because the edges of the slits reflect and add spurious radiation to the interference pattern. Although mirrors are never perfectly flat, imperfections affect lower spatial harmonics in the extreme Fresnel (or Fraunhofer) domain.

The effective distance of the two "slits" is varied either by rotating the mirrors support (changing the angle of incidence) or by raising the first mirror; raising the first mirror decreases the effective distance as seen from the source, while increases the effective distance as seen from the detector (decreases the fringes size) (Fig.).

the horizontal coherence is measured with the same setup: one of the mirror is tilted and the SR beam is deviated in the horizontal direction. The two beam are superimposed with a displacement a and the visibility decreases as the horizontal coherence at the detector plane at a distance a (Fig.).

The interference pattern can be detected either with a CCD camera or by scanning with a normal detector. With this method the whole second order correlaton function can be measured.

"holography" of a fiber [Snigirev et al.]

SR beam diffracted by the fibre interfere with the direct beam, and produces interference pattern (Fig. 1). When the distance between the projected position of the fibre and the position of the beam is bigger than the coherence distance, the interference vanish. This technique is very simple and fast. The size of the interference pattern is given by the coherence distance at the detector position; it can be recorded with a high resolution CCD camera or with a film.

Imaging with refractive lenses [P. Ellaume]

Normal lenses work if the refractive index is > 1; for X-rays the refractive index in matter is < 1, and a lens is created by drilling a hole in the material. Since the refractive index is $\simeq 1$, several of these lenses are necessary (Fig. 2).

The lenses used by P. Ellaume were made by drilling vertical (horizontal) cilindrical holes to focus in horizontal (vertical) direction on Alluminum. The focal length at 30 KeV was 12.2 m (horizontal) 10.2 m (vertical), transmission $T = 0.007$; with berillium the transmission would be 0.95. The horizontal beam size measured was in agreement with the expected value (0.11 mm), while the vertical beam size measured (0.061) was quite different from the expected value (0.010).

REFERENCES

1. L. Farvacque, The low Emittance lattice, how to win over Diffraction and reach the brightest beams, ESRF Newsletters 24 (1995), 12.
2. K. J. Kim, A new formulation of synchrotron radiation optics using the Wigner distribution, SPIE 582 (1985) 2
3. R. Coisson, Spatial coherence of Synchrotron radiation, Appl. Opt. 34 5 (1992)
4. R. Coisson et al. Gauss-Schell sources as models for Synchrotron Radiation, to be published on Journ. Of Synchr. Rad

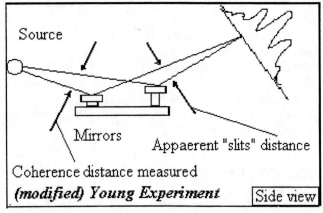

5. P. Elleaume, Two Plane focusing of 30 keV Undulator Radiation with a Refractive lens, ESRF/MACH ID 97/31, internal Report. (Also on ESRF Newsletters 28 (1997) 33)

6. A. Snigirev et al. On the requirements to the instrumentation for the new generation of the synchrotron radiation sources, NIM (1996)

7. K. Fezzaa et al. X-Ray interferometry at ESRF using two coherent beams from Fresnel mirrors, J. of X-Ray Sci. And Techn, march (1997).

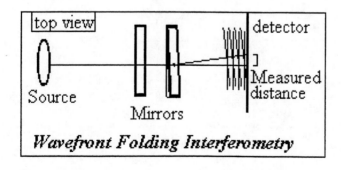

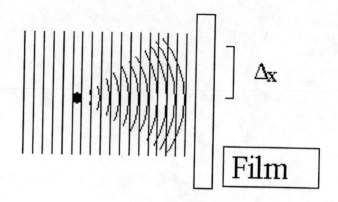

FIGURE 1. "holography" of a fiber

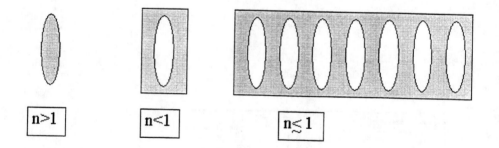

FIGURE 2. refractive lenses for hard x-ray

Suppression of the coupled-bunch instability in the SPring-8 storage ring

H. Ego *, M. Hara, Y. Kawashima, Y. Ohashi, T. Ohshima, H. Suzuki, I. Takeshita
and H. Yonehara

SPring-8, Kamigori, Ako-gun, Hyogo 678-12, Japan

* Corresponding author. E-mail ego@rikaxp.riken.go.jp

Abstract
 The bell-shaped single-cell cavities used in the SPring-8 storage ring are devised to suppress the coupled-bunch instabilities arising from the coupling impedances associated with higher-order mode resonances. The inner structures of the cavities are designed to have low longitudinal or transverse shunt impedances of the higher-order modes. The cavities are provided with a systematic modification of inner structures of the cavities and with multiple movable plungers to shift the resonant frequencies of higher-order modes thereby preventing higher-order mode resonances from being excited.

1. Introduction

 The effects of beam instabilities on the beam current and emittance of a storage ring optimized to supply highly brilliant photon beams can seriously compromise performance. In this paper, we describe a method of suppressing coupled-bunch instability arising from the narrow-band coupling impedance of higher-order mode resonance excited in RF cavities by a multi-bunch electron beam. When the instability arises, the amplitude of bunch oscillation grows exponentially and the bunch motion becomes unstable. The growth rate of the amplitude is calculated from the value of the impedance. On the other hand, in an electron ring there is radiation damping which reduces the amplitude of the bunch oscillation. Therefore, the threshold beam current for the instability is calculated by balancing the growth rate with the damping rate. The detailed formulae are described in Ref. [1] and shown in Section 2 of this paper.
 One method of suppressing the instability is to use damping couplers, which are special antennas to extract the energy of higher-order mode resonances excited in an RF cavity [2]. However, there are some difficulties of the damping coupler. The damping couplers must be devised in order not to disturb the resonant field of accelerating TM010-like mode and extract the stored energy of TM010-like mode. In addition, the cavity structure becomes complicated and cooling channels are difficult to locate effectively in the cavity body since extra ports should be prepared on the cavity for the damping couplers. Furthermore, it costs much to fabricate the damping couplers, RF absorbers and the cavities with such a complicated structure. Therefore, we don't use the damping couplers for our cavities.
 Our strategy against the coupled-bunch instability is as follows:
 (1) we design the inner structures of RF cavities so that the longitudinal or transverse shunt impedances of the dangerous higher-order modes below the cut-off frequency of a beam duct with a diameter of 100 mm are as low as possible, and
 (2) we prevent electromagnetic fields generated by an electron beam in RF cavities from resonating with a higher-order mode.

CP413, *Towards X-Ray Free Electron Lasers*
edited by R. Bonifacio and W. A. Barletta
© 1997 The American Institute of Physics 1-56396-744-8/97/$10.00

The bell-shaped single-cell cavity is used for the SPring-8 storage ring since the shunt impedances of the dangerous higher-order modes were found to be low by the URMEL [3] calculation [4-7]. The inner structure of the cavity is shown in Fig. 1.

One method of realizing the strategy (2) is to trim the length of a plunger which protrudes into the inside of an RF cavity. This method has been realized in the Photon Factory storage ring at KEK [8]. We have improved the method by replacing the plunger with a movable one. The results have been briefly reported in Ref. 7. The specification of the movable plunger and the quantitative results on the shift in resonant frequency of higher-order mode are shown in Sections 3 and 4, respectively. Another method is to modify the cavity inner structure. The correlation between the RF characteristics of higher-order mode and the inner structure parameter was examined by the URMEL calculation. The results are shown in Section 4. The method of suppressing the coupled-bunch instability both by the movable plungers and by modifying the inner structure is addressed in the Section of discussions.

2. Coupled-bunch instability in a storage ring
2.1 Growth rate and threshold current

An RF cavity has many types of electromagnetic oscillating modes such as monopole, dipole, quadrupole and sextupole. When a monopole or dipole mode resonance is excited, there is an electric or a magnetic field (or both fields) near the beam axis and the field influences the motion of electrons in a beam. The coupled-bunch instability arising from the narrow-band impedance of a monopole higher-order mode or of a dipole higher-order mode is discussed here. The theory is described in Ref. 1.

Here it is assumed that equally populated electron bunches are equally distributed in a storage ring. A higher-order mode resonance is excited by the electron bunches when the impedance of the mode overlaps the sideband or satellite frequency corresponding to a particular mode of coherent bunch motion. The electromagnetic fields of the mode drive the bunches to oscillate in longitudinal or transverse direction. This bunch oscillation is called the coupled-bunch oscillation. The frequencies of electromagnetic fields generated by electron bunches, $f^{\pm}_{\mu,n}$ are given by

$$f^{\pm}_{\mu,n} = nBf_r \pm (\mu + \delta v)f_r,$$

(1)

where n is a positive integer including zero, B the number of bunches in a storage ring, m the mode number of the bunch oscillation and f_r the beam revolution frequency [1]. δv is the fractional part of the synchrotron tune for a longitudinal bunch oscillation or the fractional part of the betatron tune for a transverse bunch oscillation. It should be noted that the plus sign in front of the parenthesis means longitudinally unstable and the minus sign transversely unstable.

When a longitudinal coupled-bunch oscillation is driven by the fields of a monopole higher-order mode resonance, the theoretical growth rate of the oscillation amplitude is given by

$$\tau^{-1} = \frac{e\alpha I f_r^2}{2Ef_s} \left\{ \sum_{n=0}^{\infty} (nB + \mu + \delta v) \operatorname{Re} Z(\omega^+_{\mu,n}) \, G^+_{\mu,n}(f_r \tau_L) - \sum_{n=1}^{\infty} (nB - \mu - \delta v) \operatorname{Re} Z(\omega^-_{\mu,n}) \, G^-_{\mu,n}(f_r \tau_L) \right\},$$

(2)

where e, E, α, I and f_s are electric charge of an electron, electron energy, momentum compaction factor, beam current and synchrotron frequency, respectively [1]. $Z(\omega)$ is the coupling impedance of a monopole mode resonance as a function of angular frequency, ω. The angular frequency, $\omega^{\pm}_{\mu,n}$ is equal to $2\pi f^{\pm}_{\mu,n}$, where $f^{\pm}_{\mu,n}$ is given by Eq.(1). $\operatorname{Re} Z(\omega)$ shows the real part of the coupling impedance. The form factor, $G^{\pm}_{\mu,n}(f_r \tau_L)$ introduced here is a function of the half bunch length, τ_L in the unit of second and is related to the Sacherer's form factor, $F_0(\Delta\phi^{\pm})$ [9] by

$$G^{\pm}_{\mu,n}(f_r \tau_L) = \frac{2F_0(\Delta\phi^{\pm})}{\Delta\phi^{\pm}}$$

with

(3)

$$\Delta\phi^{\pm} = 2\pi\,\tau_L\,f_{\mu,n}^{\pm} \quad. \tag{4}$$

The first and second terms in the right-hand side in Eq. (2) refer to contributions of anti-damping and of damping, respectively. Anti-damping means that the amplitude of the bunch motion grows and the bunch motion is unstable. Damping means that the amplitude of the bunch motion reduces and the bunch motion is stable. The amplitude of the longitudinal coupled-bunch motion grows exponentially when the growth rate exceeds the longitudinal damping rate. The longitudinal damping rate is given by

$$\frac{1}{\tau} = \frac{1}{\tau_s} + \frac{1}{\tau_{Landau}} \quad, \tag{5}$$

where τ_S and τ_{Landau} are the radiation damping time and Landau damping time, respectively. τ_S is given by

$$\tau_s = \frac{E\,T_r}{U_0} \quad, \tag{6}$$

where T_r and U_0 are the beam revolution time and the radiation loss energy, respectively. τ_{Landau} is given by

$$\tau_{Landau} \cong \frac{4}{S} \quad, \tag{7}$$

where S is a spread of the synchrotron oscillation frequency due to a non-linearity of the oscillation [1]. Since the radiation damping rate is about one hundred times larger than Landau damping rate in the SPring-8 storage ring, the effect of Landau damping is negligible. Therefore, the theoretical threshold current for the longitudinal instability is calculated from the value of τ which satisfies $\tau = \tau_S$.

TM011-like mode is a monopole higher-order mode with high longitudinal shunt impedance. Provided that only TM011-like mode resonance in RF cavities contributes to a longitudinal coupled-bunch instability, the threshold current is given by

$$I_{th} = \frac{2\,E\,f_s}{\tau_s e\alpha f_r^2\,(nB + \mu + \delta v)\,\mathrm{Re}\,Z(\omega_{\mu,n}^+)\,G_{\mu,n}^+(f_r\tau_L)} \quad. \tag{8}$$

Here, the damping by the second term in the right-hand side of Eq.(2) is neglected. The coupling impedance of TM011-like mode is given by

$$Z(\omega) = \sum_m \frac{R_{sh}^{(m)}\,\dfrac{Q_L^{(m)}}{2Q_a^{(m)}}}{1 + jQ_L^{(m)}(\dfrac{\omega}{\omega_a^{(m)}} - \dfrac{\omega_a^{(m)}}{\omega})} \quad, \tag{9}$$

where the summation is taken over all RF cavities installed in a storage ring, $Q_a^{(m}$ is the unloaded Q-value, $Q_L^{(m}$ the loaded Q-value, $R_{sh}^{(m}$ the longitudinal shunt impedance and $\omega_a^{(m)}$ the angular frequency of TM011-like mode in each cavity.

Next, we consider a transverse coupled-bunch instability arising from the high transverse impedance of a dipole higher-order mode. Provided that only the anti-damping coupled-bunch oscillation is driven, the growth rate of the oscillation amplitude is given [1] by

$$\tau^{-1} = \frac{e\beta I f_r}{2E} \sum_{n=1}^{\infty} \mathrm{Re}\,Z(\omega_{\mu,n}^-)\,F' \quad, \tag{10}$$

where F' and β are the form factor defined by Sacherer [9] and the betatron function at an RF cavity, respectively. The transverse instability arises if the growth rate exceeds the transverse radiation damping rate. The transverse radiation damping time τ_β is given by

$$\tau_\beta = \frac{2\,E\,T_r}{U_0} \, .$$

(11)

Thus, the threshold current by a single dipole higher-order mode is given by

$$I_{th} = \frac{2\,E}{\tau_\beta e\beta f_r \mathrm{Re}\ Z(\omega^-_{\mu,n})\,F'}$$

(12)

with

$$Z(\omega) = \sum_m \frac{R_t^{(m)}}{1 + jQ_L^{(m)}\left(\frac{\omega}{\omega_a^{(m)}} - \frac{\omega_a^{(m)}}{\omega}\right)} \, ,$$

(13)

where the summation is taken over all RF cavities in a storage ring, $R_t^{(m}$ is the transverse shunt impedance, $\omega_a^{(m)}$ the angular frequency, $Q_L^{(m}$ the loaded Q-value of the dipole mode in each cavity.

2.2 Cooperative coupled-bunch instability

If the resonant frequency of a monopole (or dipole) higher-order mode with high shunt impedance is the same in all RF cavities and the mode resonance is excited at the frequency, f_a by an electron beam, the coupled-bunch oscillations driven by the individual RF cavities are synchronized. We call this coupled-bunch oscillation driven in multiple RF cavities a cooperative coupled-bunch oscillation. In this case, the relations, $\omega_a^{(m)} = \omega_a$ ($\omega_a \equiv 2\pi f_a$) for all m and $\omega_{\mu,n}^+ = \omega_a$ (or $\omega_{\mu,n}^- = \omega_a$) are true in Eq. (9) (or Eq. (13)) for the cooperative longitudinal (or transverse) coupled-bunch oscillation. The real part of the coupling impedance is the maximum and given by

$$\text{Re } Z(\omega_{\mu,n}^+) = \sum_m R_{sh}^{(m)} \frac{Q_L^{(m)}}{2Q_a^{(m)}} \quad (\text{ or } \quad \text{Re } Z(\omega_{\mu,n}^-) = \sum_m R_t^{(m)} \quad). \tag{14}$$

The growth rate increases significantly and the threshold current reduces. Therefore, it is especially important to suppress the cooperative coupled-bunch instabilities.

3. Calculation and measurement of RF properties of the cavity

3.1. Calculation of the RF properties of the bell-shaped single-cell cavity

The RF characteristics such as resonant frequency, unloaded Q-value and shunt impedance of a resonant mode are obtained by the URMEL calculation. In order to express the inner structure of the bell-shaped cavity, eight elements composed of two arcs and of six straight lines are defined as shown in Fig. 2. R_0 , R_1 and Δz are the equator radius, the radius at the belly and the length of the straight section at the equator wall, respectively. The thick solid line represents the shape of the inner wall. We call a cavity with $R_0 = 255$ mm, $R_1 = 120$ mm and $\Delta z = 0$ mm as a standard-shaped cavity. We have calculated the RF characteristics of the monopole and dipole modes with a high shunt impedance as changing the length of one of the three parameters; R_0 , R_1 and Δz. The dimensions of the other elements except for both the parameter and the element 4 are kept constant. Since the element 4 must be the tangent to both the element 3 and the element 5, the length of the element 4 is changed with R_0 or R_1 .

3.2. Measurement of frequency shift by movable plungers

It is difficult to calculate exactly the shift in resonant frequency by moving a plunger. A model cavity with the structure shown in Fig. 1 was made of aluminum and used for the measurement of the shift. The cavity has four ports on the equator of the cavity at every 90 . A coaxial-loop RF input coupler was installed in the port with a diameter of 120 mm. Three movable plungers were installed in the other three ports with diameters of 73 mm. The plunger is made of an aluminum bar with a diameter of 70 mm and with length of 90 mm. The plunger is moved by a stepping motor along the center axis of the plunger. Figure 3 shows the arrangement of the movable plungers. The position of the flat top of the plunger is called the plunger position. The plunger position at which the distance between the flat top and the beam axis is equal to the value of the equator radius of the cavity is defined as the origin of the plunger. A positive value of the plunger position represents that the plunger protrudes into the cavity. A negative value represents that the plunger is drawn back into the port. The plunger 1 and the plunger 2 were set in the horizontal plunger ports and the plunger 3 was set in the vertical plunger port.

The frequency shifts of TM010-like mode and of higher-order modes by changing the position of one plunger were measured while the other two plungers were set at each origin. The frequency shifts of higher-order modes by two plungers were also measured under a specific rule as follows;

1) every plunger is set at each origin,
2) the plunger 2 is moved to a suitable position so that the frequency of TM010-like mode is 508.58 MHz,
3) the resonant frequency of a higher-order mode is measured,
4) the position of the plunger 1 or of the plunger 3 is changed,
5) the frequency of TM010-like mode changed in the preceding process is readjusted to 508.58 MHz by the plunger 2, and
6) the resonant frequency of the higher-order mode is measured again and compared with the value obtained in the process 2).

271

4. Results

4.1. Change in the RF properties of the cavity by modifying the inner structure

Figures 4, 5 and 6 show the calculated results of resonant frequency, unloaded Q-value, and longitudinal or transverse shunt impedance, respectively. ΔR_0 and ΔR_1 represent the changes in R0 and in R_1, respectively. The solid lines in Fig. 4 are obtained by the least squares method. The solid lines in Figs. 5 and 6 are only guides for eyes. The resonant frequencies are in proportion to each parameter. The proportionality constants are listed in Table 1. Unloaded Q-values are in proportion to the parameter, R_0 while there is no remarkable change in Q-value for the parameters, Δz and R_1. The shunt impedances change by less than 10 % when the absolute value of each parameter is less than 10 mm.

Figure 7 shows the frequency shifts of the higher-order modes, Δf_{HOM} as a function of the frequency shift of TM010-like mode, Δf_{TM010}. The solid lines are obtained by the least squares method. The ratios of frequency shift of higher-order mode to that of TM010-like mode, $\Delta f_{HOM}/\Delta f_{TM010}$ by modifying R_1 or Δz are larger than those by modifying R_0.

4.2. Threshold current for the coupled-bunch instability

The threshold beam current for the coupled-bunch instability was evaluated with the calculated shunt impedance of higher-order mode in the standard-shaped cavity, provided that the synchrotron radiation loss is caused only by bending magnets. When insertion devices are installed in the storage ring and the radiation loss increases, the threshold current increases as shown in Eqs.(8) and (12). It is also provided that a single higher-order mode resonance is excited in only one cavity by an electron beam. The results are shown in Table 2. The coupled-bunch instabilities don't arise in the SPring-8 storage ring at a beam current of 100 mA. However, no effect of the cooperative coupled-bunch instability is taken into account here.

4.3. Frequency shift by movable plungers

In this paper, we distinguish between the split two resonances of a dipole mode in the model cavity according to the direction of the force which electrons in a beam receive from the electromagnetic field of each resonance. When TE111-like mode is excited, electrons are kicked by the electric field. Thus, TE111-like mode with a horizontal electric field is called horizontal TE111-like mode while TE111-like mode with a vertical electric field is vertical TE111-like mode. Two TM111-like modes are also distinguished according to the direction of their electric field on the beam axis. On the other hand, the electrons are kicked by Lorentz force in the direction perpendicular to both the beam axis and the magnetic field of TM110-like mode. Thus, TM110-like mode with a horizontal magnetic field on the beam axis kicks in the vertical direction and is called vertical TM110-like mode while TM110-like mode with a large vertical magnetic field is horizontal TM110-like mode.

Figure 8 shows the frequency shifts as a function of the plunger position. The plots by open circles and by closed circles in Fig. 8 show the shifts by the horizontal plunger (the plunger 1) and by the vertical plunger (the plunger 3), respectively. Figure 8 (a) shows that the amount of frequency shift of TM010-like mode by the vertical plunger is almost the same as that by the horizontal plunger. TM011-like mode as well as TM010-like mode is an axially symmetric resonant mode. Since all movable plungers have the same structure, they are expected to induce the same amount of frequency shift. However, the vertical plunger shifts the frequency larger than the horizontal plunger as shown in Fig. 8 (b). This is because the strength of local electromagnetic field near the horizontal plunger is different from that near the vertical plunger. The electric field of TM011-like mode near the coupler port is distorted by the coupler port and by the RF input coupler. The horizontal plunger reacts on the distorted field more than the vertical plunger, since the horizontal plunger is nearer to the RF input coupler than the vertical plunger.

Figure 8 (c) shows that a negative frequency shift is induced to horizontal TE111-like mode by the horizontal plunger and a positive shift by the vertical plunger. The amount of shift of vertical TE111-like mode by the horizontal plunger is almost the same as that of horizontal TE111-like mode by the vertical plunger as shown in Fig. 8 (d). The amount of frequency shift of vertical TE111-like mode by the vertical plunger is almost the same as that of the horizontal TE111-like mode by the horizontal plunger. The frequencies of TM110-like modes and of TM111-like modes are sensitive to the position of either horizontal or vertical plunger as shown in Figs. 8 (e), (f), (g)

272

and (h). The horizontal plunger shifts the frequencies of horizontal TM110-like and TM111-like modes largely. The vertical plunger shifts the frequencies of vertical TM110-like and TM111-like modes largely.

Figure 9 shows the frequency shifts of the higher-order modes by two movable plungers. There are two types of plunger combination; (I) the plunger 1 and the plunger 2 and (II) the plunger 2 and the plunger 3. The frequency of TM010-like mode was kept at 508.58 MHz by the plunger 2 wherever the position of the other plunger was set. The plunger 1 or the plunger 3 was used for shifting the frequencies of the higher-order modes. The results of the combination (I) and of the combination (II) are shown in the left-hand column and in the right-hand column in Fig. 9, respectively. The abscissa represents the position of the plunger 1 in (I), or of the plunger 3 in (II). The solid lines are obtained by the least squares method. The frequency shifts of the higher-order modes are small in (I) but large in (II). Since the amount of frequency shift by the plunger 1 is almost the same as that by the plunger 2 for all the resonant modes, the frequency shift by the plunger 1 is reduced by the plunger 2 in (I). On the other hand, the frequency shift by the plunger 3 is not canceled out by the plunger 2 in (II), since the plunger 3 shifts the frequencies of the higher-order modes differently from the plunger 2 as shown in Fig. 8. In results, the frequencies of the higher-order modes are shifted in a rate between 30 kHz/mm and 100 kHz/mm by using both the horizontal plunger and the vertical plunger while TM010-like mode is kept resonating at 508.58 MHz by the horizontal plunger.

5. Discussions

The results in Section 4.2 indicate that the coupled-bunch instabilities don't arise at a beam current of 100 mA. If all RF cavities are fabricated with the same inner structure, they have the same RF properties. In case that the resonant frequency of a higher-order mode with high shunt impedance is equal to the frequency given by Eq.(1), the higher-order mode resonance is excited in all RF cavities by an electron beam and contributes to the cooperative coupled-bunch instability as discussed in Section 2.2. A specific higher-order mode resonance should not be excited at the same frequency in all cavities. Therefore, we modify the inner structure of each cavity slightly and systematically to change the resonant frequencies of the higher-order modes. Since the frequency of TM010-like mode is changed by modifying the inner structure, the frequency change of TM010-like mode must be as small as possible while the Q-values and the shunt impedances of all the modes must not deteriorate remarkably.

As shown in Section 4.4, R_1 or Δz is better than R_0 as a parameter to be modified. Here, we select R_1 as the parameter. Let the fundamental length of R_1 be 120 mm and ΔR_1 represent a small change in R_1 ; R_1 is given by $R_1 = 120 + \Delta R_1$ [mm]. Since the frequency of TM010-like mode is changed by modifying R_1, the movable plungers are used to adjust the frequency of TM010-like mode. The frequency of higher-order mode is also shifted by the plungers as described in Section 4.3. Therefore, the total frequency change of higher-order mode is obtained by taking account of the frequency shift by the plungers. We evaluate the total frequency change, using the results in Section 4 and assuming that the same amount of frequency change is caused to the two resonances of a dipole higher-order mode by modifying R_1 and the amount of frequency shift by the plungers is independent of ΔR_1.

At first, we evaluate the frequency change in the case that the frequency of TM010-like mode is adjusted by one horizontal or vertical plunger. Let all plungers installed for the standard-shaped cavity with $R_1 = 120$ mm be set in a position of x_0 and the frequency of TM010-like mode be 508.58 MHz. The condition, in which the frequency of TM010-like mode in a cavity with ΔR_1 is adjusted to 508.58 MHz by one plunger, is represented by the following equation.

$$\Delta f_{TM010}(\Delta R_1) + F(x_a) - F(x_0) = 0 \ , \quad (15)$$

where $\Delta f_{TM010}(\Delta R_1)$ is the frequency change of TM010-like mode as a function of ΔR_1, $F(x)$ the resonant frequency of TM010-like mode as a function of the plunger position and x_a the plunger position at which the frequency of TM010-like mode in the cavity with ΔR_1 is 508.58 MHz. By transforming Eq. (15), x_a is given by

273

$$x_a = F^{-1}(F(x_0) - \Delta f_{TM010} (\Delta R_1)) . \quad (16)$$

Therefore, the total frequency change of a higher-order mode, $\Delta f^\wedge_{HOM}$ is given by

$$\Delta f^\wedge_{HOM} = \Delta f_{HOM} (\Delta R_1) + G_{HOM} (x_a) - G_{HOM} (x_0) , \quad (17)$$

where $\Delta f_{HOM} (\Delta R_1)$ is the frequency change of the higher-order mode as a function of ΔR_1, and $G_{HOM} (x)$ the frequency of the higher-order mode as a function of the plunger position. $F(x)$ and $G_{HOM} (x)$ obtained by measurement are used.

In case that both the horizontal plunger and the vertical plunger are used for adjusting the frequency of TM010-like mode, $F(x)$ in Eq. (15) is replaced with

$$F^\wedge(x) = F^{(h)}(x) + F^{(v)}(x) , \quad (18)$$

where $F^{(h)}(x)$ is the resonant frequency of TM010-like mode as a function of the horizontal plunger position, and $F^{(v)}(x)$ as a function of the vertical plunger position. Here, let the value of the horizontal plunger position be the same as that of the vertical plunger position. $G_{HOM}(x)$ in Eq. (17) is also replaced with

$$G^\wedge_{HOM} (x) = G^{(h)}_{HOM} (x) + G^{(v)}_{HOM} (x) , (19)$$

where $G^{(h)}_{HOM} (x)$ is the frequency of a higher-order mode as a function of the horizontal plunger position, and $G^{(v)}_{HOM}(x)$ as a function of the vertical plunger position.

The results are shown in Fig. 10; (a) the single horizontal plunger, (b) the single vertical plunger and (c) both the horizontal plunger and the vertical plunger. Here, the value of x_0 is +6 mm. The left-hand and right-hand vertical axes show the frequency change of a higher-order mode, $\Delta f^\wedge_{HOM}$ and the plunger position x_a, respectively. The thick solid line and the thick dotted line represent the frequency changes of a horizontal resonance of a dipole higher-order mode and of a vertical one, respectively. The thin solid line represents the plunger position, x_a. When only one plunger is used, the frequency of either horizontal or vertical resonance is changed largely as shown in Figs. 10 (a) and (b). TM110-like and TM111-like modes are remarkable for this phenomena. The negative frequency change of TM111-like mode caused by the plunger is larger than that by the increase in R1. The frequency of TM110-like mode is changed largely by the plunger while the frequency slightly increases by the increase in R1. The frequency of TM011-like mode is changed largely by the increase in R1 and by the plunger. Figure 10 (c) shows that the resonant frequencies of the horizontal and vertical dipole modes are changed more when both the horizontal plunger and the vertical plunger are used. Therefore, the use of the two plungers is effectual for changing the frequencies of the higher-order modes in the cavities with ΔR_1.

Taking into account that the vertical oscillation of electron bunch is executed independently of the horizontal oscillation, we determine the length of R_1 for thirty-two cavities. If a dipole higher-order mode resonance with a horizontal electric field is excited in two RF cavities at the same frequency, electron bunches are kicked in the horizontal direction by the electric field. However, when one cavity is rotated around the beam axis by 90 , the electric field in the cavity becomes vertical and the electron bunches are kicked vertically in the cavity but horizontally in the other cavity. As a result, the cooperative coupled-bunch oscillation by the higher-order mode resonance is not caused in the cavities. Thus, we design sixteen types of cavity with different length of R_1 and fabricate each type of cavity two by two. We rotate one of two cavities with the same length of R_1 around the beam axis by 90 , when we arrange the cavities in the storage ring.

As shown by Eq. (1), electromagnetic fields oscillating at integral multiples of the beam revolution frequency are generated by the electron beam in a storage ring. In order to prevent a higher-order mode resonance from being excited at the same frequency in all cavities by the beam, the resonant frequency of a higher-order mode in one cavity is desirably different from that in any other types of cavity by more than the revolution frequency, 209 kHz. Figure 10 (c) shows that the difference between a resonant frequency in the cavity with $\Delta R_1 = -1$ mm and that in the cavity with $\Delta R_1 = +2$ mm is 4.6 MHz for TE111-like modes, 2.9 MHz for TM110-like modes, 12.6 MHz for TM011-like mode, 13.9 MHz for horizontal TM111-like mode and 11.4 MHz for vertical TM111-like mode. The frequency change of TM110-like mode is the smallest among the higher-order modes. When R1 is systematically increased by a step of 0.22 mm, the frequency of TM110-like mode increases by about 200 kHz for every step while the frequency of TM010-like mode is kept at 508.58 MHz by the two plungers. Since the frequency difference of about 200 kHz is larger than the half-power band width of TM110-like mode resonance, TM110-like mode resonance is not simultaneously excited at a monochromatic frequency in all the systematically modified cavities. Since the frequency changes of the other higher-order modes are larger than that of TM110-like mode, the cooperative coupled-bunch instabilities by the other mode resonances also don't arise.

However, there are some factors disturbing the above method of suppressing the instabilities; change in temperature of cooling water for RF cavities and errors in manufacturing cavities. In order to prevent the change in the temperature, we prepare a control system keeping the RF power loss at cavity inner wall and the temperature of cooling water constant even if the load of beam current changes. Since fabricated RF cavities usually have errors in their dimensions, the frequencies of the higher-order modes will not be perfectly manipulated by modifying R_1. In order to make up for the defect, the method of changing the resonant frequencies of the higher-order modes described in Section 4.3 is used. By changing the positions of the horizontal plunger and of the vertical plunger appropriately, we can prevent each cavity from resonating at the frequencies given by Eq.(1). Figure 11 shows the resonant frequencies of the higher-order modes in the model cavity as a function of the vertical plunger position. The frequency of TM010-like mode is adjusted to 508.58 MHz by the horizontal plunger wherever the position of the vertical plunger is set. The notation of the abscissa is the same as that of Fig. 9 (II). Open circles on each curve indicate the values given by Eq.(1). The thin solid and the thick solid lines represent the resonant frequencies of horizontal dipole higher-order mode and of vertical mode, respectively. A coupled-bunch oscillation is caused when the resonant frequency of a higher-order mode is close to one of the values given by Eq.(1). However, when we set the vertical plunger in a suitable position and adjust the frequency of TM010-like mode by the horizontal plunger, the frequencies of the higher-order modes are different from any frequencies given by Eq. (1) and the higher-order mode resonances are not excited in the cavity by an electron beam. The position at -2 mm is an example position. A lot of such positions are easily found in the diagram of Fig. 11.

6. Conclusions

The evaluated threshold currents for the coupled-bunch instabilities by the higher-order mode resonances in one bell-shaped single-cell cavity exceed a value of 100 mA. The cooperative coupled-bunch instabilities by multiple RF cavities can be suppressed both by modifying the radius at belly of each cavity systematically and by using the horizontal plunger and the vertical plunger. The resonant frequencies of the higher-order modes can be changed by the two plungers as keeping the frequency of TM010-like mode at 508.58 MHz. Therefore, the electromagnetic fields generated by an electron beam are prevented from resonating at the frequencies of the higher-order modes.

Acknowledgements
The authors would like to express their gratitude to Mr. K. Inoue, Mr. T. Kusaka of KOBE STEEL, LTD., and Mr. T. Yoshiyuki of TOSHIBA Co. for their many contributions to the SPring-8 project. Especially, they wish to thank Prof. Y. Yamazaki of the KEK for his encouragement and useful discussions throughout the present work.

Needle Cathodes as Sources of High-Brightness Electron Beams

C. A. Brau

Department of Physics, Vanderbilt University, Nashville, TN 37235, USA

Abstract. At the present time, the normalized brightness of electron beams available for use in free-electron lasers is of the order of 10^{11} A/m²-steradian. Needle cathodes emitting by field emission have demonstrated high current density, low electron temperature, and small emittance at peak currents exceeding 1 A. Photoelectric field emission can be used to control the pulse length. Measurements are underway to measure the emittance and electron energy spread of beams produced by photoelectric field emission.

Using beams with very high brightness, as much as five orders of magnitude beyond those in use now, it is possible to construct compact free-electron lasers at wavelengths from the far infrared to the ultraviolet. Cerenkov and Compton lasers are discussed in detail.

INTRODUCTION

The development of newer, shorter-wavelength and more powerful free-electron lasers has been paced by the development of better electron beams, that is electron beams with higher current and smaller emittance. There are two fundamental reasons for the need for smaller emittance. In the first place, the electron beam must be focused inside the laser beam for the interaction to take place. For the purposes of this discussion, we define the effective emittance ε by the formula(1)

$$\varepsilon = 4\pi\sqrt{\left\langle x^2 \right\rangle\left\langle x'^2 \right\rangle - \left\langle xx' \right\rangle^2} \,, \tag{1}$$

where x is the transverse position of an electron at the point z along the beamline, $x' = dx/dz$, and the brackets $\langle \; \rangle$ indicate an average over all the electrons in the beam. The edge of the beam ($1/e$ point of a Gaussian beam or edge of a top-hat beam) then satisfies the envelope equation(2)

$$\frac{d^2 w_e}{dz^2} = \frac{K}{w_e} + \frac{\varepsilon^2}{\pi^2 w_e^3} \,, \tag{2}$$

where the dimensionless perveance is(3)

CP413, *Towards X-Ray Free Electron Lasers*
edited by R. Bonifacio and W. A. Barletta
© 1997 The American Institute of Physics 1-56396-744-8/97/$10.00

$$K = \frac{1}{2\pi\beta^3\gamma^3}\frac{I_e}{I_0}, \tag{3}$$

$\beta = v/c$ is the velocity of the electrons normalized to that of light, $\gamma = 1/\sqrt{1-\beta^2}$ is the electron energy normalized to its rest energy, I_e is the total current, and $I_0 = \varepsilon_0 mc^3/e = 1356\,\mathrm{A}$ is a characteristic current, in which ε_0 is the permeability of free space (SI units are used throughout), m the electron mass and e the electron charge. If we ignore, for the moment, the space-charge term, the envelope of the focused beam has the form(4)

$$w_e = \overline{w}_e\sqrt{1+z^2/z_e^2}, \tag{4}$$

where the Rayleigh range z_e and beam waist $\overline{w}_e$ satisfy the relation

$$\pi\overline{w}_e^2 = \varepsilon z_e. \tag{5}$$

In the absence of gain guiding of the laser beam by the electron beam, the radius of the focused laser beam has the same form except that the wavelength λ_L plays the role of the emittance. Thus, to focus the electron beam inside the laser beam, we require that

$$\varepsilon < \lambda_L. \tag{6}$$

A second reason for needing small emittance follows from the longitudinal motions of the electrons. Due to the transverse motions of the electrons, they move as though they had an equivalent energy spread proportional to the emittance.(5) When this is compared with the energy acceptance of the FEL, it is found that the ratio depends, again, on the quotient ε/λ_L. The requirements change somewhat in the case of high-gain lasers, where gain guiding is generally important, but still apply in modified form.

Of course, the gain and power increase with the electron-beam current I_e, so a useful figure of merit for electron beams is the so-called brightness(6)

$$B = d^2 I_e/d\Omega dA, \tag{7}$$

where A is the area and Ω the solid angle through which the electrons pass. In terms of the effective emittance and total current, the brightness of a top-hat beam is approximately(7)

$$B = 2I_e/\varepsilon^2. \tag{8}$$

For beams in a time-independent field, the emittance decreases with increasing energy. Thus, it is useful to define the normalized effective emittance and the normalized brightness

$$\varepsilon_N = \beta\gamma\varepsilon, \tag{9}$$

$$B_N = \beta\gamma B = 2I_e / \varepsilon_N^2 , \tag{10}$$

which are generally only weakly dependent on the electron energy.

The earliest sources of electrons were dc thermionic cathodes. For a thermionic cathode of radius R and temperature T_e, the normalized brightness is(8)

$$B_N = mc^2 J_e / 2\pi k T_e , \tag{11}$$

where J_e is the current density at the cathode, k Boltzmann's constant, and T_e the electron temperature. Since the current density is generally limited to $J_e \sim 10^5$ A/m^2, and the temperature is typically about $T_e \sim 0.1$ eV, the brightness is limited to about $B_N \sim 10^{11}$ A/m^2 – steradian, although higher brightness can be achieved from small sources.

In rf linacs, the process of bunching a thermionic beam to higher peak current degrades the emittance. To avoid this, rf guns have been developed. With thermionic cathodes the beam must still be magnetically bunched to achieve high peak current, but peak brightness as high as 10^{11} A/m^2-steradian has been achieved.(9,10) Photocathodes avoid the bunching process, at some price in terms of system complexity and reliability, but brightness as high as 10^{11} A/m^2-steradian has been achieved at somewhat higher current.(11,12)

Storage rings scale differently than accelerators. Whereas the emittance of accelerators decreases as the energy increases, so that the normalized brightness is essentially frozen, the emittance of storage rings is formed by radiative processes whose photon energy increases with the electron energy, causing the energy spread and emittance of the electron beam to increase.(13) It is not difficult to show that the normalized emittance and relative energy spread of a conventional isomagnetic storage ring are(14)

$$\varepsilon_N = \frac{495\sqrt{6}}{864} \frac{\lambda_C}{v_x^3} \frac{R}{\rho} \gamma^3 , \tag{12}$$

$$\frac{\langle \Delta E^2 \rangle}{\langle E \rangle^2} = \frac{495\sqrt{3}}{3456\pi} \frac{\lambda_C}{\rho} \gamma^2 , \tag{13}$$

where R is the mean radius of the ring, ρ the orbit radius in the bend magnets, v_x the betatron number, and $\lambda_C = h/mc$ the Compton wavelength, in which h is Planck's constant. Typically $R/\rho = O(1)$, and the betatron number (the number of transverse oscillations about the nominal orbit described by an electron in a single circuit around the ring) is a small number, $1 < v_x < 10$, so for $\gamma \sim 2000$ the normalized emittance is $\varepsilon_N \sim 10^{-4}$ π m - radians, and increases with γ^2. The emittance in the vertical direction is generally somewhat smaller, and depends on the coupling between the horizontal and vertical motions. The brightness depends on the peak current, which depends on a variety of nonlinear effects such as the Touschek

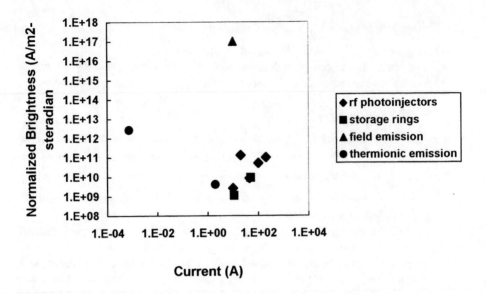

FIGURE 1. The normalized brightness of various sources of electron beams. Conventional sources group around a brightness of 10^9 to 10^{11} A/m^2-steradian.

effect and a host of instabilities. In general, the peak current improves with increasing energy, and peak current $I_e \sim 100$ A has been achieved with a normalized brightness $B_N \sim 10^{10}$ A/m^2 – steradian .(15)

The status of these three technologies is summarized in Fig. 1. We see there that the highest brightness among conventional sources is still achieved by dc thermionic technology, albeit at low current. Storage rings were once the best technology for achieving high brightness at high current, and still offer the most graceful means of obtaining high energy. Nevertheless, rf linacs now provide higher brightness, and their performance is constantly improving.

However, the most interesting point shown in Fig. 1 is that of the field-emission cathode near the top of the graph. As discussed below, field-emission cathodes have demonstrated high current density, small energy spread, and (in pulses) high peak current. Recently discovered photoelectric field-emission cathodes have produced comparable current in well-controlled short pulses. The prospect, therefore, is for an increase of five or more orders of magnitude in the normalized brightness of electron sources for free-electron lasers.

FIELD EMISSION

Near the tips of sharp needles, the electric field can take on very high values even for modest voltages. For an elliptical needle protruding from an infinite plane

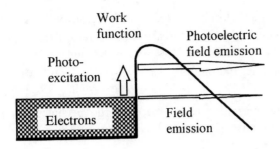

FIGURE 2. Schematic diagram illustrating the mechanisms for field emission and photoelectric field emission.

cathode into a uniform electric field E_0, the field can be computed using elliptical coordinates.(16) When the tip radius R_{tip} is small compared with the distance to the anode, the equipotential corresponding to the anode is nearly flat, and the field at the tip of the needle is

$$E_{tip} = E_0 \frac{a^2}{b^2} \frac{\eta^3}{\arctan\eta - \eta} \approx E_0 \frac{2a/R_{tip}}{\ln(4a/R_{tip})}, \qquad (14)$$

where a is the length of the needle, b the diameter of the needle at its base, $\eta = \sqrt{1 - b^2/a^2}$, and $R_{tip} = b^2/a$ is the radius of the tip of the needle. The approximation is valid for sharp needles, $a/b \gg 1$. For example, for a tip radius $R_{tip} \sim 1\,\mu m$, a needle length $a \sim 1\,cm$, and an anode-cathode separation of 3.3 cm, at 50 kV ($E_0 \sim 1.5 \times 10^6$ V/m), the electric field at the tip of the needle is $E_{tip} \sim 3 \times 10^9$ V/m.

Around the turn of the century, anomalous emission of electrons was observed under such conditions. A satisfactory explanation was finally developed when the concept of quantum mechanical tunneling was introduced, as shown in Fig. 2.(17) When the electric field is very strong, the potential barrier at the surface of the metal becomes thin enough for the electrons near the Fermi level to penetrate through the classically forbidden region. The quantum mechanical theory was first derived by Fowler and Nordheim, who showed that the current density at low temperatures (room temperature and below) is given by the formula bearing their names:(18)

$$J_e = 1.54 \times 10^{-6} \frac{E^2}{\varphi} \exp\left[-6.83 \times 10^9 \frac{\varphi^{3/2}}{E} f(y)\right], \qquad (15)$$

where E is the electric field at the surface, φ the work function of the metal (in eV), and $y = 3.79 \times 10^{-5} \sqrt{\varphi}/E$. The function $f(y)$ is a dimensionless elliptical function introduced to account for image forces near the surface. It varies from 1 at $y = 0$ to

0 at $y = 1$, and is actually closely approximated by $\cos(\pi y / 2)$. Current is typically emitted over the tip of the needle out to about 30 degrees.(19) Using as an example the needle introduced above, we find that for a work function $\varphi \sim 4.5\,\mathrm{eV}$, which is characteristic of tungsten, the current density is $J_e \sim 7 \times 10^6\,\mathrm{A/m}^2$, and the total current is $I_e \sim 5\,\mu\mathrm{A}$

At sufficiently large current density, space charge becomes important. For a spherical geometry, the space-charge limit is(20)

$$ J_e = \frac{4\varepsilon_0}{9a^2} \sqrt{\frac{2e}{m}} \frac{V^{3/2}}{r_{cathode}}, \tag{16} $$

where V is the voltage, and a^2 is a slowly-varying function of the ratio of the anode and cathode radii. For $10^3 < r_{anode}/r_{cathode} < 10^4$, it is found that $5 < a^2 < 10$. However, the field-emission current density is very sensitive to the electric field at the surface, so space-charge effects become noticeable at current densities about an order of magnitude below this limit. The energy spread of the electrons is predicted to be of the order of a volt in the longitudinal direction, and smaller in the transverse direction.(21) At higher temperatures, the emission increases due to faster tunneling by electrons thermally excited to higher energy.(22)

The experimental results confirm the theoretical predictions in remarkable detail. The current density has been observed to follow the Fowler-Nordheim relation over more than six orders of magnitude, up to a current density $J_e \sim 10^{11}\,\mathrm{A/m}^2$.(23) Above this value, space-charge effects reduce the field at the surface. Nevertheless, a current density of $3 \times 10^{11}\,\mathrm{A/m}^2$ has been observed in microsecond pulses. The longitudinal energy distribution has been carefully measured, and confirms both the predicted shape and the predicted width of about 1 eV.(24) The largest total current which has been observed is 6.5 A, obtained in microsecond pulses from a cathode with a tip radius of 3 μm.(25) The emittance has not been measured directly. However, the resolving power of field-emission microscopes, about 3 nm, supports the predicted transverse electron temperature of less than 1 eV at low currents. If this persists to higher currents, the corresponding normalized brightness would exceed $10^{17}\,\mathrm{A/m}^2$-steradian.

Recently, it has been observed that shining a laser on the tip of a needle will turn on field emission at surface electric fields well below those otherwise required for significant emission.(26,27) Although no theoretical analysis has been developed, it is believed that the electrons are photoexcited by the laser to energy levels above the Fermi level where they can more easily tunnel through the potential barrier. This is illustrated in Fig. 2. The results are very promising for laser applications. The largest total current observed so far is $I_e \sim 2\,\mathrm{A}$, extracted in nanosecond pulses from needles with $R_{tip} \sim 50\,\mathrm{nm}$. The corresponding current density is $J_e \sim 10^{14}\,\mathrm{A/m}^2$, which is near the space-charge limit. The quantum efficiency for this process, based on the photons absorbed by the shiny metallic surface, is about unity for photons in the visible part of the spectrum, and near unity

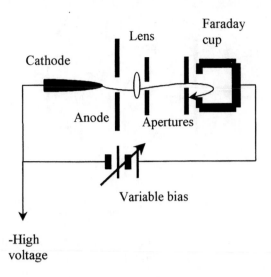

FIGURE 3. Schematic diagram of the energy-spread measurement showing the apertures used to restrict the beam focused into the Faraday cup.

for 1-μm photons. Similar results were obtained with picosecond laser pulses. However, no measurements were made of the electron temperature or the emittance at either pulse length.

To determine the electron temperature and emittance of beams produced by photo-field emission, experiments are being undertaken at Vanderbilt. A schematic diagram of the energy-spread measurement is shown in Fig. 3. The electrons are collected in a Faraday cup connected to the cathode through a variable bias of the order of a few volts. As the bias is increased, electrons with increasing energy are reflected from the cup. By differentiating the current collected as a function of the bias voltage, the electron energy distribution function is obtained. The resolution of the measurement is limited by the focusing of the electrons into the Faraday cup. This introduces a transverse component of the energy

$$E_\perp / E \approx p_\perp^2 / p^2 \approx \vartheta^2, \tag{17}$$

where $p_\perp$ and p are the transverse and total momenta, respectively, and ϑ is the angle of the electron trajectory from the axis. To achieve an energy resolution better than 100 meV, which corresponds to about ten parts per million at a total energy of 10 keV, the diameter of the aperture at the focusing lens (1 m from the focus) must be less than 6 mm. The electrons are also deflected transversely by the fields at the entrance of the Faraday cup. Simulations indicate that the energy resolution is roughly 0.2 eV out of 10 keV for a 50-μm (radius) electron beam entering the Faraday cup. The electron beam size is determined principally by the spherical aberration of the solenoidal lens, for which the minimum circle of confusion is(28)

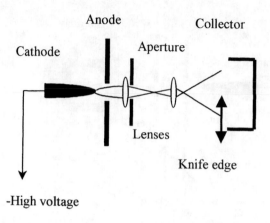

Anode

Collector

Cathode

Aperture

Lenses

Knife edge

-High voltage

FIGURE 4. Schematic diagram of the emittance measurement showing the magnification of the focal spot to achieve adequate resolution at the knife edge.

$$W_{aberration} = \frac{5}{128} \left(\frac{w_{lens}}{R_{lens}} \right)^2 \frac{z_i}{f_{lens}} \left(1 + \frac{224}{45\pi} \frac{R_{lens}}{f_{lens}} \right) w_{lens},$$ (18)

where w_{lens} is the beam radius at the lens, R_{lens} the lens winding radius, z_i the image distance (distance from the lens to the focus), and f_{lens} the focal length of the solenoidal lens The lens has a a winding radius $R_{lens} \sim 5\,cm$, and brings the electron beam from an effective source 30 cm in front of the lens to a focus at $z_i \sim 1\,m$. For an aperture at the lens $w_{lens} \sim 3\,mm$, the circle of confusion is $w_{aberration} \sim 2\,\mu m$. This is larger than the spot size based on the emittance alone, which is a problem for the emittance measurements, but well within the 50-μm acceptance of the Faraday cup. The total current transmitted by the aperture at the lens is expected to be about 100 nA out of a total emitted current of 30 μA.

A schematic diagram of the emittance measurement is shown in Fig. 4. Again, the solenoidal lens is used to focus the beam at a point $z_i \sim 1\,m$. The emittance-limited spot radius here is $w_e \sim 0.2\,\mu m$. The beam is then magnified by a second solenoidal lens 1 cm in radius, to a spot radius $w_e \sim 10\,\mu m$ at a point 1 m beyond the second lens. At this point, the beam is large enough to be scanned across a knife edge and the current measured by the Faraday cup. The emittance is determined in the usual way, by varying the current in the second lens and observing the spot size. The principle source of error is spherical aberration in the first lens. To avoid this, the radius of the aperture at the first lens must be reduced to $w_e \sim 1\,mm$. The total current at the Faraday cup is correspondingly reduced to about 10 nA.

The initial experiments are being conducted using a 4-W cw argon laser to generate total currents up to 30 µA. This avoids problems due to space charge. Later measurements will use a Q-switched Nd:Yag laser to generate higher peak currents. The effect of space charge at the cathode is discussed above, and may alter the actual brightness and energy spread of the electron beam. It may also limit the resolution of the energy-spread and emittance measurements. To avoid reflections from the Faraday cup in the energy-spread measurements, the beam must be focused to a spot diameter smaller than 100 µm, as discussed above. By integrating Equation (2), ignoring the emittance term, we find that the envelope has the form(29)

$$z = w_e \sqrt{2/KF} \left[\sqrt{\ln(w_e / \overline{w}_e)} \right], \tag{19}$$

where

$$F(x) = \exp(-x^2) \int_0^x \exp(u^2) du . \tag{20}$$

Thus, to form a 50-µm spot from a 3-mm aperture at a distance of 1 m, the focused current is limited to less than 10 µA. The corresponding total emitted current is of the order of 3 mA.

In the emittance measurement, the spherical aberration is reduced by restricting the aperture at the lens to 1 mm. The spot size based on the emittance is then about 0.5 µm. In this case we find that the focused current is limited to less than 0.4 µA, which coresponds to a total emitted current of about 1 mA.

LASER APPLICATIONS

With a five-order-of-magnitude increase int e brightness of the electron beam, it is possible to conceive of many new laser systems. We discuss here the Cherenkov free-electron laser and the Compton free-electron laser.

The principle of a Cherenkov free-electron laser is illustrated in Fig. 4. The optical wave is guided by a thin dielectric film on a conducting surface, and the index of refraction of the dielectric slows the phase velocity of the wave to synchronism with the electrons. The electron beam interacts with the evanescent part of the wave which, since it is not a plane wave, has a longitudinal component of the electric field. When the gain is very large, it may be expressed by the formula(30)

$$G_0 = \tfrac{1}{9} \exp \left[(\tfrac{1}{2} j_e)^{1/3} \sqrt{3} \right], \tag{21}$$

where j_e is the dimensionless current density. In the one-dimensional approximation this is given by the expression(31)

$$j_e = \frac{2}{1+\chi} \frac{eI_e}{\varepsilon_0 mc^3} \frac{k_L^2 L^3}{\beta^5 \gamma^6 \sigma_x} e^{-2k_L h/\beta\gamma}, \tag{22}$$

in which $k_L = 2\pi / \lambda_L$ is the laser wavenumber in vacuum, L the length of the laser, h the height of the electron beam above the surface of the dielectric, and σ_x the width of the laser beam, presuming that the electron beam is narrower than the laser beam. The factor χ is the ratio of the energy propagating through the dielectric to that propagating through the evanescent wave, and is given by the formula

$$\chi = \frac{\tan\theta}{n^4}\frac{2\theta + \sin 2\theta}{1 + \cos(2\theta)}, \tag{23}$$

where

$$\theta = \arctan\left(\frac{n^2}{\gamma\sqrt{n^2\beta^2 - 1}}\right). \tag{24}$$

To reach saturation in a single pass, beginning from noise, the overall gain must be very large. For conventional free-electron lasers, Kim finds that saturation occurs when $j_e \sim 2(4\pi)^3 \approx 4000$, which corresponds to $G \approx e^{20}$.(32) In the high-gain regime, diffraction spreading of the laser beam in the transverse direction is controlled by gain guiding.(33) In this case, the effective width of the optical beam corresponds to a Rayleigh range of the order of the gain length, $L_{gain} = L/\ln G_0$, so that $\sigma_x \sim \sqrt{2\beta L / k_L \ln G_0}$. It is, however, important to focus the electron beam inside the evanescent part of the laser beam, which falls off exponentially with a scale height $s = \beta\gamma\lambda_L / 4\pi$. The radius of the electron beam at the focal point is $w_e = \sqrt{\varepsilon L / 2\pi}$, which must be smaller than the scale height. Thus, the Cherenkov free-electron laser is most useful at long wavelengths, typically in the far infrared and beyond.

To illustrate these ideas with an example, we examine a laser $L \sim 5$ cm in length, operating at a wavelength $\lambda_L \sim 250\mu$m with an electron energy of 50 keV. The scale height of the evanescent wave is $s \sim 10\mu$m. Using an e-beam with a total current of $I_e \sim 10$ mA and a normalized brightness $B_N \sim 10^{15}$ A/m^2 − steradian, the focused spot radius of the electron beam is $w_e \sim 10$ μm, which satisfactorily matches the scale height. The dimensionless current density is then $j_e > 4 \times 10^3$, which satisfies the requirement for reaching saturation in a single pass.

As another application, we consider the possibility of operating in the ultraviolet regime. If we use a laser as the wiggler, in which case the FEL is really using stimulated Compton backscatter, then the laser wavelength is given by the double Doppler shift(34)

$$\lambda_L = \frac{1-\beta}{1+\beta}\lambda_P, \tag{25}$$

where λ_P is the pump wavelength. The gain is again given by (21), but the dimensionless current density is(35)

$$j_e = 4\pi\frac{e}{\varepsilon_0 mc^3}\frac{a_P^2 L^3 J_e}{\gamma^3\beta^3\lambda_P}, \tag{26}$$

in which L is the length of the interaction region, and

$$a_P^2 = \frac{e^2}{4\pi^2 \varepsilon_0 m^2 c^5} \lambda_P^2 I_P \qquad (27)$$

is the dimensionless wiggler vector potential, where I_P is the pump laser intensity. Although these expressions are derived using the one-dimensional approximation, they should be satisfactory for the case of interest here. The diffraction-limited spot size of the UV beam in the high-gain regime is controlled, again, by gain guiding. In this case, the diffraction-limited spot corresponds to a Rayleigh range of the order of the gain length, that is, to a radius $w_0 \sim \sqrt{\lambda_x L / \pi \ln G_0}$. But from the envelope equation we see that the radius of the beam at the focus is $\overline{w}_e \sim L\sqrt{K}/2$, which is generally larger than the diffraction-limited laser beam. Under these circumstances, the actual laser spot size approximates that of the electron beam, validating the one-dimensional approximation. The current density is then $J_e = I_e / \pi w_e^2 \sim 8\beta^3 \gamma^3 I_0 / L^2$, independent of the total current. Unfortunately, the length of the interaction region cannot be made arbitrarily long because quantum effects become important and lower the gain.(36) Electron bunching occurs on a length scale of the order of $L\sqrt{3}/\ln(9G_0)$.(37) If we argue that the electron phase shift in one bunching length caused by the electron recoil must be less than unity, then we obtain the restriction

$$\frac{4\pi\sqrt{3}}{\ln(9G_0)} \frac{1+\beta}{\beta\gamma} \frac{\lambda_c L}{\lambda_P \lambda_L} \le 1. \qquad (28)$$

Combining these ideas, we arrive at the formula

$$j_e \sim 16 a_P^3 \left(\frac{\lambda_L}{\lambda_c} \frac{\beta\gamma}{1+\beta} \right)^{3/2} \qquad (29)$$

for the optimized performance of a Compton free-electron laser. To reach saturation, we again require that $j_e \sim 2(4\pi)^3$. For an electron energy of 150 keV ($\beta \sim 0.6$), and a pump wavelength $\lambda_P = 1.064 \ \mu m$, the laser wavelength is $\lambda_L \sim 0.24 \ \mu m$, which is in the ultraviolet. The minimum pump laser intensity is $I_e = 2 \times 10^{19} \ W/m^2$, but the electron-beam current is only $I_e \sim 2 \ mA$. Pump laser intensities of this magnitude are available from high-power glass laser systems. The difficulty is that the intensity must be constant in time and space. Although the dimensionless wiggler intensity is only $a_P^2 \sim 10^{-3}$, the wiggler is quite long ($N_W \sim 10^5$ wiggler periods), so the laser intensity must be constant to about 10 percent, even allowing for the expansion of the laser bandwidth by the high gain.

CONCLUSION

Due to the enormous current density possible by using field emission from needle cathodes, it appears possible to extend the brightness of electron sources for free-electron lasers by five orders of magnitude. Photoelectric field emission has demonstrated nanosecond control of electron current in excess of 1 A, but the detailed properties of these sources (energy spread and emittance) have not been measured. Experiments are underway to determine these properties at total currents up to about 1 mA. Beyond this point, space-charge makes it impossible to focus the beams as required to resolve the expected energy spread and emittance.

Using beams with normalized brightness as high as 10^{15} to 10^{17} A/m²-steradian, it is possible to build compact lasers at wavelengths from the infrared to the ultraviolet. For example, a Cerenkov free-electron laser operating at a wavelength of 250 μm can reach saturation in a single pass at a total electron-beam current of only 10 mA. Similarly, a Compton free-electron laser operating at a wavelength of 240 nm can reach saturation in a single pass at a total electron-beam current of only 2 mA.

REFERENCES

1. Lawson, J. D., *The Physics of Charged-Particle Beams*, Oxford: Clarendon Press, 1988, p. 192. I include the factor of π, making this the phase-space area. Then the emittance becomes equivalent to the wavelength of a light beam.

2. Lawson, J. D. ., *The Physics of Charged-Particle Beams*, Oxford: Clarendon Press, 1988, p. 176.

3. Lawson, J. D. ., *The Physics of Charged-Particle Beams*, Oxford: Clarendon Press, 1988, p. 117.

4. Brau, C.A., *Free-Electron Lasers*, Boston: Academic Press, 1989, p. 173.

5. Brau, C.A., *Free-Electron Lasers*, Boston: Academic Press, 1989, p. 184.

6. Lawson, J. D. ., *The Physics of Charged-Particle Beams*, Oxford: Clarendon Press, 1988, p. 160.

7. Brau, C.A., *Free-Electron Lasers*, Boston: Academic Press, 1989, p. 164.

8. Brau, C.A., *Free-Electron Lasers*, Boston: Academic Press, 1989, p. 381.

9. Westenkow, G. A. and Madey, J. M. J., *Lasers and Part. Beams* **2**, 223 (1984).

10. Madey, J. M. J., private communication.

11. Fraser, J. S. and Sheffield, R. L., *IEEE J. Quant. Electron.* **QE-23**, 1489 (1987).

12. O'Shea, P. G., *et al.*, *Nucl. Inst. Meth.* **A341**, 7-11 (1994).

13. Sands, M., "The Physics of Electron Storage Rings. An Introduction", in Touschek, B, *Physics with Intersecting Storage Rings*, New York: Academic Press, 1971, pp. 257-411.

14. Brau, C. A., unpublished.

15. Litvenenko, V. N., *et al.*, *Nucl. Inst. Meth.* **A375**, 46-52 (1996).

16. Landau, L. D. Lifshitz, and Pitaevskii, L. P., *Electrodynamics of Continuous Media*, Oxford: Pergamon Press, 1984, p. 27.

17. Fowler, R. H., and Nordheim, L. W., *Proc. Roy. Soc.* **A119**, 173 (1928).

18. Dyke, W. P., and Dolan, W. W., "Field Emission", in Marton, L., *Advances in Electronics and Electron Physics, Volume III,*New York: Academic Press, 1956, pp. 89-185.

19. Dyke, W. P., and Dolan, W. W., "Field Emission", in Marton, L., *Advances in Electronics and Electron Physics, Volume III,*New York: Academic Press, 1956, p. 104.

20. Langmuir, I., and Blodgett, K, *Phys Rev.* **24**, 49 (1924).

21. Mueller, E. W., *Z. Physik* **120**, 261 (1943).

22. Dolan, W. W., and Dyke, W. P., *Phys. Rev.* **95**, 327 (1954).

23. Dyke, W. P., and Dolan, W. W., "Field Emission", in Marton, L., *Advances in Electronics and Electron Physics, Volume III,* New York: Academic Press, 1956, p. 106.

24. Mueller, E. W., *Z. Physik* **120**, 261 (1943).

25. Dyke, W. P., and Dolan, W. W., "Field Emission", in Marton, L., *Advances in Electronics and Electron Physics, Volume III,* New York: Academic Press, 1956, p. 168.

26. Bousoukaya, M., *et al., Nucl. Inst. Meth.* **A279**, 405-409 (1989).

27. Ramian, G., and Garate, E., presented at the 16th International Free-Electron Laser Conference, Stanford, CA, August 23, 1994.

28. Hawkes, P. W., and Kasper, E., *Principles of Electron Optics, Volume 1*, London: Academic Press, 1989, Section 24.3.

29. Lawson, J. D. ., *The Physics of Charged-Particle Beams*, Oxford: Clarendon Press, 1988, Section 3.2.8.

30. Colson, W. B.,"Classical Free-Electron Laser Theory," in Colson, W. B., *et al., Laser Handbook Volume 6*, Amsterdam: North-Holland, 1990, p. 143.

31. Brau, C. A., unpublished .

32. Kim, K.-J., *Nucl. Inst. Meth.* **A250**, 396 (1986).

33. Moore, G. T., *Opt. Commun.* **52**, 46 (1984).

34. Landau, L. D., and Lifshitz, E. M., *The Classical Theory of Fields*, Oxford: Pergamon Press, 1975, Section 48.

35. Brau, C. A., unpublished.

36. Brau, C. A., *Free-Electron Lasers*, Boston: Academic Press, 1989, p. 55.

37. Colson, W. B.,"Classical Free-Electron Laser Theory," in Colson, W. B., *et al., Laser Handbook Volume 6*, Amsterdam: North-Holland, 1990, p. 143.

Coherent Radiation of an Electron Bunch Moving in an Arc of a Circle

E.L. Saldin[1], E.A. Schneidmiller[1], M.V. Yurkov[2]

[1] *Automatic Systems Corporation, 443050 Samara, Russia*
[2] *Joint Institute for Nuclear Research, Dubna, 141980 Moscow Region, Russia*

Abstract.

The paper presents some analytical results of the theory of coherent synchrotron radiation (CSR) describing the case of finite curved track length.

INTRODUCTION

Analysis of project parameters of linear colliders [1,2] and short-wavelength FELs [1,3,4] shows that the effects of coherent synchrotron radiation of short electron bunches passing bending magnets influence significantly on the beam dynamics (see, e.g., [5,6]). The first investigations in the theory of coherent synchrotron have been performed about fifty years ago [7–9]. In these papers the main emphasis was put on the calculations in far zone of CSR produced by a bunch of relativistic electrons moving on a circular orbit. Another part of the problem, namely that of the radiative interaction of the electrons inside a bunch has been studied for the first time in refs. [10,11] and later in refs. [5,12] where the energy loss along the bunch has been calculated. The results of the above mentioned CSR theories are valid for a model situation of the motion of an electron bunch on a circular orbit and do not describe the case of an isolated bending magnet. The first analytical results describing this case have been presented in ref. [13]. In particular, analytical expressions have been obtained for the radiative interaction force, for the energy loss distribution along the bunch and for the total energy loss of the bunch. The criterium for the applicability region of the previous theories to the case of a finite magnet length has been derived. In this report some analytical results of ref. [13] are presented.

CP413, *Towards X-Ray Free Electron Lasers*
edited by R. Bonifacio and W. A. Barletta
© 1997 The American Institute of Physics 1-56396-744-8/97/$10.00

RESULTS FOR A RECTANGULAR BUNCH

Let us consider a rectangular bunch of the length l_b passing a magnet with the bending angle ϕ_m and the bending radius R. We use the model of ultrarelativistic electron bunch with a linear distribution of the charge (zero transverse dimensions) and assume the bending angle to be small, $\phi_m \ll 1$. We neglect the interaction of the bunch with the chamber walls assuming the electrons to move in free space. The total number of particles in the bunch is equal to N and the linear density is equal to $\lambda = N/l_b$.

When the electron bunch passes the magnet, the electromagnetic field slips over the electrons due to the curvature and the difference between the electron's velocity and velocity of light c. The slippage length L_{sl} is given by the expression:

$$L_{sl} \simeq \frac{R\phi_m}{2\gamma^2} + \frac{R\phi_m^3}{24} , \tag{1}$$

where γ is relativistic factor. When applying the results of steady-state CSR theory (periodical circular motion) to the case of isolated magnet it is assumed usually that the bunch length is much shorter than the slippage length. To obtain more correct criterium for the applicability region one has to develop more general theory including transient effects when the bunch enters and leaves the magnet. Such an investigation has been performed in ref. [13]. In particular, it has been stressed that the radiation formation length of the order of $l_b\gamma^2$ before and after magnet plays an important role in CSR effects. In practically important case when the conditions $\gamma\phi_m \gg 1$ and $R/\gamma^3 \ll l_b < L_{sl}$ are satisfied, the expression for the total energy loss of the bunch can be written in the following form [13]:

$$\Delta\mathcal{E}_{tot} = -\left(\frac{3^{2/3}e^2N^2}{l_b^{4/3}R^{2/3}}\right)(R\phi_m)$$

$$\times \left\{1 + \frac{3^{1/3}4}{9}\frac{l_b^{1/3}}{R^{1/3}\phi_m}\left[\ln\left(\frac{l_b\gamma^3}{R}\right) + C\right]\right\}, \tag{2}$$

where e is the charge of the particle and

$$C = 2\ln 2 - \frac{1}{2}\ln 3 - \frac{11}{2} \simeq -4 .$$

The first term in Eq. (2) is the solution obtained in the framework of steady-state approach (see, e.g., refs. [9,11,5]). Therefore, with logarithmical accuracy we can set the applicability region of the results of the steady-state theory for the case of a finite curved track length:

$$\frac{l_b^{1/3}}{R^{1/3}\phi_m}\ln\left(\frac{l_b\gamma^3}{R}\right) \ll 1 . \tag{3}$$

In particular, the steady-state theory provides completely incorrect results for the case of the electron bunch much longer than the slippage length, $l_b \gg L_{sl}$. In this case the energy losses of the particles in the bunch due to CSR are proportional to the local linear density and take place mainly after the magnet [13]. For a "short" magnet, $\gamma \phi_m \ll 1$, the total energy loss of rectangular bunch is equal to

$$\Delta \mathcal{E}_{tot} = -\frac{2}{3} \frac{e^2 N^2}{l_b} \gamma^2 \phi_m^2 . \tag{4}$$

The energy loss of the rectangular bunch passing a "long" magnet, $\gamma \phi_m \gg 1$, is equal to:

$$\Delta \mathcal{E}_{tot} = -\frac{N^2 e^2}{l_b} [4 \ln(\gamma \phi_m) - 2] . \tag{5}$$

These results has been obtained in ref. [13] by means of calculation the radiative interaction of the electrons in the bunch. It is interesting to compare the total energy loss of the bunch with the energy of coherent radiation in far zone. The radiation energy in far zone can be calculated as an integral over frequency of the spectral density of the radiation energy:

$$\frac{dW_{coh}}{d\omega} = N^2 \eta(\omega) \frac{dW}{d\omega} , \tag{6}$$

where $\eta(\omega)$ is the bunch form factor (squared module of the Fourier transform of the linear density distribution). The form factor for the rectangular bunch of the length l_b is given by the expression:

$$\eta(\omega) = \left(\sin \frac{\omega l_b}{2c} \right)^2 \left(\frac{\omega l_b}{2c} \right)^{-2} . \tag{7}$$

Function $dW/d\omega$ entering Eq. (6) is the spectral density of the radiation energy of a single electron. The angular and the spectral characteristics of the radiation of an electron moving in an arc of a circle have been studied in ref. [14][1]. It has been shown that the spectrum of the radiation emitted by an electron moving in an arc of a circle differs significantly from that of conventional synchrotron radiation of an electron executing periodical circular motion. In the latter case the spectral density at low frequencies is proportional to $\omega^{1/3}$ [16]. In the case of a finite curved track length the spectral density is constant at $\omega \to 0$. When the bending angle is small, $\phi_m \ll 1$, the spectral density of the radiation energy emitted by ultrarelativistic electron is function of the only parameter $\gamma \phi_m$ [14]:

---

[1] The same problem has been considered later in ref. [15], but the results of this paper are incorrect.

$$\frac{dW}{d\omega} = \frac{e^2}{\pi c} f_m \ ,$$

(8)

where

$$f_m = \left(\mu + \frac{1}{\mu}\right) \ln \frac{1+\mu}{1-\mu} - 2 \ ,$$

and

$$\mu = \frac{\gamma \phi_m / 2}{\sqrt{1 + (\gamma \phi_m / 2)^2}} \ .$$

Formula (8) is valid in the frequency range $\omega \ll c/L_{sl}$. Taking into account formula (7) we can estimate that typical frequencies of the coherent radiation are below the frequency $\omega \sim c/l_b$. It means that we can use the asymptotical expression (8) in the case when $l_b \gg L_{sl}$. Integrating Eq. (6) over the frequency, we obtain:

$$W_{coh} = \frac{e^2 N^2}{l_b} f_m \ .$$

(9)

It is easy to obtain that in the case of a "long" magnet, $\gamma \phi_m \gg 1$, the energy of coherent radiation (9) coincides exactly with the bunch energy loss given by Eq. (5) taken with opposite sign. In the limit of a "short" magnet, $\gamma \phi_m \ll 1$, there is also complete agreement between formulae (9) and (4).

BUNCH WITH AN ARBITRARY DENSITY PROFILE

The solutions obtained in ref. [13] for the rectangular bunch can be generalized for the case of an arbitrary linear charge density. We present here the results of the calculation of the transition process when the bunch enters the magnet. Let the bunch have the density distribution $\lambda(s)$ which satisfies the condition

$$\frac{R}{\gamma^3} \frac{d\lambda(s)}{ds} \ll \lambda(s) \ .$$

(10)

Under this condition the rate of the energy change of an electron is given by the expression [13]:

$$\frac{d\mathcal{E}(s,\phi)}{d(ct)} = -\frac{2e^2}{3^{1/3} R^{2/3}} \left\{ \left(\frac{24}{R\phi^3}\right)^{1/3} \left[\lambda\left(s - \frac{R\phi^3}{24}\right) - \right. \right.$$

$$\left. \lambda\left(s - \frac{R\phi^3}{6}\right) \right] + \int_{s-R\phi^3/24}^{s} \frac{ds'}{(s-s')^{1/3}} \frac{d\lambda(s')}{ds'} \right\} \ ,$$

(11)

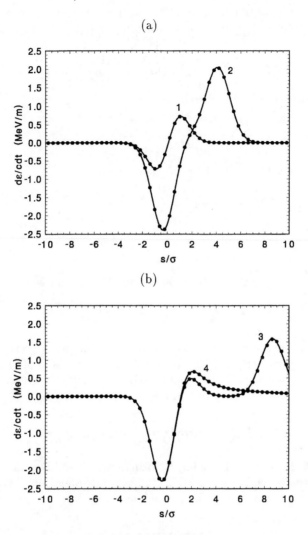

(a)

(b)

FIGURE 1. The rate of an electron energy change as a function of its position along the Gaussian bunch entering the magnet. Curve (1) in the graph (a): 5 cm after the begin of the magnet. Curve (2) in the graph (a): 14 cm after the begin of the magnet. Curve (3) in the graph (b): 18 cm after the begin of the magnet. Curve (4) in the graph (b): steady state. The curves are the results of calculations with formula (13) and the circles are the results of numerical simulations presented in ref. [6]. The parameters are as follows: $R = 1.5$ m, $\sigma = 50$ μm, $q = 1$ nC.

where s is the position of the electron in the bunch and ϕ is azimuthal angle. For the Gaussian density distribution:

$$\lambda(s) = \frac{N}{(2\pi)^{1/2}\sigma}\, exp\left[-\frac{s^2}{2\sigma^2}\right] ,\qquad(12)$$

expression (11) takes the form:

$$\frac{d\mathcal{E}}{d(ct)} = -\frac{2e^2 N}{3^{1/3}(2\pi)^{1/2}R^{2/3}\sigma^{4/3}}G(\xi,\rho) ,\qquad(13)$$

where function $G(\xi,\rho)$ is given by the expression:

$$\begin{aligned}
G(\xi,\rho) = \rho^{-1/3}\left[e^{-(\xi-\rho)^2/2} - e^{-(\xi-4\rho)^2/2}\right] \\
+ \int_{\xi-\rho}^{\xi}\frac{d\xi'}{(\xi-\xi')^{1/3}}\frac{d}{d\xi'}e^{-(\xi')^2/2} .
\end{aligned}\qquad(14)$$

Here $\xi = s/\sigma$ and $\rho = R\phi^3/24\sigma$. Function $G(\xi,\rho)$ reduces to

$$G(\xi,\rho) \simeq -\frac{9}{2}\xi\exp(-\xi^2/2)\rho^{2/3}$$

at $\rho \ll 1$. In the opposite case, at $\rho \to \infty$, expression (14) tends to the steady-state solution [11,5,12]. In Fig.1 we present the plot of function (13). One can see that there is excellent agreement of analytical results [13] and the results obtained by means of numerical simulation code [6].

ACKNOWLEDGMENTS

We wish to thank C.L. Bohn, R. Brinkmann, Ya.S. Derbenev, M. Dohlus, K. Floettmann, D. Jaroszynski, T. Limberg, H. Mais, A. Piwinski, J. Rossbach, V.D. Shiltsev and D. Trines for useful discussions.

REFERENCES

1. *Conceptual Design of a 500 GeV e+e− Linear Collider with Integrated X-ray Laser Facility* (Editors Brinkmann R., Materlik G., Rossbach J., Wagner A.), **DESY 97-048**, Hamburg, 1997.
2. *Zeroth-Order Design Report for the Next Linear Collider*, **LBNL-PUB-5424, SLAC Report 474, UCRL-ID-124161**, 1996.
3. Tatchyn R. et al., *Nucl. Instr. and Methods* **A375**, 274 (1996).
4. Rossbach J., *Nucl. Instr. and Methods* **A375**, 269 (1996).
5. Derbenev Ya.S., Rossbach J., Saldin E.L., Shiltsev V.D., *DESY Print* **TESLA-FEL 95-05**, Hamburg (1995).

6. Dohlus M., Limberg T., *DESY print* **TESLA-FEL 96-13**, Hamburg (1996).

7. Schiff L.I., *Rev. Sci. Instr.* **17**, 6 (1946).

8. Schwinger J., *On radiation by electrons in a betatron*, 1945 (unpublished), see references in [9].

9. Nodvick J.S., and Saxon D.S., *Phys. Rev.* **96**, 180 (1954).

10. Tamm I.E., *Report of the Institute of Physics, Acad. of Sciences*, USSR, 1948 (in Russian).

11. Iogansen L.V., and Rabinovich M.S., *Sov. Phys. JETP* **37(10)**, 83 (1960).

12. Murphy J.B., Krinsky S., and Glukstern R.L., "Longitudinal Wakefield for Synchrotron Radiation", Proc. of IEEE PAC 1995, Dallas(1995).

13. Saldin E.L., Schneidmiller E.A., and Yurkov M.V., *DESY Print* **TESLA-FEL 96-14**, Hamburg (1996).

14. Bagrov V.G., Ternov I.M., and Fedosov N.I., *Phys. Rev.* **D28**, 2464 (1983).

15. Wingham D.J., *Phys. Rev.* **D35**, 2584 (1987).

16. Schwinger J., *Phys. Rev.* **75**, 1912 (1949).

Bunch Compression and Beam Transport for SPring-8 Linac SASE

Kenichi Yanagida, Shinsuke Suzuki, Tsutomu Taniuchi,
Hiroshi Yoshikawa, Akihiko Mizuno, Hiroshi Abe,
Toshihiko Hori, Hironao Sakaki,
Takao Asaka and Hideaki Yokomizo

SPring-8, Kamigori, Hyogo, 678-12, Japan

Abstract. For SASE, a bunch compression and a beam transport system which follow the layout of the present SPring-8 linac are designed. This is the first order evaluation trying to preserve the emittances at an RF photocathode gun. There are two bunch compressors in the injector section. There are one energy compressor, 90° bending system and two bunch compressors in the final compression section. The RF gun generates electron beam which has the normalized transverse emittances of 1.5πmm·mrad, and the longitudinal emittance of 0.16πps·MeV. The final emittances before a undulator become 3.5πmm·mrad for x direction and 0.38πps·MeV respectively.

INTRODUCTION

The 1.3GeV SPring-8 linac was completed in August 1996 as the injector for the large synchrotron radiation facility and now provides the electron beam to the booster synchrotron. In October 1997 the use of public beamline of the storage ring will be started. Then the linac will be operated twice a day for the beam injection. In the rest of the time there are some plans of linac utilization for various applications. For example, a New SUBARU beam injection, an inverse Compton scattering, a parametrix X-ray generation and a slow positron generation are planned. Especially a single pass FEL operating in the Self Amplified Spontaneous Emission (SASE) mode is planned as the most interesting application. Fortunately there will be a large assembly hall and a long beam transport tunnel beside the linac for this experiment. These buildings will be completed in 1998.

A target wavelength is determined as 20nm at the beam energy of 690MeV. Because the linac was optimized only for the injector of the booster synchrotron, its beam characteristics are not adequate for the single pass FEL

CP413, *Towards X-Ray Free Electron Lasers*
edited by R. Bonifacio and W. A. Barletta
© 1997 The American Institute of Physics 1-56396-744-8/97/$10.00

without improvement. For example the conventional thermionic cathode, Y796 EIMAC, is used for the electron gun. After the 20nm FEL is realized, the shorter wavelength FEL, 4nm at the beam energy of 1.55GeV, will be challenged.

In this paper the beam transport system, including bunch or energy compressors, is presented as a first order evaluation using TRACE 3-D code. In order to minimize the change of the present linac, the main acceleration section should be preserved. On the other hand an injector section that contains an RF photocatode gun and bunch compressors will be added in the upstream of the main acceleration section. And a final compression section that containes an energy compressor, a 90° bending system and bunch compressors will be added in the downstream of the main acceleration section as shown in Fig. 1. Of course the present thermionic gun will be settled in parallel with the RF gun injector, and used by switching. In the final section of this paper, a laser system is also described.

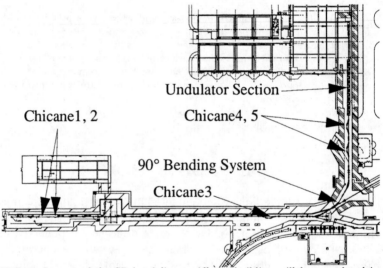

FIGURE 1. Layout of the SPring-8 linac. All the building will be completed in 1998. The the main accelerator section will be preserved. An injector section contains an RF gun and two bunch compressors. A final compression section containes an energy compressor, a 90° bending system and two bunch compressors.

I INJECTOR SECTION

In order to obtain a low emittance such as $\sim 1\pi$mm·mrad an RF photocathode gun is indispensable. For research and development the single cell

RF photocathode gun was designed. The designed parameters and calculated beam characteristics are shown in Table 1. This RF gun cavity is under construction and will be examined from April 1998. These values are not the final ones. However the calculation of transport was done using these values. In this paper value of bunch length and energy deviation are defined as ones of full width at half maximum.

TABLE 1. RF gun parameters and calculated beam characteristics.

Parameter or Characteristics	Value	Dimension
Number of Cells	1	
Field Gradient	150	MV/m
Input Power	18	MW
Beam Energy	3.4	MeV
Charge per Bunch	1.0	nC
Transverse Emittance	1.5	πmm·mrad
Bunch Length	10	ps
Energy Deviation	1.8	%

After the electron beam is generated at the RF gun, electrons must be accelerated rapidly. A regular acceleration tube with solenoid coils, denoted ACC-I1, succeeds the RF gun. Its distance between the gun and the ACC-I1 is designed as short as possible. Through the ACC-I1 the electron beam is accelerated up to 70MeV.

In the injector section there are two bunch compressors. One consists of ACC-I1 and Chicane1 and the other consists of ACC-H0 and Chicane2. Table 2 shows the elements and beam characteristics in the injector section.

TABLE 2. Elements and beam characteristics in the injector section.

Element	Beam Energy (MeV)	Norm. Emittance x (πmm·mrad)	Emittance z (πps·MeV)	Bunch Length (ps)
RF Gun	3.4	1.5	0.16	10
ACC-I1	70	1.5	0.40	10
Chicane1	70	3.0	0.40	4.6
ACC-I2	120	3.0	0.38	4.6
ACC-H0	170	3.0	0.39	4.6
Chicane2	170	3.3	0.38	2.3

The authors think this low energy section is very important. In the first acceleration tube, ACC-I1, longitudinal emittance growth occurs, because the nonlinearity of the acceleration voltage is relatively large to the initial energy deviation at the RF gun. Therefore bunch must be compressed and its energy deviation must be enlarged rapidly. For the preservation of the longitudinal emittance, the authors are studying the use of subharmonic, for example 1428MHz or 714MHz, acceleration tubes.

II MAIN ACCELERATION SECTION

Because the linac is a part of the SPring-8 accelerator complex, the change of the linac should be as small as possible. In order to minimize the change, the main accelerator section should be preserved. Especially the positions of acceleration tubes will not be changed. There are twenty five acceleration tubes. These are denoted as ACC-H1~H6 and ACC-M1~M19. In the main acceleration section, electron beam is accelerated up to the final energy, 690MeV.

III FINAL COMPRESSION SECTION

The final compression section consists of an energy compressor, a 90° bending system and two bunch compressors. In order to transport the electron beam to the assembly hall, 90° bending system of triple bend achromat is designed. The advantage of the triple bend achromat is the tunability of compression ratio, if we use it as a bunch compressor. However, in the initial stage, the 90° bending system woks as an isochronous transport system. The 90° bending system will be completed in 1998 for the other experiment in the assembly hall.

The energy compressor will be used to enlarge the bunch length. It is composed of Chicane3 and ACC-M20. Because it is difficult to transport the high current or high density beam, the peak current should be decreased using this energy compressor. The Chicane3 will be installed in January 1998 for measurement of energy stability.

After the beam is bent by 90°, the bunch is recompressed by two bunch compressors (ACC-L1, Chicane4 and ACC-L2, Chicane5). Using these bunch compressors beam characteristics can be processed to the required ones. In the case of 20nm SASE the energy deviation will be required as 0.5% FWHM ($\pm 0.25\%$). And corresponding bunch length and peak current become 0.44ps and ~2kA. Table 3 shows the elements and beam characteristics in the final compression section. Figures 2, 3 and 4 show the beam characteristics for all over the linac.

IV LASER SYSTEM

The RF photocathode gun system is now under construction. The first experiment of RF gun will be started from April 1998. In this section the laser system is described as a part of accelerator. The seed laser is the model 131 LIGHTWAVE. It is a 178.5MHz CW laser and has the nominal pulse width of 10ps FWHM. These pulse trains are sliced to the single pulse in the regenerative amplifier. Its energy per pulse will be 20mJ at 1047nm (fundamental),

302

TABLE 3. Elements and beam characteristics in the final compression section.

Element	Beam Energy (MeV)	Norm. Emittance x (πmm·mrad)	Emittance z (πps·MeV)	Bunch Length (ps)
Main Accel. Section	690	3.3	0.38	2.3
Chicane3	690	3.3	0.38	4.8
ACC-M20	690	3.3	0.38	4.8
90° Bend.	690	3.3	0.38	4.8
ACC-L1	690	3.3	0.38	4.8
Chicane4	690	3.3	0.38	2.2
ACC-L2	690	3.3	0.38	2.2
Chicane5	690	3.5	0.38	0.44

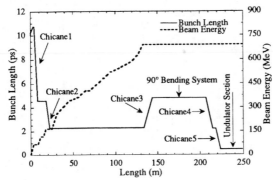

FIGURE 2. Beam energy and bunch length for all over the linac.

2mJ at 261nm (fourth harmonic) and 0.5mJ at 209nm (fifth harmonic) respectively. The fourth harmonic light will be used for copper cathode. The fifth harmonic light will be used for the research of new cathode material. For example diamond crystal has a negative work function with the excitation energy of 5.5eV.

The scheme of synchronization is shown in Fig. 5. The model 131 work as a master oscillator. The model 131 outputs the 178.5MHz sinusoidal signal. The signal is processed to the pulse of 175ps width, which is the half period of 2856MHz, triggered at the positive slope zero crossing point. These generated pulse trains are filtered by the band pass filter of 2856MHz, and then the acceleration frequency can be obtained.

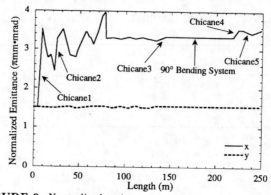

FIGURE 3. Normalized emittance x and y for all over the linac.

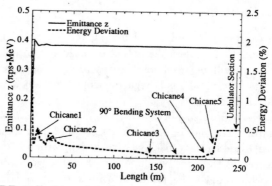

FIGURE 4. Emittance z and energy deviation for all over the linac.

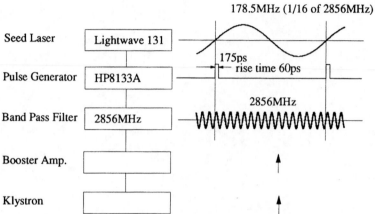

FIGURE 5. Scheme of synchronization between laser pulse and acceleration frequency.

Short Pulse Electron Beam Characteristics in an RF Gun with a Photocathode

Kai Masuda, Takashi Inamasu, Masaaki Sobajima,
Jiro Kitagaki, Kiyoshi Yoshikawa, Masami Ohnishi,
Yasushi Yamamoto and Hisayuki Toku

Institute of Advanced Energy, Kyoto University, Gokasho, Uji, Kyoto 611, Japan

Abstract. Electron trajectories in a 4½-cavity RF gun were calculated by a 2-dimensional simulation code we have developed with full Maxwellian equations with space charge-effects taken into account self-consistently, to evaluate short pulse electron beam charasteristics for FEL application. Especially, the beam emittance evolution was evaluated and, to reduce emittance growth in the RF gun, design of focusing magnetic field was carried out by use of the 2-D simulation code.

INTRODUCTION

As is well known, RF guns have the advantages over conventional electrostatic guns on such as (a) more rapid acceleration in excess of 100 MV/m, resulting in

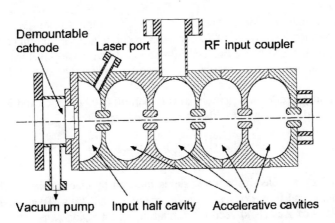

FIGURE 1. Schematic of an S-band 4½-cavity RF gun.

CP413, *Towards X-Ray Free Electron Lasers*
edited by R. Bonifacio and W. A. Barletta
© 1997 The American Institute of Physics 1-56396-744-8/97/$10.00

appreciable reduction of emittance growth due to space charge effects, (b) very compact structure in producing high current, low emittance beam, and (c) pre-bunched beam and longitudinal phase-space lending itself to magnetic compression. On the other hands, with thermionic cathodes, there are some disadvantages as well such as (a) electrons emitted from the cathode suffer RF defocusing effects in the RF gun, (b) wide energy spectrum extending from the peak energy down to zero, and (c) short operation period and short cathode life time due to backstreaming electron-induced heat.

To overcome these disadvantages, a various kinds of photo cathode RF guns have been developed successfully. However, in the application to short wavelength FELs, they still suffer from high emittance compared with the required normalized rms emittance of around $3\pi\text{mm} \cdot \text{mrad}(1)$, due to as high peak current as 2500A.

In this study, electron trajectories in a 4½ cavities RF gun with photo cathode were calculated by a 2-D simulation code we have developed, to evaluate electron beam charasteristics, especially, emittance growth in the gun. To reduce the emittance growth, design of focusing magnetic field was carried out by use of the simulation code.

PARAMETERS OF THE RF GUN

An S-band RF gun was developed by AET Associates Inc., having 4½ side-coupled cavities with a thermionic cathode as is shown in Fig.1, to provide simple method without additional accelerator tubes for IR FEL generation as was first made for Far-IR at Stanford (2). It was originally designed to provide about 4 MeV electron beam with 5 MW RF input.

SIMULATION CODE

To fully simulate transient behavior of the electron trajectories even for higher RF input power which are envisaged in our experiment, a 2-D simulation code was newly developed (3), by modifying the 2-D klystron code we have developed (4,5), which takes into account space charge-effects self-consistently with full Maxwellian equations.

The simulation code uses particle-in-cell method (2-D in space and 3-D in velocity coordinates) in time domain. Fields are calculated separately,

$$E = E_b + E_c, B = B_b + B_c + B_f, \tag{1}$$

where E_b and B_b are electromagnetic fields induced by electron beam, E_c and B_c are RF cavity fields and B_f is external focusing magnetostatic field. To calculate E_b and B_b, wave equations of scalar and vector potential in Lorentz gauge are fully solved at each time step by use of the finite element method (FEM) within quadratic rectangular cells.

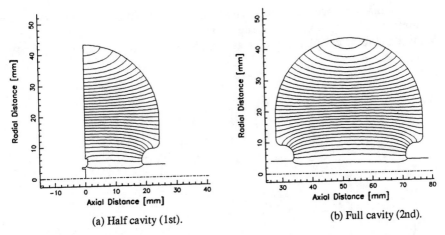

| (a) Half cavity (1st). | (b) Full cavity (2nd). |

FIGURE 2. Calculated electric field lines of eigenmodes in the RF gun cavities.

E_c and B_c are given by,

$$E_c(r,t) = E_{c0}(r) \operatorname{Re}\left[V_c e^{j\omega t}\right]$$

$$B_c(r,t) = \mu_0 H_{c0}(r) \operatorname{Re}\left[j\sqrt{\frac{\varepsilon_0}{\mu_0}} V_c e^{j\omega t}\right], \tag{2}$$

where $k_0 E_{c0} = \nabla \times H_{c0}$ is the eigenmode pattern of each cavity precalculated by use of FEM within quadratic triangular cells (see Fig. 2), and V_c is the cavity voltage which are given based on the electric fields measured by the bead-drop method. The electron emission current density is assumed to be constant over the cathode surface of a 6mm diameter.

SIMULATION RESULTS

Beam Characteristics with a Thermionic Cathode

The simulation code was applied to the 4½ RF gun with thermionic cathode, to evaluate performance characteristics of the originally designed gun for comparison with the photo cathode gun. The current density on the cathode surface is assumed to be 0.3A/mm² from the experimentally measured value.

The output beam is found to have a wide energy spectrum from the peak energy of 5.80MeV down to zero. The peak current and normalized rms emittance corresponding to $\beta\gamma > 0.9(\beta\gamma)_{peak}$ is calculated to be 21.5A (0.265nC, 12.3psec) and

4.58πmm·mrad respectively. It is also found that the high energy backstreaming electrons onto the cathode originate mainly in the 4th cavity, and backstreaming averaged power is calculated as high as 210 kW for total beam output of 7.34 MW.

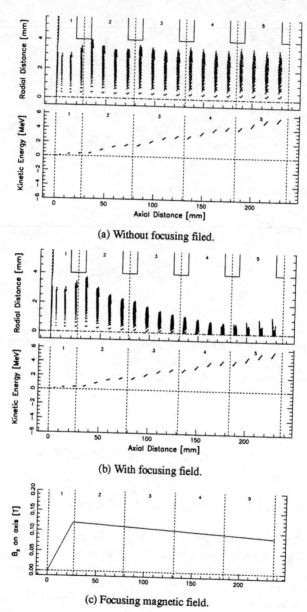

(a) Without focusing filed.

(b) With focusing field.

(c) Focusing magnetic field.

FIGURE 3. Beam snapshot of 1nC beam with 2psec laser pulse, comparing (a) without and (b) with focusing field shown in (c).

Beam Characteristics with a Photocathode

Electron trajectories in the RF gun were simulated with short pulse photo cathode which provides 1nC pulse beam with 2psec laser pulse, leading to peak current of 500A, which is to be compressed up to around 2500A by magnetic compression.

Fig. 3(a) shows beam snapshots in the RF gun with 2psec laser pulse. The emittance of output beam is calculated to be 13.3πmm·mrad which is bad compared with the final goal of around 3πmm·mrad and 4.58πmm·mrad for the lower current beam with thermionic cathode. It is also found that the bunch length of the output beam is 10.7psec, which is much longer than the laser pulse period of 2psec. All these result from large space charge effects due to high beam current.

Reduction of Emittance Growth with Focusing Magnetic Field

To reduce emittance growth in the RF gun, external focusing magnetic field was induced as shown in Fig. 3(c). Simulation results (Fig. 3(b)) indicate a reduced normalized rms emittance of 11.0πmm·mrad while it is still not low enough. Fig. 4 shows the emittance evolution with focusing field and without focusing. It is clearly seen that, without focusing, the emittance increases (and also decreases) near the cavity noses, where the cavity field E_r is not proportional to the radial distance r as shown in Fig. 5(a). On the other hand, with focusing field induced, the beam is not affected by the nonlinear field in the 3rd, 4th and 5th cavity, since it is focused within the region $r<2$mm (see Fig.3(b)), where $E_r \propto r$ as shown in Fig. 5(a).

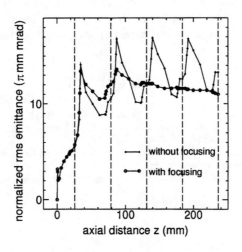

FIGURE 4. Emittance evolution in the RF gun.

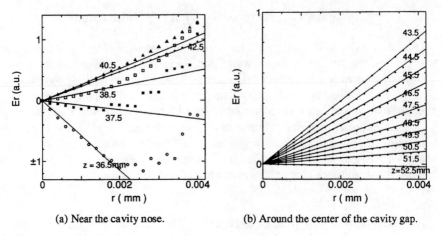

(a) Near the cavity nose.

(b) Around the center of the cavity gap.

FIGURE 5. Radial cavity field E_r as function of radial distance r (a) near the nose and (b) near the gap center (z=52.5mm).

SUMMARY

Electron trajectories in the 4½ cavity RF gun were calculated using newly developed 2-D simulation code, to evaluate beam characteristics, especially emittance growth in the gun. It is found that, without focusing magnetic field, 1nC, 10.7psec and 13.3πmm·mrad beam can be achieved. With focusing field induced, the normalized rms emittance is reducesd to 11.0πmm·mrad which results from the fact that the focused beam avoids being affected by the nonlinear cavity field. Furether improvement will be required to achieve the design goal of around 3πmm·mrad

REFERENCES

1. Kim, K. J., and Xie, M., *Nucl. Instr. Methods in Phys. Res.* **A331**, 359-364 (1993).
2. Huang, Y. V., Schmerge, J., Harris, J., Gallerano, G. P., Pantell, R. H., Feinstein, and J., *Nucl. Instr. Methods in Phys. Res.* **A318**, 765-771 (1992).
3. Inamasu, T., "Analysis of Electron Beam Characteristics in a Microwave Gun", BS thesis, Dept. of Electrical Engeneering, Kyoto University (1996), in Japanese.
4. Masuda, K., Yoshikawa, K., Ohnishi, M., Yamamoto, Y., and Sobajima, M., *Fusion Technology* **30**-3, 805-809 (1996).
5. Shintake, T., KEK Report **90**-3 (1990)

TREDI: A Self consistent three-dimensional integration scheme for RF-gun dynamics based on the Lienard-Wiechert potentials formalism

Luca Giannessi* and Marcello Quattromini*

*ENEA, Dipartimento Innovazione,
Divisione Fisica Applicata,
C.R. Frascati
Via E. Fermi 27, 00044 Frascati, Rome, Italy

Abstract. We describe the model for the simulation of charged beam dynamics in radiofrequency injectors used in the three dimensional code TREDI, where the inclusion of space charge fields is obtained by means of the Lienard-Wiechert retarded potentials. The problem of charge screening is analysed in covariant form and some general recipes for charge assignment and noise reduction are given.

INTRODUCTION

TREDI is a numerical code devoted to the simulation of charged electron beam dynamics in three spatial dimensions, in situations in which the space charge and radiative effects give a significant contribution to the dynamics of the beam itself and cannot be treated perturbatively. The development of TREDI was originally motivated by the necessity of simulating radiofrequency (rf) injectors in not-axisymmetric conditions. Deviation from the axial symmetry in rf injectors is due to inhomogeneity of photocathodes quantum efficiency, to laser misalignements or to multipolar rf fields in the accelerating cavities. These effects introduce a distortion from the axial symmetry that can be amplified by the nonlinear behavior of the system and that affects emittance compensation schemes inducing an emittance growth. Interest has been furthermore shown in asymmetric injectors with large transverse aspect ratio for future linear colliders applications. When the wavelength associated to the self consistent fields is of the same order of magnitude of the beam size itself, the code may in principle provide interesting informations even in other situations not strictly related to injectors dynamics. The code may be

CP413, *Towards X-Ray Free Electron Lasers*
edited by R. Bonifacio and W. A. Barletta

indeed used in the simulation of emittance diluition in beam compressors, in long wavelength ultrashort optical pulses production, coherent harmonic emission and superradiance in FELs. In this note we describe the code structure, we analyse the problem of fields regularization and finally we show some results obtained comparing the code predictions with those of existing and well established programs.

DESCRIPTION OF THE CODE

TREDI is a particles pushing code, based on the solution of the Lorentz force equation for each simulated macroelectron. The electromagnetic (em) fields are the superposition of two terms: the radiofrequency (rf) external fields, and the space charge (sc) fields produced by the beam itself. This approach is particularly efficient in the simulation of beam dynamics in three dimensions, due to the fact that the sc fields evaluation is obtained by means of the phase space particles coordinates, i.e. through the solution of ordinary differential equations, and are not the solution of partial differential equation (Maxwell's equations) requiring a set of constraints on the mesh size and time step to ensure numerical stability. Although the em fields are evaluated in both cases on a mesh, the mesh in the first approach covers a portion of space surrounding the beam, while in the other case it fills the whole resonating cavity, with a comparable discretization to ensure a correct reproduction of the fields on the region occupied by the electron bunch. In three spatial dimensions this leads, especially in the simulations of very short bunches where the high frequency fields contributions are more significant, to a huge number of mesh points and to very large memory requirements and computation time. In fig. 1 a flowchart of the program is shown. We assume a starting time t in which the particles positions and the relevant rf and sc fields are known.

- First: an adaptive stepsize method is used on a number of test particles to determine the time step length Dt ensuring the evaluation of the beam position at time t+Dt within a given accuracy. Different integration schemes can be used: 4th order Runge Kutta, the Burlish Stoer Method [1] or a third order symmetric split operator technique [2]. This method opposes to the disadvantage of a low order convergence in the stepsize, the advantage of being a phase space area invariant scheme thus preserving the beam emittance from numerical distortions.

- Second: The whole beam is advanced at time t+Dt according to the same method as that used to determine the step length. The space and time scale of variation of the force arising from either the rf and the sc fields is substantially different. During a single step, each particle is advanced independently from all the others. Typically, rf fields are characterized by a precise functional dependence both on time and space coordinates,

Tredi Flowchart

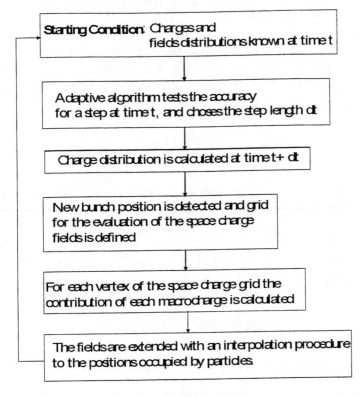

FIGURE 1. TREDI flowchart

while sc fields concern mainly the "internal dynamics" of the beam, which is almost invariant under displacements. These considerations allow the integration with considerably long steps (a large fraction of the bunch length itself).

- Third: The beam position in space is determined and a grid for the evaluation of the sc fields is defined. The definition of the grid allows to reduce the size of the problem by considering the interaction of the macroparticle with the average fields produced by the whole beam. As it will be shown below, starting from the field values on the mesh point, a regularization procedure can be applied to settle upper bounds on the frequency spectra of the self-consistent fields.

- Fourth: The fields produced by each particle on each grid point is evaluated. The number of evaluation can be quite large and this is the main cpu time consuming step in the program. The evaluation of each link particle-grid point is independent from all the others. This in principle allows a parallelization of the program with a consistent reduction of the global computation time.

In an rf injector, during extraction, we face a situation in which the beam head is nearly relativistic, while the tail is still almost at rest nearby the cathode. To take into account for the finite propagation velocity of light the fields must be evaluated according to the Lienard Wiechert retarded potentials. Each macroelectron contributes to the fields at a given grid point $\vec{X}$ when the retarded condition is satisfied:

$$t - t' = \frac{|\vec{X} - \vec{r}(t')|}{c} \qquad (1)$$

i.e., when its trajectory $\vec{r}(t')$ crosses the light cone of space-time event $(t, \vec{X})$. This way of evaluating the fields makes not trivial the inclusion of the effects of wake fields from cavity walls or other boundaries. In the case of rf injectors the only boundary taken into account is the cathode wall, assumed as planar and included with the method of the image charges.

- Fifth: the sc fields on the grid vertices are known. A fitting procedure is used to interpolate the fields values on the particles positions.

The situation at step one is reproduced.

Covariant formulation of the charge screening problem

The regularization of fields is a well known topic to deal with when modelling the dynamics of a large assembly of charged particles interacting with their self-consistent electric and magnetic fields [3]. Because of practical computational limitations, the simulations are usually bounded to a few thousand or a few million particles, whereas for typical values of the beam charge the number of electrons actually involved is of the order of 10^{10} or even 10^{11}. Thus, the amount of physical elementary charges each particle in the model ("macroparticle" or "macroelectron") can be thought as representing of is huge ($10^4 - 10^6$). This leads to space charge fields unphysically large when particles pass close to each other, a situation that harms either the stability of the simulations or - by introducing a fictious collisional contribution to the physics of the model - the accuracy of the results.

Various techniques have been considered to reduce the harmonic content of the space charge fields. For example, when the fields are to be computed on the vertices of a regular mesh, the size of the cell itself defines a cutoff on the upper value of the fields spectra, even though the eventuality of a macroparticle passing very close to a mesh point must be considered and accounted for (usually by weighted summation of contributions to the fields from the

neighbour vertices or through other sharing algorithms). Here, the classical approach of giving the macroparticle a size and shape has been followed.
Since the fields are evaluated through the Lienard-Wiechert retarded potentials, much emphasis has been devoted to the goal of deriving a smoothing
procedure manifestly covariant. A fully covariant approach to noise reduction in fields evaluation presents three valuable aspects. First, the effects of
charge screening can be examined where the contributing particle is at rest.
Second, in each reference frame, the smoothing procedure yields values of the
fields that are connected through Lorentz transformations, as they should be.
As a consequence, the reduction of harmonic content of fields spectra turns
to be *naturally* independent from the dynamical regime of the beam. Third,
the calculation can be actually carried out wherever the dynamical variables
of the particle are directly available, usually the accelerator reference frame.
To do this, a relativistic generalization of purely geometrical form-factor need
to be introduced: for a gaussian-shaped particle, for example, a 4-dimensional
inverse covariance tensor must be defined, both reproducing the relativistic
distortions of the fields in an arbitrary reference system and reducing to the
usual purely spatial inverse covariance matrix in the frame where the particle
is at rest. In the particle's reference frame, where an observer at rest experi-
ences a purely static electric field, an "effective charge" can be defined as the
total charge included in the iso-density surface associated to the value at the
observer point. It can be proved that for the effective charge the following
expression holds:

$$Q_{\text{eff}} = Q \left[\text{erf} \left(\frac{R}{\sqrt{2}} \right) - \sqrt{\frac{2}{\pi}} \cdot R \cdot \exp \left(-\frac{R^2}{2} \right) \right] \qquad (2)$$

where

$$R = x^T \hat{\Sigma}^{-1} x = x'^T \hat{\Sigma}'^{-1} x' \qquad (3)$$

and

$$\hat{\Sigma}^{-1} = \begin{pmatrix} 0 & \vec{0}^T \\ \vec{0} & \hat{\sigma}^{-1} \end{pmatrix} \qquad (4)$$

$$\hat{\Sigma}'^{-1} = \begin{pmatrix} \vec{\beta}^T \hat{\sigma}'^{-1} \vec{\beta} & -\vec{\beta}^T \hat{\sigma}'^{-1} \\ -\hat{\sigma}'^{-1} \vec{\beta} & \hat{\sigma}'^{-1} \end{pmatrix} \qquad (5)$$

are the generalized inverse covariance matrices in the particle's and accelerator
frames, respectively. The (2) can be used to approximate the electric field by
assuming that:

- it is produced by a pointlike charge Q_{eff} at a distance $\vec{R}$;

- it is spherically symmetric (see fig. (2)):

$$\vec{E}_{\text{eff}}(\vec{R}) = \frac{Q_{\text{eff}}}{\vec{R}^3} \cdot \vec{R} \tag{6}$$

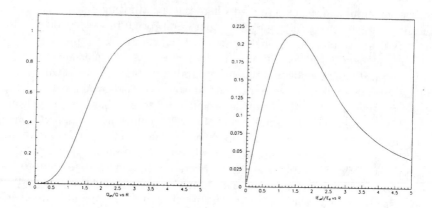

FIGURE 2. Normalized effective charge (left) and static electric field strenght (right) as a function of distance.

It is immediately seen that between the effective electric field and the un-screened one there is only a scale factor Q_{eff}/Q that remains unchanged when applying the Lorentz transformation to recover the values of electromagnetic fields in the acclerator frame. This picture is approximated in the sense that for a non spherically-symmetric charge distributions (in its own rest frame) a different scale factor should be introduced for each component of the electric field. One can be led to such a distribution because the most natural choice is to assign the macroparticle in the injector frame the same shape of the whole beam, suitably downscaled, as a remedy against the possibility of overscreening the fields along the "thin" directions of the beam, or vice versa. A covariant formulation of the screening procedure in this case is not a trivial task as before and will be treated in a forthcoming paper [4].

RESULTS

Comparison with Itaca.

The program is entirely written in C language, and has been successfully ported to a wide number of platforms. We have considered the following test configuration resembling the original S-Band Brookhaven injector [5] (the parameters relevant to the simulation are listed in table .

Charge	Q	$1\,nC$
Acc. Gradient	E_0	$100\,MV/m$
Frequency	ν_{RF}	$2.856\,GHz$
Bunch width (radial)	σ_r	$3\,mm\ (rms)$
Bunch length	σ_z	$2\,ps\ (rms)$
Avg. injection phase	ϕ_0	65°
No. of cells	N	$2+1/2$

TABLE 1. Parameters used for simulations.

The results have been compared with those obtained from the PIC code ITACA [6]. In figs. (3a) and (3b) are shown the behaviours of the energy (γ) and energy spread ($\Delta\gamma/\gamma$) as a function of the longitudinal coordinate. In figs. (4a) and (4b) are shown respectively the normalized transverse emittances and RMS values. The agreement between the two codes in this test case is remarkably good.

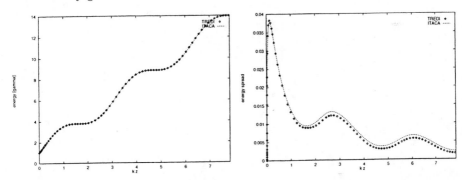

FIGURE 3. Energy (γ) and energy spread ($\Delta\gamma/\gamma$) vs. longitudinal coordinate (kz).

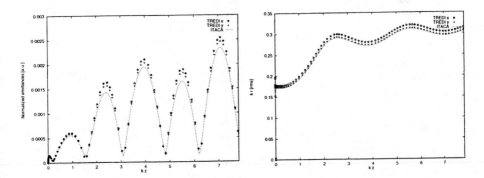

FIGURE 4. Transverse emittances (left) and RMS beam sizes (right).

REFERENCES

1. Press W. H., Teukolsky S. A., Vetterling W. T., Flannery B. P., *Numerical Recipes in C: the art of scientific computing*, 2nd ed., Cambridge: Cambridge University Press, 1992, p. 707ff.

2. Dattoli G., Giannessi L., Quattromini M. and Torre A., *to be published in Physica D.* Singapore: McGraw-Hill, 1985, p. 159.

3. Dawson J. M., *Rev. Mod. Phys.* **55**, 403 (1983). New York: John Wiley and Sons, 1975, p. 552ff.

4. Giannessi L. and Quattromini M., *in preparation.*

5. K. Bachelor et al., *Nucl. Instr. and Meth.*, **A318**, 372, (1992).

6. L. Serafini and C. Pagani, *Proceedings of the I European Particle Accelerator Conference (EPAC) Rome 7-11 1988*, 866, (1988).

THE SHORT BUNCH BLOW-OUT REGIME
IN RF PHOTOINJECTORS

Luca Serafini

INFN and Universita' di Milano
Via Celoria 16, 20133 Milano, Italy

Abstract. A new beam dynamics regime of RF Photoinjectors is presented here, dealing with a violent bunch elongation under the action of longitudinal space charge forces. It is shown that such a blow-out expansion of the electron bunch can lead to highly linear behaviors of both the longitudinal and the transverse space charge field, a well known prerequisite to achieve minimum emittance dilution in photoinjectors. If operated in the ultra-short pancake-like bunch regime, such an effect can be very beneficial to the emittance correction mechanism, making it effective also for ultra-short pancake like bunches. The anticipated performances are presented: kA peak current beams can be generated directly out of the photoinjector (10 to 20 MeV exit energy) with rms normalized emittances below 1 mm·mrad.

I. INTRODUCTION

The theoretical understanding of transverse beam dynamics in RF Photoinjectors, which basically deals with relativistic quasi-laminar beams, has recently become a quite mature subject, thanks to several recent publications presented in the literature[1], which actually provide a satisfactory and general framework to understand the mechanism underlying the emittance correction process[2]. After a pioneering work by K.J. Kim[3], capable to describe quantitatively the emittance dilution effects using a ballistic model, but still unable to provide possible explanations for correcting such effects, making the emittance dilution process a reversible one[4], the coherent cold plasma oscillations associated with this process were recently analyzed and described[5] in the framework of the envelope equation treatment by Lapostolle and Sacherer[6].

The success of this new model, based on the concept of invariant envelope, an equilibrium state for accelerated laminar relativistic beams, is reported in various references[1,5], where the predictions of optimum injector operating conditions to achieve maximum beam brightness are compared to measurements and simulation results.

On the other hand, the longitudinal dynamics has been studied in quite less details, although various authors analyzed the phase focusing effect[7] in a single particle description and others[8] evaluated the space charge induced bunch lengthening under a perturbative approximation. A complete understanding of

CP413, *Towards X-Ray Free Electron Lasers*
edited by R. Bonifacio and W. A. Barletta
© 1997 The American Institute of Physics 1-56396-744-8/97/$10.00

longitudinal cold plasma oscillations in a photoinjector is however still missing. This might be due to the initial approach followed in conceiving the laser-driven Radio-Frequency gun: the longitudinal space charge effects were expected to be weak, possibly perturbative, so to obtain time profiles for the electron bunch current basically replicating the intensity time distribution of the laser pulse illuminating the photocathode - this has been in the past the basic idea associated with RF guns.

Recent experimental results from BNL[9], together with some less recent - but well reported in the literature - results from simulations, pointed out how longitudinal space charge effects can be dramatic even in usual regimes where they were thought to be negligible. Main concern is their possible impact on the emittance dilution mechanisms due to longitudinal correlations in transverse RF and space charge forces.

Although a severe bunch lengthening is usually thought to be a parasitic effect in RF guns - mainly because this typically occurs close to the maximum charge limit where a huge correlated energy spread is induced - we will show that the same effect, if occurring in the ultra-short bunch regime and far from the maximum charge limit, can have great beneficial impact on the beam quality, preventing the formation of bifurcated halos, typical of the emittance correction mechanism, and producing density distribution close to the uniform ellipsoidal distribution, well known for the linearity of the self-field in both the longitudinal and transverse planes. In other words, we will show that in this particular regime one can have severe debunching associated with modest energy spread.

At the same time, simulations show that the emittance correction technique is applicable successfully to ultra-short bunches, leading to unprecedented values of beam brightness achieved right out of the injector without the need of a further magnetic compression.

II. LONGITUDINAL DYNAMICS OF DISK-LIKE BUNCHES IN RF PHOTOINJECTORS

The maximum charge limit is set by

$$Q_{MAX} = \frac{I_A}{c} \gamma' \sigma^2 \tag{1}$$

($I_A \cong 17$ kA), which gives the maximum amount of charge that can be extracted from a photocathode illuminated by a laser pulse having gaussian transverse intensity distribution of width σ, with a peak applied RF field (normally directed to the cathode surface) E_0 [10]: γ' comes out to be the rate of increase of the electron normalized energy $\gamma' \equiv \dfrac{d\gamma}{dz}$. This has a secular linear growth for a standing wave mode, as usual for RF gun cavities, for which $\gamma' \cong \dfrac{eE_0}{2mc^2}$.

A more detailed analysis, as the one performed in ref.11, shows that this limit can be actually overcome to some extent, because eq.1 corresponds to a space charge saturation reached only at the centre of the gaussian laser pulse. Using a 1 mm rms laser spot size at the cathode, in a 100 MV/m peak RF field, eq.1 gives a maximum of 5.6 nC that can be extracted from the gun: however, the closer the effective bunch charge is to this limit, the higher is the induced rms energy spread on the electron bunch, which can be evaluated[8], at first order, as

$$\frac{\Delta\gamma}{\gamma} \cong \frac{2Qc}{\sqrt{3}I_A\gamma'\sigma^2} - \sigma_\phi \cos\overline{\varphi}, \tag{2}$$

where Q is the electron bunch charge ($Q \le Q_{MAX}$), σ_ϕ the bunch rms length (in radians of the RF cycle) and $\overline{\varphi}$ the (average) bunch RF phase ($\overline{\varphi} = \frac{\pi}{2}$ is the maximum acceleration phase). One should note that the second term on the r.h.s. of eq.2 is due to the RF field in the gun: this contribution may mitigate the space charge induced energy spread (for $\overline{\varphi} < \frac{\pi}{2}$) by means of an opposite energy-phase correlation. When the bunch charge is close to Q_{MAX} (see eq.1) the space charge induced energy spread (first term on the r.h.s. of eq.2) cannot be any longer corrected by the RF field (being $\frac{2Q_{MAX}c}{\sqrt{3}I_A\gamma'\sigma^2} = O(1)$), and the bunch time structure is lost, leading to unacceptable bunch lengthening in the gun, which is typically accompanied by severe non-linear behaviors in the longitudinal phase space, giving in turns huge emittance dilutions.

As discussed in ref.10, eq.1 (as well as eq.2) is valid in the short bunch regime, i.e. under the condition

$$\sigma_\phi[°RF] << \frac{21}{\alpha} \tag{3}$$

where α is the fundamental gun parameter, defined as the normalized RF vector potential amplitude $\alpha \equiv \frac{eE_0}{2mc^2k}$ (note $\gamma' = \alpha k$): since almost all photoinjectors fall within $\frac{1}{2} < \alpha < 4$, eq.3 is easily satisfied in any typical RF gun operating range. This is true because RF guns are designed to produce bunches much shorter than the accelerating gap (i.e. in a transient regime), which in this case can be considered as the first half cell (usually 1/4 of RF wavelength in length) of the RF gun cavity. One of the main differences between RF guns and thermoionic injectors (based on DC guns) is that the latter work in a steady state regime where the bunch is much longer than the accelerating gap, well described by the Child-Langmuir law. In case of RF guns the limitations to the maximum charge is therefore expressed in terms of surface charge density more than current. The intermediate regime has been analyzed by various authors, for example J.L. Coacolo and M. Dolique[12], who derived the correct relativistic Child-Langmuir law for RF guns, or J. Girardeau-Montaut[13], for the short bunch case in DC guns.

One should recall that eq.2 is valid for low aspect ratio bunches, as typically in use in RF guns, *i.e.* under the condition $A \ll \gamma$. Since the beam normalized energy γ at the gun exit is usually in excess of 5, typical aspect ratios $A \equiv \dfrac{\sigma}{\sigma_z} = \dfrac{k\sigma}{\sigma_\phi}$ in use satisfy this condition. We want to investigate now the domain of high aspect ratio bunches, *i.e.* those characterized by being much radially larger than axially long, the so-called pancakes: we will see that these pancake bunches may undergo violent longitudinal de-bunching even far below the maximum charge threshold, keeping at the same time a highly linear behavior in the fields, as well as in phase space distributions.

The transverse $\omega_\perp$ and longitudinal $\omega_\parallel$ plasma frequencies for a water-bag uniform ellipsoid scale like[14]

$$\omega_\perp^2 \propto \frac{Q}{\gamma^3 \sigma^2 \sigma_z} \left[\frac{1}{4A_r^2 + 1} \right]^{1/4} \qquad \omega_\parallel^2 \propto \frac{Q}{\gamma^3 \sigma^2 \sigma_z} \left\{ 1 - \left[\frac{1}{4A_r^2 + 1} \right]^{1/4} \right\} \qquad (4)$$

where $A_r \equiv \dfrac{\sigma}{\gamma \sigma_z}$ is the rest aspect ratio, *i.e.* the aspect ratio of the electron bunch in its rest reference frame (here σ and σ_z are again the rms sizes). In case the laser pulse aspect ratio $A_l = \dfrac{\sigma}{\sigma_{las}}$ is quite large (and the debunching moderate so that $\sigma_z \cong \sigma_{las}$), A_r will remain large up to the gun exit, or at least in the first cells, where the longitudinal space charge effects are important. We then have, in such a regime of pancake bunches,

$$\omega_\parallel^2 \propto \frac{Q}{\gamma^3 \sigma^2 \sigma_z} \qquad\qquad \omega_\perp^2 \propto \omega_\parallel^2 \sqrt{\frac{\gamma}{A_l}} \qquad (5)$$

stating that the longitudinal plasma frequency is much larger that the transverse one (*i.e.* the space charge field is mainly longitudinal, as expected for the case of a sheet of charge). Moreover, since the longitudinal frequency scales like $1/\sigma_z$ the perveance term in the longitudinal rms envelope equation will be independent on σ_z

$$\sigma_z'' - \frac{Qc}{2I_A \beta^2 \gamma^3 \sigma^2} = 0 \qquad (6)$$

324

So we are in a particular situation in which the self-force does not depend on the moment of the distribution (*i.e.* σ_z) in that degree of freedom, unlike the transverse envelope equation,

$$\sigma'' + \sigma'\frac{\gamma'}{\gamma} + K_r\sigma - \frac{Qc}{2I_A\gamma^3\sigma_z\sigma} = 0$$

which always retains (for round beams!) the non-linear dependence of the perveance on the rms beam size σ (note that both equations are written under the hypothesis of beam laminarity, as discussed in ref.1,14). The bunch lengthening (blow-out) described by eq.6 will occur within an ideal linear force distribution inside the bunch for the case of uniform laser pulses (in time): this will in turns favour the formation of a water-bag distribution, after the blow-out, with optimum linear behavior for the transverse fields too.

An useful expression for the bunch lengthening in RF guns was derived in ref.8 under the approximation of perturbative longitudinal space charge field:

$$\frac{\sigma_z}{\sigma_{las}} = \frac{I_p}{I} = 1+\Delta_{SC}-\Delta_{RF} \tag{7}$$

This solution takes into account the transition from high rest frame aspect ratios A_r close to the cathode down to very small A_r at high energies at $\gamma \gg 1$ the relativistic elongation of the bunch in its reference frame makes the aspect ratio A_r progressively small); in other words eq.7 is a result of an integration from the pancake regime to the cigar-like regime under perturbative approximation, *i.e.* $\sigma_z \cong \sigma_{las}$. In case of eq.6 (no dependence on σ_z, *i.e.* on the bunch length) the approximation is valid even for violent expansion (blow-out, *i.e.* $\sigma_z \gg \sigma_{las}$). The bunch expansion is given in terms of the nominal peak current $I_p \propto \dfrac{Q}{\sigma_{las}}$, the RF phase focusing term

$$\Delta_{RF} = \frac{1}{2\alpha\sin^2\overline{\varphi}}\left(\cos\overline{\varphi}+\frac{2}{3}\cot\overline{\varphi}\right) \tag{7'}$$

and the space charge debunching term

$$\Delta_{SC} = \frac{I_p}{I_A(\gamma'\sigma\sin\overline{\varphi})^2}f(A_l,\gamma) , \tag{7''}$$

where the form factor $f(A_l,\gamma)$ is given by

$$f(A_l,\gamma) = \left\{ \begin{array}{l} A_l\left[1 - \dfrac{1}{\gamma} + \sinh^{-1}(A_l) - \sinh^{-1}\left(\dfrac{A_l}{\gamma}\right)\right] + (1-A_l)\ln\left[\dfrac{2\gamma}{1+\gamma}\right] + \\[4mm] \sqrt{1 + \dfrac{A_l^2}{\gamma^2}} - \sqrt{1 + A_l^2}\left(1 + \ln\left[\dfrac{A_l^2(1+\gamma)}{A_l^2 - \gamma + \sqrt{1+A_l^2}\sqrt{\gamma^2 + A_l^2}}\right]\right) \end{array} \right\}$$

$f(A_l,\gamma)$ has the following asymptotic expression $f(A_l \gg 1, \gamma \gg 1) \cong Log[2]$, while it can be well approximated by the simple expression $f(A_l, \gamma \gg 1) \cong 0.534\sqrt{A_l} - 0.072A_l - 0.24A_l^{1/4}$ in the region $0 \le A_l \le 20$.

At zero charge ($\Delta_{SC} = 0$) and $\overline{\varphi} = \dfrac{\pi}{2}$ (hence $\Delta_{RF} = 0$) one has a bunch rms length at the gun exit σ_z actually equal to the laser pulse rms length σ_{las}, as expected. The intermediate regime, where a significant but small space charge debunching $\Delta_{SC} \ge 0$ can be eventually compensated by the RF phase focusing contribution (with $\Delta_{RF} \ge 0$ at $\overline{\varphi} \le \dfrac{\pi}{2}$) is widely discussed elsewhere[8].

Here we are interested in the so-called blow-out regime, defined as the domain where $\Delta_{SC} \gg 1$, so that Δ_{RF}, being $O(1)$, can induce only a small correction that can be neglected. This regime can be explored by using very high I_p, i.e. by illuminating the cathode with very short laser pulses and by extracting moderate charges, so to stay far below the charge limit: a particular example of interest is presented below in the next section. In this way Δ_{SC} can be large but $\dfrac{\Delta\gamma_{SC}}{\gamma} = \dfrac{2Qc}{\sqrt{3}I_A\gamma'\sigma^2}$ (see eq.2) small. Therefore, in the blow-out regime the actual beam current $I = \dfrac{I_p}{1 + \Delta_{SC} - \Delta_{RF}} \cong \dfrac{I_p}{\Delta_{SC}}$ at the gun exit (at a beam energy γ) will be almost independent on the nominal peak current I_p , being

$$I = \dfrac{I_A(\gamma'\sigma\sin\overline{\varphi})^2}{f(A_l,\gamma)} \qquad (8)$$

giving a current density ($J \equiv \dfrac{I}{2\pi\sigma^2}$) scaling like the square of the accelerating field: taking e.g. a laser pulse with aspect ratio $A_l = 10$ and rms spot size $\sigma = 1.5$ mm , with a gun exit energy of 10 MeV ($\gamma = 21$) one obtains a beam current $I = 1.6$ kA at 150 MV/m RF peak field at the cathode.

The independence of the beam current I at the gun exit on the laser pulse length σ_{las} (i.e. on the nominal peak current I_p) comes out to be a great factor of stability versus the laser pulse length fluctuations: the current produced by the gun is no longer set by the laser pulse characteristics but by the equilibrium reached between the blow-out space charge expansion on one side and the relativistic freezing given by the fast acceleration in the gun, plus a possible weak RF phase focusing term, on the other side.

The condition to operate the gun in the blow-out regime can be cast in terms of the peak field E_0 and laser rms spot size σ, once the desired beam current and laser aspect ratio are chosen. This is expressed, in practical units, by

$$E_0[\text{MV/m}]\sigma[\text{mm}] \geq \frac{10.8}{\sin\overline{\varphi}}\sqrt{f(A_l,\gamma)I[A]} \tag{9}$$

Since we want a bunch charge much smaller than the charge limit, a second condition to be satisfied will be, recalling eq.1 and $I_p = \dfrac{Qc}{\sqrt{2\pi}\sigma_{las}} = \dfrac{QcA_l}{\sqrt{2\pi}\sigma}$, $\gamma'\sigma \gg \dfrac{\sqrt{2\pi}I_p}{I_A A_l} = \dfrac{\sqrt{2\pi}I\Delta_{SC}}{I_A A_l}$. In practical units we have

$$E_0[\text{MV/m}]\sigma[\text{mm}] \gg 0.15\frac{I[A]\Delta_{SC}}{A_l} \tag{10}$$

which, being typically $\dfrac{\Delta_{SC}}{A_l} \leq 1$, seems easier to be satisfied than eq.9.

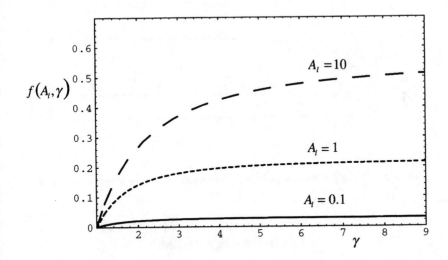

FIGURE 1. $f(A_l,\gamma)$ as function of beam energy γ, at some values of the aspect ratio A_l

As a last remark on the space charge term Δ_{SC}, we note that its expression is valid and correct also for non relativistic energies, *i.e.* $\gamma \cong 1$, in which case eq.8 predicts the equilibrium current of a DC gun operated in the transient short bunch regime (*i.e.* $\sigma_z \ll$ gap length), once $\overline{\varphi} = \dfrac{\pi}{2}$ and $\gamma' = \dfrac{eE_0}{mc^2}$ are defined ((as proper for a DC field of amplitude E_0).

The behavior of the form factor $f(A_l, \gamma)$ is plotted in Fig.1 for fixed laser aspect ratios A_l, while the same quantity is plotted for fixed values of the beam energy γ in Fig.2: note that for $\gamma = 1$ (bunch at the cathode) we always have $f(A_l, \gamma = 1) = 0$ as expected.

Again, we see that high aspect ratio bunches undergo larger expansions than cigar-like bunches having the same spot size σ and nominal current I_p (see eq.7'). It is only entering the pancake domain ($A_l > 10$) that makes the bunch lengthening no longer dependent on the laser aspect ratio.

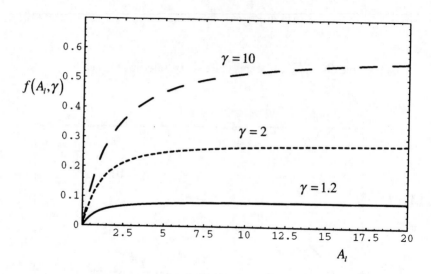

FIGURE 2. $f(A_l, \gamma)$ as function of the laser aspect ratio A_l, at some beam energies γ

III. PRELIMINARY SIMULATION RESULTS IN THE BLOW-OUT REGIME

In order to investigate the blow-out regime we performed some preliminary simulations using the P.I.C. (particle in cell) electro-magnetic code

ITACA[15], able to push particles in the time domain from the cathode surface up to relativistic energies in their own self-consistent field. As a test structure, we adopted a 3+1/2 cell RF gun (first half cell 1/4 of wavelength long) operated at S-band with 150 MV/m peak field at the cathode, injecting into a booster linac made of a 4 cell Standing Wave structure operated at the same RF frequency (2.856 GHz) with 120 MV/m peak field on axis.

Illuminating the photocathode with a 1.2 mm rms spot size laser pulse would yield a 12 nC maximum charge limit, therefore we chose to extract a bunch charge $Q = 0.5$ nC.

Since we want to minimize the emittance dilution in transverse phase space one has to apply the emittance correction technique, which properly corrects linear space charge correlations through the bunch. So far, several studies based on simulations as well as analytical models have shown that this technique is less effective in the domain of high aspect ratio bunches because of the high non-linear behaviors of transverse space charge forces in this regime[1,5]. Indeed, it seems that the technique is less effective (though good) in correcting non-linear effects. Cigar-like bunches (*i.e.* low aspect ratio) have fairly linear transverse fields: the non-linearities are typically confined in the bunch tails, fact that gives rise to bifurcations, as discussed extensively elsewhere[1,14]. However, time ago an optimum charge density distribution was proposed to minimize field non-linearities in pancake bunches[16]: this requires longitudinally uniform density distributions which have an inverse parabolic radial profile (*i.e.* denser in the core than in the radial outer edge)

For this reason we chose to illuminate the cathode with a uniform time profile laser pulse, 0.3 ps total length, having a radial intensity distribution

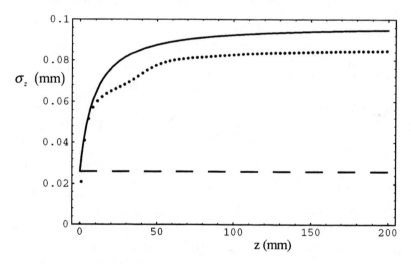

FIGURE 3. Rms bunch length σ_z from the cathode (z=0) up to the gun exit (z=184 mm) as predicted analytically (solid line), compared to simulation results (dots). The dashed line gives the laser pulse rms length σ_{las} =26 μm (for comparison).

scaling like $I(r) = I_0(1 - 0.6r^2)$, as predicted to be the optimum profile in ref.16 . The rms laser pulse length for such a distribution is $\sigma_{las} = 0.3c/2\sqrt{3} = 26$ µm, giving a laser aspect ratio $A_l = 46$ and a nominal current $I_p = 2.3$ kA.

The results of the simulations are shown in Figures 3-5: the beam energy at the gun exit (located at z=184 mm far from the cathode) is about 12.5 MeV with a rms energy spread of 0.5 %. The energy is raised up to 24.5 MeV by the booster linac, located between z=750 mm and z=960 mm. In Fig.3 the bunch rms length behavior along the gun (dots) is compared to the analytical prediction of eq.7 (solid line): since the electron bunch is formed in time at the cathode surface, the simulation properly reproduce this effect giving an initial bunch length (first dot) close to zero. Due to the assumption done by the analytical model[8] of a bunch instantaneously formed at the cathode location with rms length σ_z equal to the laser rms length σ_{las}, the two curves are initially mismatched, but they show thereafter a remarkable agreement. At the gun exit the analytically predicted bunch lengthening factor σ_z/σ_{las} is 3.6 versus a simulation value of 3.3. The difference is in part due to the radial expansion the bunch undergoes in the first cells of the gun, as shown in Fig.4, which lowers the bunch density hence the bunch lengthening: this radial expansion is caused by combined RF and space charge transverse forces which are not taken into account by the analytical model, which therefore tends to over-estimate the bunch lengthening.

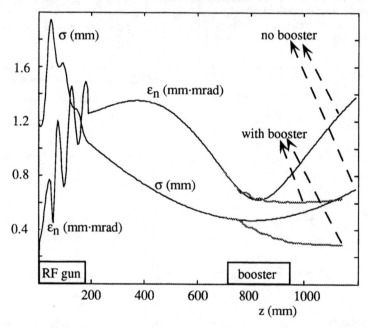

FIGURE 4. Rms bunch radius σ and normalized transverse emittance ε_n plotted along the photoinjector with and without booster acceleration

The performances of such a configuration in terms of beam quality are reported in Fig.4, where the rms normalized emittance is plotted versus the bunch position z from the cathode, together with the rms envelope σ. Under the focusing action of a solenoid lens placed around the first cell of the gun, set at a field amplitude of 3.9 kGauss, the rms envelope decreases down to a gentle space charge dominated laminar waist in order to be injected in the booster linac matched to the invariant envelope so to perform emittance correction, as widely discussed in ref.1. The rms normalized emittance is actually damped down to a value of 0.6 mm·mrad, therefore proving that the emittance correction process is effective also in the pancake bunch regime. The matching to the invariant envelope also assures a parallel beam at the exit of the booster (see Fig.4), as also shown by the transverse phase space distribution of the bunch plotted in Fig.5 (right diagram) displaying an average beam divergence around 0.1 mrad.

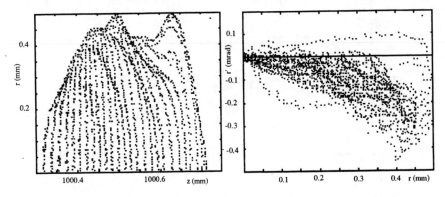

FIGURE 5. Phase space (r,r') distribution (right diagram) of the bunch particles at the exit of the booster: the left diagram shows the particle distribution in the configuration space (z,r)

For sake of comparison the envelope and emittance behaviors in absence of booster acceleration are plotted in Fig.4, showing that the transverse space charge field is still effective in blowing up the rms beam spot and, at the same time, the normalized emittance - a clear signature that the beam is evolving under a laminar space charge dominated flow, *i.e.* a cold plasma coherent oscillation mode, instead of the usual emittance dominated betatron incoherent flow, where the normalized emittance is supposed to stay invariant.

Looking at the (r,r') particle distribution of Fig.5 we can notice at a glance the absence of bifurcated halos typical of the emittance correction process, as those reported in ref.14, which are mainly caused by the over focusing applied by the solenoid field to the bunch head and tail regions being subject to lower defocusing space charge forces. These outer slices of the bunch, because of the longitudinal laminarity (no mixing between slices), go through a crossover instead of a laminar waist, hence developing a halo downstream the waist itself. We believe the reason why halo is absent in pancake bunches is due to the blow-out longitudinal effect which brings the bunch distribution toward a uniformly filled ellipsoid, as clearly depicted (a part from some local fluctuations) by the particle distribution in the configuration (r,z) space (see left diagram in Fig.5). It is well known that in such a distribution the transverse space charge field is not only linear versus the radial coordinate but also independent on the longitudinal (slice) coordinate along the bunch - each slice has almost the same transverse plasma frequency.

As a further confirmation of the evolution of the initial uniform charge density distribution (in time) into that of a water-bag distribution, we plotted in Fig.6 the current distribution along the bunch at 4 different locations during the acceleration.

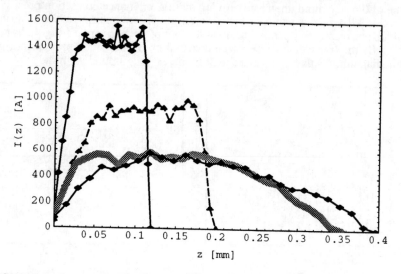

FIGURE 6. Current time profiles within the bunch at 4 different positions along the photoinjector

The first current profile is taken as a snapshot of the bunch when it is 2.6 mm far from the cathode surface, 190 keV average kinetic energy, and clearly displays the uniform time profile of the laser pulse: some lengthening has already taken place, since the total electron bunch length is 120 μm versus the laser pulse length of 90 μm (0.3 ps), so that the peak current is about 1500 A. The second snapshot is taken at 6.2 mm from the cathode surface, 650 keV kinetic energy: it shows still a fairly uniform distribution (peak current 900 A) with begins to develop smooth tails. At the gun exit (430 mm from the cathode, 12.5 MeV, bolded line) the current profile is no longer uniform: under the action of the longitudinal space charge field causing the blow-out the hard edge central core of the bunch is smoothed out, so to reach progressively a sort of parabolic distribution in time as shown by the last current profile, displaying a 550 peak current with 0.95 rms bunch length - this has been taken at the booster exit (1000 mm from the cathode, 24.5 MeV energy).

IV. CONCLUSIONS

These preliminary results show the quite relevant interest of the blow-out regime in producing high brightness beams: at the moment, we found a better optimization for the solenoid and booster position leading to a 700 A peak current beam at 0.5 mm·mrad rms normalized emittance. Unfortunately, due to the very short laser pulses in use (on the scale of the RF wavelength - tens of microns versus hundreds of mm), the time and space discretizations to be applied in the PIC simulations are so small (mesh steps around 3-6 microns are used) that the required CPU times to accomplish a whole photoinjector run are huge, preventing an easy scan over the operating diagram to find out the optimum operating points.

Nevertheless, noting that emittances are calculated in this case all over the bunch, *i.e.* no tail clipping is performed in the calculation, as is customary in cigar-like bunch simulations to get rid of bifurcated halos, we can say that these results give beam brightness almost one order of magnitude higher than typically obtained with best simulations in the cigar-like domain. This makes possible to conceive beam generation for short wavelength FEL based on photoinjectors without the need of magnetic compression to raise up the peak current beyond the kA level.

As anticipated, such performances on beam quality are possible due to the unique behavior of the blow-out longitudinal dynamics, which leads toward uniformly filled ellipsoidal distributions, an optimum from the point of view of space charge field linearity.

The promising potentialities of such a new regime in RF gun beam dynamics are likely to be of interest also for DC gun applications: namely, future analysis should address the possibility to take advantage of the ideal linear behavior in the space charge fields of bunched electron beams produced in laser-driven DC guns, so to evaluate the enhancement of beam brightness achievable by this technique.

Finally, future works on this subject should involve a detailed analysis on modal expansion of the envelope equation to evaluate the sensitivity of the final density distribution on the initial one, including cathode time response effects which might produce tails in the distribution due to delayed photo-emission processes.

ACKNOWLEDGEMENTS

The author acknowledges many useful discussions with J. Rosenzweig, D.T. Palmer, R. Miller, M. Cornacchia, J. Clendenin and D. Yeremian. In particular, I am indebted to Claudio Pellegrini who invited me to study again the beam dynamics of short bunches in RF guns and continuously supported me throughout the work.

REFERENCES

1. L. Serafini, J.B. Rosenzweig, *Phys. Rev. E* **55**, 7565 (1997)
   J.B. Rosenzweig *et al.*, "Beam Dynamics in an Integrated Plane Wave Transformer Photoinjector at S and X band",*Proc. Part. Acc. Conf. 1997*, Vancouver, Canada
   L. Serafini, J.B. Rosenzweig, "Optimum Operation of Split RF Photoinjectors",*Proc. Part. Acc. Conf. 1997*, Vancouver, Canada
2. B.E. Carlsten, *Nucl. Instr. Methods A* **285**, 313 (1989)
3. K.J. Kim, *Nucl. Instr. Methods A* **275**, 201 (1989)
4. P. O'Shea, "Entropy and Emittance Growth", *Proc. 1995 IEEE Particle Accel. Conf.* 3478 (1996)
5. J.L. Coacolo *et al.*, "TTF-FEL Photoinjector Simulation giving a High Quality Beam", *Proc. of FEL'96 Conf.*, August 1996, Rome
   D.T. Palmer *et al., Proc. 1995 IEEE Particle Accel. Conf.* 2432 (1996)
   D. Yeremian, "A Proposed Injector for LCLS", *these proceedings*
6. P. Lapostolle, *Proton Linear Accelerators* (Los Alamos, 1980)
7. B. Carlsten, D. Nguyen, *Particle Accelerator* **56**, 127 (1997)
   X.J. Wang et al., *Phys. Rev. E* **54**, 3121 (1996)
8. L. Serafini, *IEEE Trans. on Plasma Science*, Vol. **24**, 421 (1996)
   L. Serafini, R. Zhang, C. Pellegrini, *Nucl. Instr. Meth. A* **387**, 305 (1997)
9. D.T. Palmer, "Photocathode Electron Sources for Single Pass X-Ray FEL", *these proceedings*
10. C. Travier, *Ph.D. Thesis*, Universite' De Paris Sud, Orsay-LAL, Dec. 1995
    C. Travier, *Nucl. Instr. Methods A* **304**, 285 (1991)
11. J.B. Rosenzweig *et al.*, *Nucl. Instr. Methods A* **341**, 379 (1994)
12. J.M. Dolique and J.C. Coacolo, *Nucl. Instr. Methods A* **340**, 231 (1994)
13. J.P. & C. Girardeau-Montaut, *J. Applied Physics* **65**, 2889 (1989)
14. L. Serafini, *AIP CP* **395**, 47 (1997)
15. L. Serafini *et al.*, *Proc. 1st European Particle Accelerator Conf.*, World Scientific, Singapore, 1988, p.866
16. L. Serafini, *Nucl. Instr. Methods A* **340**, 40 (1994)

AUTHOR INDEX

P

Palmer, D. T., 155
Pellegrini, C., 49
Petrillo, V., 45
Piovella, N., 205
Prazeres, R., 81

Q

Quattromini, M., 313

R

Reiche, S., 29, 49
Rosenzweig, J., 49

S

Sakaki, H., 299
Saldin, E. L., 29, 195, 219, 291
Scharlemann, E. T., 171
Schneider, J. R., 219
Schneidmiller, E. A., 29, 195, 219, 291
Serafini, L., 179, 321
Shimada, S., 241
Sobajima, M., 241, 307
Suzuki, H., 267
Suzuki, S., 121, 299

T

Takeshita, I., 267
Takii, T., 127
Taniuchi, T., 121, 299
Toku, H., 241, 307
Tomimasu, T., 127
Travish, G., 37, 49
Tremaine, A., 49

V

Varfolomeev, A., 49

W

Weits, H. H., 231

Y

Yamamoto, Y., 241, 307
Yanagida, K., 121, 299
Yokomizo, H., 121, 299
Yonehara, H., 267
Yoshikawa, H., 121, 299
Yoshikawa, K., 241, 307
Yurkov, M. V., 29, 195, 219, 291